AF678960

From Milk By-Products to Milk Ingredients

From Milk By-Products to Milk Ingredients

Upgrading the Cycle

Ruud de Boer
Wageningen UR (University & Research centre), The Netherlands

WILEY Blackwell

This edition first published 2014 © 2014 by John Wiley & Sons, Ltd

Registered office: John Wiley & Sons Ltd, The Atrium, Southern Gate, Chichester, West Sussex, PO19 8SQ, UK

Editorial offices: 9600 Garsington Road, Oxford, OX4 2DQ, UK
The Atrium, Southern Gate, Chichester, West Sussex, PO19 8SQ, UK
111 River Street, Hoboken, NJ 07030-5774, USA

For details of our global editorial offices, for customer services and for information about how to apply for permission to reuse the copyright material in this book please see our website at www.wiley.com/wiley-blackwell.

The right of the author to be identified as the author of this work has been asserted in accordance with the UK Copyright, Designs and Patents Act 1988.

All rights reserved. No part of this publication may be reproduced, stored in a retrieval system, or transmitted, in any form or by any means, electronic, mechanical, photocopying, recording or otherwise, except as permitted by the UK Copyright, Designs and Patents Act 1988, without the prior permission of the publisher.

Designations used by companies to distinguish their products are often claimed as trademarks. All brand names and product names used in this book are trade names, service marks, trademarks or registered trademarks of their respective owners. The publisher is not associated with any product or vendor mentioned in this book.

Limit of Liability/Disclaimer of Warranty: While the publisher and author(s) have used their best efforts in preparing this book, they make no representations or warranties with respect to the accuracy or completeness of the contents of this book and specifically disclaim any implied warranties of merchantability or fitness for a particular purpose. It is sold on the understanding that the publisher is not engaged in rendering professional services and neither the publisher nor the author shall be liable for damages arising herefrom. If professional advice or other expert assistance is required, the services of a competent professional should be sought.

Library of Congress Cataloging-in-Publication Data

Boer, Ruud de.
Ingredients from milk : principles and processes / Ruud de Boer.
pages cm
Includes bibliographical references and index.
ISBN 978-0-470-67222-8 (cloth : alk. paper) 1. Milk. 2. Dairy products. I. Title.
SF251.B625 2014
636.2′142–dc23

2013046821

A catalogue record for this book is available from the British Library.

Wiley also publishes its books in a variety of electronic formats. Some content that appears in print may not be available in electronic books.

Cover design by www.hisandhersdesign.co.uk

Set in 10.5/13pt Times by Aptara Inc., New Delhi, India
Printed and bound in Singapore by Markono Print Media Pte Ltd

1 2014

Dedicated to
my wife Fiene and our daughter Hilde
for their encouragement and support

Ingredients
Milk specifics
Patents
Reuse
Outside constraints
Vital membrane processes
End users

Contents

Foreword

The Netherlands is one of the leading dairy countries in the world. Education of professionals is of vital importance for the future of the dairy industry. Given the interest in milk and milk products worldwide, this is important not only for The Netherlands but also for other countries. The developments in the emerging countries can be accelerated by strengthening the cooperation between universities, international dairy companies and national governments. The book *Dairy Science and Technology* by Pieter Walstra *et al.* is in this respect an excellent source of specific knowledge. It focuses particularly on the properties and technology of the more traditional dairy products.

This book, *From Milk By-Products to Milk Ingredients – Upgrading the Cycle*, highlights a lesser known part of the dairy industry. Milk, being a liquid, is a convenient source for the processing of ingredients. Due to their nutritional value and specific properties, most of the ingredients from milk are positioned at the high end of the ingredient market. More and more the dairy industry recognizes the importance of these ingredients for adding value to milk. The value of the individual components has to be in line with the strategy of a company. Milk protein and milk fat are commonly used for calculating the farm milk price. Up to now lactose, being a major component of the milk, has not been directly involved. In the new milk price system of FrieslandCampina lactose has been recently introduced as a third component. Protein : fat : lactose are valued in the ratio 10 : 5 : 1 – the first time the contribution of the dairy ingredient, lactose, is recognized.

The background, quality, manufacture and application of numerous dairy ingredients in a variety of finished products are described in this book. It shows that innovation of ingredients from milk has been very successful during recent decades. An addendum of more than 50 information sheets for dairy ingredients illustrates this development and increases accessibility to the information. The book is a recommended source for the education of undergraduate students in food science, graduate students in dairy science and technology and refresher training of dairy staff. The knowledge of the dairy industry might be also valuable for other parts of the food industry, which may include product development activities. This book is therefore recommended also as a reference source for professionals with a miscellaneous background.

Ruud de Boer's diverse and extensive background in the dairy industry guarantees a wide and interesting coverage of this important topic. *From Milk By-Products to*

Milk Ingredients – Upgrading the Cycle also greatly benefits from his experience within the International Dairy Federation and the USA.

Toon van Hooijdonk
Professor of Dairy Science and Technology,
Wageningen UR (University & Research centre)

Preface

Milk is a fascinating fluid. It has a complex composition with thousands of constituents. It is an honour to share this fascination with my readers and highlight diverse aspects of my professional career. I will take you with me on my long-lasting journey in the dairy industry and follow my 'Milky Way'. My path crossed both the industrial and academic worlds. It included aspects of the whole dairy chain from farm to consumer, with topics ranging from milk hygiene on farms, research in membrane processes and development of consumer products to biotechnology of ingredients and the protection of knowledge. My journey led me to the USA as well, where a stay of 5 years enriched my professional experience in ingredients.

Milk is traditionally consumed in well-known dairy products such as liquid milk, cheese and butter. Its application as an ingredient in many kinds of other products is less apparent. Nevertheless, a major part of milk is used in non-specific dairy products. I will introduce you to the ever-extending range of applications. In doing so I will highlight the scientific and industrial aspects of a wide range of ingredients from milk. Dairy Science and Technology has a long history. Past events will also be used to place current developments in perspective.

The subtitle of this book, 'Upgrading the Cycle', reflects the background of milk ingredients – the manner in which they are extracted from milk and their application, while taking care of environmental issues. The word 'cycle' refers to a situation in which all constituents of milk are used in an economical way. In our global society goods that do not deplete the earth are vital. As will be explained, sustainability is a challenge for the dairy industry. The word 'grade' has been used in the dairy industry for decades. Grading ensures that the specific ingredients have a uniform quality. 'Upgrading' adds the aspect of adding value.

Ruud de Boer

Acknowledgements

Milk seems to be a very restricted field of knowledge. However, due to the numerous applications of ingredients from milk, it has a wide scope. For this reason I was grateful to receive support from a long list of experts. I wish to acknowledge Professor Toon van Hooijdonk (Wageningen UR), who gave me the opportunity and support to write this book. His critical and valuable comments were highly appreciated. Professor Hans de Jong (Wageningen UR) helped me in the early stages of my book project; it was an interesting learning experience.

I am very thankful for support on patents from Rene van Duyvenbode (European Patent Attorney – NLO) and support on kosher from Gad Buchbinder (Orthodox Union). The cooperation and suggestions of my former colleagues at Friesland-Campina were a great help too. In this respect I refer to Eef van Arem, Frank Burghout, Toine Hendrickx, Armand Jansen, Jacques Koenraads, Johan Lemmen, Hans Maas, Jan Steijns, Albert Verver, George van der Werff and Tonnie Westerlaken. The experience of Klaas Kussendrager (DFE Pharma) was very beneficial for understanding the nature and properties of pharmaceutical lactose. I also received valuable information from Frans Boer (Royal VIV Buisman), Sana Leppäjoki (Valio), Jerry van Loon (GEA Filtration), Ron McKernan (Whitehall Specialities), Ben Rehwinkel (Arla Foods Ingredients), Paul Roeleveld (Royal HaskoningDHV), Marleen van Russelt (Corman), Bill Schmitz (MSI) and Corina Sneek (Kaasfabriek Eyssen).

Introduction

It never ceases to amaze me that hundreds of ingredients are produced from the single source 'milk'. These ingredients are used in thousands of applications. This is an incredible achievement, which is the result of numerous investigations carried out over the course of time. In the title the word 'upgrading' is mentioned; it is a synonym for improvement. 'Improve' is a verb and it usually goes hand in hand with considerable investigations. The acronym 'improve' is reflected in the structure of the book. The first letters of the titles of the chapters form the word 'IMPROVE'. By selecting the chapters I wish to underline the value of an integrated approach. Such an approach is vital for successful production and use of ingredients. The main chapters of this book are set out below:

Ingredients

Milk specifics

Patents

Reuse

Outside constraints

Vital membrane processes

End users

The knowledge per ingredient is presented in an extensive appendix of information sheets. The ingredients are categorized into milk, lipids, cheese and whey ingredients. These sheets provide a comprehensive reference to more than 50 ingredients from milk. Each information sheet gives a description of the process, composition, applications, properties and relevant literature.

1 Ingredients

An ingredient is a part of a recipe that gives valuable properties to the final product. Food ingredients are used, for instance, to enhance flavour, nutrition and texture. Thousands of recipes can be found in a wide variety of cookbooks. Some of them have a long history. Cookbooks have the great advantage that the information is properly recorded and is not dependent on oral transmission. In antiquity, milk or a part of it, such as cheese, was already applied in various recipes. Ingredients from milk were born. In the course of time food habits were developed; for instance, a combination of cheese and fish was already disliked in Italy in the fifth century BC. The origin of this lies in the difference in diet between the farmers who prepared the cheese, and the fishermen, who were active on the coast (Sancisi-Weerdenburg, 1995).

Preparation of food is a time-consuming activity. Nowadays, processed foods are commonly used and the time spent in the kitchen is reduced considerably. A lot of work in the kitchen has been moved from home to process plants. For example, sauces and dressings containing various ingredients are sold in huge variety in supermarkets and are rarely prepared at home today. Advantages are that the outcome in the meal is predictable and its quality is meticulously controlled. The reproducibility of the processed food relies on the properties of the individual ingredients. They are produced in specialized ingredient plants. Large-scale manufacture is not only responsible for a consistent quality, but also the shelf life is improved and the cost of the ingredients is reduced. Consistency is crucial in processing. Deviations in quality result in defective products, which is expensive due to loss of raw materials, energy and process time.

Food labels in today's supermarkets are an interesting source of information. The ingredient list reflects the choice of ingredients of the manufacturer. Sometimes up to 50 ingredients are mentioned on the label of the package, drawn from a family of tens of thousands of food ingredients. Important categories of food ingredients, including additives, are carbohydrates and sweeteners, oil and fats, vegetable and animal proteins, water, spices, fruits and flavours, stabilizers, emulsifiers, vitamins,

From Milk By-Products to Milk Ingredients: Upgrading the Cycle, First Edition. Ruud de Boer.
© 2014 John Wiley & Sons, Ltd. Published 2014 by John Wiley & Sons, Ltd.

minerals, colours and preservatives. The numbers are still growing over time. Milk ingredients represent approximately 5% of the total number of ingredients, based on the author's own estimation and various buyers' guides (Food Product Design, 2007–2008).

The proportion of ingredients from milk used in any given product will depend on the type of industry considered. In case of one of the largest dairy companies, FrieslandCampina, the average proportion of milk ingredients in the overall ingredients list is about 25% (GRA van der Werff, 2010, personal communication). However, there can be big differences between the various segments of the dairy industry itself. Large users of all kinds of ingredients are processed cheese, imitation cheese and liquid milk products. The latter can be complex products, which include, for instance, fruit yoghurts, drinks and desserts. In summary it can be concluded that the dairy industry is an important customer of its own ingredients.

Since the publication of the well-known book entitled *Byproducts from Milk* (Whittier & Webb, 1950), enormous development has taken place in the dairy industry. At that time the basis of this industry was dominated by butter, cheese, liquid milk and cream. The remaining by-products were used as liquid animal feed or applied as fertilizer. The farmers' milk price was based on milk fat only; the value of milk protein was underestimated, while lactose was totally out of the picture. Changes in consumer preference took place particularly after World War II. The gradual switch from traditional products to convenience foods posed a new challenge for the food industry. The dairy industry also took part in this development, for instance by the introduction of coffee whiteners and desserts. Milk proteins in the form of caseinates were developed and introduced in all kinds of food products such as sausages. After 1970 new technologies emerged, of which various membrane processes were a major breakthrough. Separation of undenatured whey proteins became a reality. The foaming, gelling and emulsifying properties of undenatured whey proteins were favourable for new applications. The end uses of this type of ingredient increased substantially. In selecting an ingredient, adding value became an important topic for both end users and suppliers. Addition of value can be related to several properties of the finished product such as texture, flavour, nutritional value and colour, and also to yield.

At the same time the cooling of raw milk on the farm became more common. The scientific knowledge that cooling of milk (<4°C) retarded bacterial growth considerably was translated into the design and adoption of bulk storage tanks. The prolonged storage of milk made the distance of the farm to the process plant less critical. It favoured the development of larger plants and kept costs at a reasonable level. In areas with a large dairy industry, for instance in The Netherlands, ingredient plants with an enormous capacity could be constructed.

After 1970 environmental issues became more critical than ever as the industry had to pay for its own pollution. This forced the dairy industry to use all of the milk constituents and close the cycle as much as feasible. All these developments helped to create a situation in which by-products such as whey became actual ingredients.

An ingredient is a word that can be applied in various ways. In case of ingredients from milk their application is not limited to food. As will be shown in this chapter and in Chapter 7 ('End User'), these include applications in feed, pharmaceuticals and personal care products. The following definition will be used in this book:

> An ingredient is a substance of dairy origin, which is used in the manufacture or preparation of a variety of food and non-food finished products.

Ingredient is the key concept. A food ingredient distinguishes itself from a food additive. Milk and milk derivatives are normally considered characteristic ingredients of a food (EU, 2008). Dairy ingredients are safe to use and in general do not require extensive permission procedures. The description is simpler than the statutory definition of a food ingredient (EU, 2011) because of the limitation to dairy origin. Milk powder, which is applied in finished products such as reconstituted liquid milk and yoghurts, is also included in the definition. Besides food ingredients the field of application is much wider as non-food finished products are included too. Manufacture as well as preparation is mentioned, which reflects the possible scale of operation of the end user. Indeed, it may take some years before an application reaches maturity.

In the following sections the focus will be on ingredient list, ingredient and cycle, and ingredient and adding value.

1.1 Ingredient list

Based on various food laws, for instance in the USA and the European Union (EU), food labels have to contain information about ingredients in order to allow customers to make more informed purchasing decisions. In most countries it is a meticulously regulated area. The food label includes a list of ingredients, nutrition information and shelf life, but it should also hold warnings against allergens. In this chapter we will restrict ourselves to the list of ingredients as it fits best in the context of this book. The list of ingredients mostly includes also small percentages of additives, which are applied for purposes such as texture, flavour, colour and shelf life. Additives, being approved substances, are not considered to have nutritive value. This is in contrast to natural products like milk, eggs and honey. In the Middle Ages the whey/serum part of the milk was even believed to have value as a medicine. It was considered beneficial for purification of the human body. Nowadays, a variety of substances with biological activity have been discovered that have more specific properties (Hettinga *et al.*, 2011).

Table 1.1 lists the ingredients of a butter pastry with almond paste, apple and cinnamon. The list includes all the ingredients in descending order of weight. In Europe food additives are in the vast majority not indicated by their names but with an E-number. E-numbers are codes for food additives that have been assessed for use in the EU. Dry milk solids (whey powder and milk protein) and butter fat / oil are

Table 1.1 Ingredient list of butter pastry with apple and cinnamon.

Wheat flour (contains gluten)
Raisins
Almond paste (15%)
  [almonds, sugar, free range eggs, stabilizer (E420), water, thickening agent (E407)]
Water
Dried apples (9%)
Bread improver
  [dextrose, wheat gluten, *whey powder* (contains milk) *butterfat* (contains milk) emulsifiers (vegetable origin: E471, E481, E472e), salt with iodine, rye flour (puffed, contains gluten), pea flour, glucose syrup, *butter oil* (contains milk), *milk protein*, colour agent (E100)]
Baker's yeast
Cinnamon (0.3%)
Powdered sugar
  [dextrose, wheat starch, vegetable fat (palm)]
Vegetable oil (palm)
Preservative (E220)

Source: Courtesy of 'Lekker Vers Bakkerij', Obdam (The Netherlands).

included as ingredients from milk. 'Contains milk' is added for allergy reasons. The milk constituents are incorporated in butter pastry at a modest level. Traditionally they provide flavour, texture, structure and colour to bakery products. Although butterfat is an expensive ingredient it is well-liked by bakers particularly due to its flavour. To comply with European food laws, labels should list the percentage of ingredients that appear in the name under which the foodstuff is sold (almond paste, apple, cinnamon). More details regarding legislation are provided in the *'European Food Law'* (O'Rourke, 2005).

As was shown by the butter pastry example, long lists of ingredients are not unusual nowadays. Such complex recipes sometimes have a long history. Ice cream is given as an example in Figure 1.1. Cold desserts – a mixture of snow and fruit

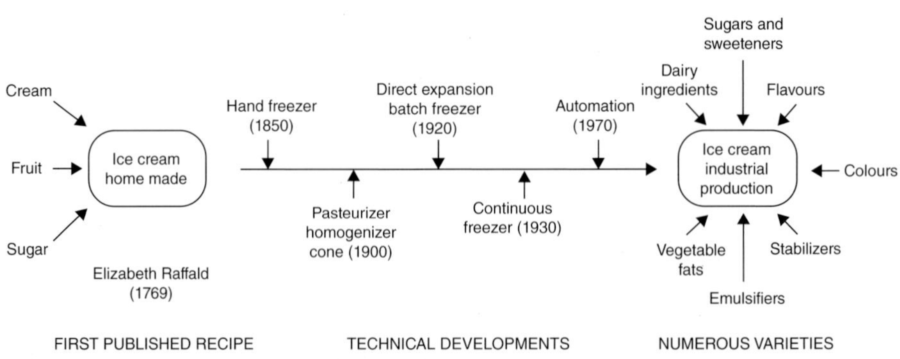

Figure 1.1 Ice cream – from simple to complex: recipes and numerous varieties.

juices – were already known in China thousands of years ago (Blomdahl, 1982). The art of ice cream making was brought by Marco Polo from China to Italy in the 14th century. From there it spread to other parts of the world. However, it was not until 1769 that the first recipe was published (Raffald, 1769). This ice cream was homemade and contained three ingredients: dairy cream, fruit and sugar. Nowadays, most ice creams are made in industrial plants. The USA in particular became the pioneer country for industrial ice cream manufacturing. In the mid-1800s a hand-operated freezer was invented (Johnson, 1843). The ingredients were poured into a container with a stirrer and freezing took place by using a mixture of ice and salt. In subsequent developments the pasteurizer, homogenizer and the ice cream cone were introduced. Around 1920 it became possible to replace the ice-salt mixture by a direct expansion batch freezer. However, the batches varied in quality and the process was expensive and time consuming. A major step forward in technical development was the introduction of a continuous freezer (Marshall & Arbuckle, 1995). It took many efforts, but finally resulted in a simplified production and allowed for the mass production of ice cream. The process consists of continuously feeding a metered amount of an ingredient mix and air into one end of the freezing chamber. The mix is agitated and partially frozen while it passes through the chamber. At the other end of the freezing chamber it is discharged in a continuous stream. Due to the incorporation of air the volume of the mix is, for instance, doubled (overrun of 100%); it is dispensed into packages and then placed in a hardening unit to complete the freezing process. The first plant was opened in Detroit around 1930 and was based on a cluster of patents (Vogt, 1930). We shall see the importance of patents in Chapter 3 ('Patents'). Automation completed the technical development. In the meantime product development was more focused on increasing the popularity of ice cream. Important drivers for improving the recipes were consumer preferences (diversification in flavour, structure and appearance, reduction in calories), cost (various vegetable fats, whey ingredients) and stability against temperature changes during storage (heat shock resistance). This resulted in numerous varieties and complex recipes. Ingredients from milk have an important contribution to the texture and flavour of this complex product, which consists of ice crystals, air bubbles and unfrozen liquid (Udabage & Augustin, 2003). New developments have come forward such as an ice-structuring peptide derived from milk protein. This peptide improves heat shock resistance (Nestec, 2008).

In contrast to the label for butter pastry the list of an infant formula is led by ingredients from milk. In the early months of life, milk is the only food that infants can easily digest, so it is critical for survival. As an example the ingredient list of the infant formula produced by the US company Abbott is included in Table 1.2. The list of this powder is far more complicated than Nestlé's infant formula that went on sale in 1868. The latter contained two main components: sweetened condensed cow's milk and a specially prepared wheat flour. Nowadays the ingredient list is full of technical feats. Of course, the 'gold standard' for infant food is human breast milk. Because cow's milk – the basis of most infant formulas – is quite different in composition, numerous efforts have been made to mimic human milk. The list in

Table 1.2 Ingredient list of an infant formula.

Nonfat milk	**Less than 2% of the following:**	D-alpha-tocopheryl acetate
Lactose	*C. cohnii* oil (source of docosahexaenoic acid, DHA)	Niacinamide
High oleic safflower oil	*M. alpina* oil (source of arachidonic acid, ARA)	Calcium pantothenate
Soy oil	Beta-carotene	L-Carnitine
Coconut oil	Lutein	Vitamin A palmitate
Galacto-oligosaccharides	Lycopene	Cupric sulfate
Whey protein concentrate	Potassium citrate	Thiamine chloride hydrochloride
	Calcium carbonate	Riboflavin
	Ascorbic acid	Pyridoxine hydrochloride
	Soy lecithin	Folic acid
	Potassium chloride	Manganese sulfate
	Magnesium chloride	Phylloquinone
	Ferrous sulfate	Biotin
	Choline bitartrate	Sodium selenate
	Choline chloride	Vitamin D3
	Ascorbyl palmitate	Cyanocobalamin
	Sodium chloride	Calcium phosphate
	Taurine	Potassium phosphate
	m-Inositol	Potassium hydroxide
	Zinc sulfate	Nucleotides (adenosine 5′-monophosphate, cytidine 5′-monophosphate, disodium guanosine 5′-monophosphate, disodium uridine 5′-monophosphate)
	Mixed tocopherols	

Source: Courtesy of Abbott BV.

Table 1.2 reflects these attempts. The main ingredients of the formula, except the fat source, are ingredients from milk such as nonfat dry milk, whey protein, lactose and the lactose derived oligosaccharides. They mirror the increase in knowledge regarding ‘humanized’ infant food as well as the progress made in technology. The use of oligosaccharides is a hot topic for many food scientists. They are a source of nutrients for a beneficial microbial gut flora. This supports the health of the infant and goes a step further than fulfilling nutritional needs only. In some formulas even specific bacteria are added in a dry form (Thompkinson & Suman Kharb, 2007). In the list of Table 1.2, some 50 ingredients are mentioned by name, all of which have specific functions. Abbreviations such as E-numbers are not in use in the USA. The minor compounds are given in the category ‘Less than 2% of the following’, which makes reading more easy.

The milk ingredients in Tables 1.1 and 1.2 can be replaced by other ingredients, for reasons of cost for instance. Figure 1.2 shows the alternatives. Milk chocolate, which is the most popular type of chocolate, will be given as an example. Its history goes back to 1875, and its inventor was Daniel Peter, a Swiss like Henri Nestlé. Until then only a bitter-tasting solid chocolate was consumed. Daniel Peter was well aware of the work of Henri Nestlé, who used sweetened condensed milk for the production

Figure 1.2 Various dairy ingredients applied in milk chocolate. AMF, anhydrous milk fat; SMP, skim milk powder; WMP, whole milk powder.

of infant food. The sweetened condensed milk was also the key for the development of milk chocolate with a good shelf life (Anonymous, 1975). In combination with cocoa substances and sugar a good-tasting new product was introduced. Since then knowledge of the properties of chocolate and its manufacture has increased tremendously (Afoakwa *et al.*, 2007). Also, information about the application of various ingredients from milk has grown considerably (Haylock & Dodds, 2009). In Figure 1.2, 12 choices are shown. Sweetened condensed milk/evaporated milk, being a liquid, contains some water, which has to be removed during chocolate processing. An alternative to sweetened condensed milk/evaporated milk in a dried form is milk crumb. This combination of milk solids and sugar also gives a caramel flavour to the milk chocolate. The remaining ingredients are based on the main milk constituents. Three clusters can be distinguished. They are sources of:

- milk fat – AMF (anhydrous milk fat), fat fractions, high-fat powder, WMP (whole milk powder);
- milk protein – SMP (skim milk powder), buttermilk powder and yoghurt powder;
- milk sugar (lactose, demineralized whey powder and whey powder).

Within the food rules and regulations milk chocolate manufacturers may consider the following factors in selecting ingredients: cost savings, product texture, flavour and process flexibility. Milk fat has a great influence on the quality. It provides a favourable flavour to the chocolate. The texture is influenced by the amount of free fat and its melting point. Milk proteins add to the perceived creaminess and nutritional value. Milk sugar can be used to replace part of the sugar and results

Table 1.3 Ingredient list of a toothpaste.

Water	Sodium benzoate
Hydrated silica	Sodium saccharin
Sorbitol	Potassium thiocyanate
Glycerin	Zinc gluconate
Surfactant (Steareth-30)	*Colostrum*
Carrageen (Chondrus Crispus Extract)	Lysozyme
Aroma	*Lactoferrin*
Titanium dioxide	*Lactoperoxidase*
Disodium phosphate	Amyloglucosidase
Sodium fluoride	Glucose oxidase
Citric acid	

Source: Courtesy of Unilever Benelux.

in a reduction of sweetness. From a cost view the latter is a less expensive dairy ingredient. A combination of AMF and SMP can sometimes be found on the same ingredient list. Such a combination gives more flexibility in formulating recipes than WMP alone.

A non-food application of ingredients from milk is rendered in Table 1.3. It is a personal care application with toothpaste as an example. In contrast to the ingredients of infant food, here the ingredients of milk are found at the end of the ingredient list. These ingredients are shown in italics. Thus far only minor amounts are used. Together with potassium thiocyanate and lysozyme they strengthen the antimicrobial system of the saliva. Colostrum is the first of the dairy ingredients listed. Colostrum is the milk obtained from the cow in the first few days post partum. It is collected separately and is not allowed to be mixed with the bulk milk at the farm. Colostrum is, for instance, rich in antimicrobial components such as lactoferrin, lysozyme, lactoperoxidase and immunoglobulins (Tripathi & Vashishtha, 2006). Besides colostrum per se, lactoferrin and the enzyme lactoperoxidase are also incorporated in the toothpaste. In particular, lactoperoxidase plays an interesting role. It stimulates the production of hypothiocyanite starting with hydrogen peroxide and potassium thiocyanate. Hypothiocyanite counteracts bacterial growth. The inhibition takes place according to the following chemical reaction:

$$H_2O_2 + \underset{\text{Thiocyanate}}{SCN^-} \xrightarrow{\text{Lactoperoxidase}} \underset{\text{Hypothiocyanite}}{OSCN^-} + H_2O$$

With the help of two other enzymes, amyloglucosidase and glucose oxidase, hydrogen peroxide is produced from dietary fermentable carbohydrates. Both enzymes are mentioned at the end of the ingredient list. They boost the natural lactoperoxidase system and reduce gingival inflammation (Midda & Cooksey, 1986). The toothpaste list is an example of a complex ingredient list, which is common these days. It is the result of growing demands of consumers and the greater knowledge of the various ingredients.

Table 1.4 Ingredient list of imitation mozzarella cheese.

Water
Partially hydrogenated soybean oil
Rennet casein
Modified food starch
Non-fat dry milk
Salt
Sodium aluminum phosphate
Natural flavour
Lactic acid
Sodium citrate
Sorbic acid (as a preservative)
Artificial colour

Source: Courtesy of Whitehall Specialities.

The most versatile ingredients from milk are derived from cheese. These cheese ingredients are, for instance, used in convenience foods. Convenience foods are prepared on a large scale and designed for ease of consumption. They were developed after World War II. Of the ingredients from milk the cheese ingredients are a complex group, and are able to provide a diversity of savoury flavours and textures. Their main constituents are fat and protein as well as substances developed during cheese ripening. Sources of cheese ingredients are natural cheese, processed cheese, imitation cheese and cheese powders. Mozzarella, Cheddar, Gouda, Emmentaler and feta cheeses are often used as a source of natural cheese. They are available in size reduced forms such as crumbled, diced, shredded and sliced. Processed cheese contains a variety of cheeses. It is made heat stable (i.e. fat and water do not separate during heating) through the addition of emulsifiers, and has a smoother texture and longer shelf life than natural cheese. Both natural and processed cheeses are consumed by themselves, but their use as ingredients is also substantial (Guinee & Kilcawley, 2004).

Imitation cheeses, or analogues, which are particularly popular in the USA, are a less expensive version of processed cheese. The milk fat is replaced by vegetable fat, and even part of the milk protein is sometimes substituted by vegetable protein. In many cases the amount of dairy ingredients is low and therefore its flavour contribution is insufficient compared to real cheese (Bachmann, 2001). Table 1.4 gives an example of the ingredient list of an imitation mozzarella cheese. This imitation cheese is, for instance, available in sizes of about 10 kg or greater. It is an intermediate product and its ingredient list can be found on the package in the warehouses of processors. The main application is for pizzas, but it can be applied in frozen entrées and salads too. The product contains a fair amount of rennet casein (see the information sheet 'Unmatured cheese non-fat') and it belongs therefore to the upper segment of the imitation cheeses. Rennet casein is also present in natural cheese, where it is called paracasein. Rennet casein or paracasein is important for

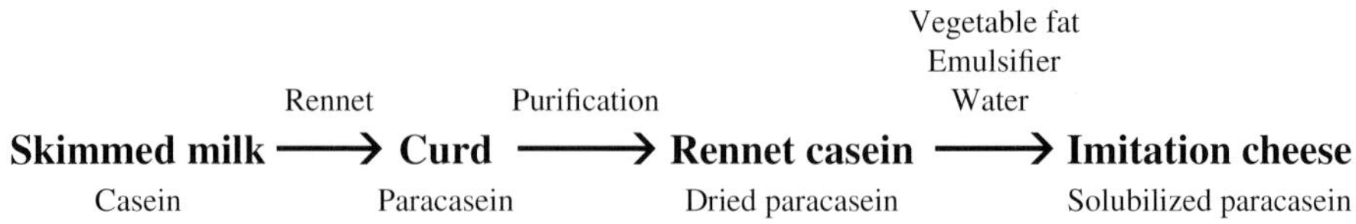

Figure 1.3 Processes involved in the production of imitation cheese. *Source:* Nägeli, 1941.

the good stretch and melt properties that are characteristic of mozzarella cheese (Jana & Upadhyay, 2003). A survey of the processes is set out in Figure 1.3.

Besides rennet casein, the process used is also crucial. In a twin-screw cooker the kneading and stretching of the conventional mozzarella cheese-making process is mimicked. Emulsifiers – here sodium aluminium phosphate and sodium citrate – are applied to solubilize the paracasein (Lucey *et al.*, 2011). In this way a stable emulsion is obtained and oiling-off can be avoided. Modified starch is used to prevent the separation of whey during storage. In the ingredient list non-fat dry milk is also mentioned; it contributes to the desired colour and flavour.

As opposed to the aforementioned cheeses, which contain a substantial amount of water, cheese powders are delivered in a dried form. These are a concentrated source of flavour and are, for instance, used as a coating on snacks (Johnson, 2000).

To summarize, cheese ingredients provide nutrition value, texture and flavour and are available in several forms and qualities.

1.2 Ingredient and cycle

Decisions in dairy companies are based on which markets add the highest value to the milk. For this reason, dairy consumer products generally have the highest priority. As the composition and the quantity of these finished products usually do not match those of raw milk the remaining by-products have to be processed. One of the key problems is how to create a situation in which effluent is avoided. Ingredients play a crucial role in solving this problem. Before an ingredient is even introduced into the market many studies have to be carried out. Nowadays, extended evaluations take place based on technological possibilities, market expectations, environmental issues and detailed cost calculations. Upgrading adds to this complexity. By-products have to be processed even if they have a negative value, such as some mineral fractions. In this section we will look at progress in upgrading the production cycle. Past experiences will be used to place current developments into perspective.

The by-product of cheese manufacture is whey, which contains about half of the solids from the cheese milk. In the ideal situation cheese and whey are all used at the same location. The use of whey as pig feed on farms has a long history. In the 18th century some of the whey was upgraded, because of the belief that consumption of whey had a beneficial and wholesome effect on people's health (Zollikofer, 1974).

Figure 1.4 Whey as health drink Switzerland 18th century (Zollikofer 1974, WUR Library-special collections).

An impression of a whey cure is shown in Figure 1.4, in which regular drinking of fresh whey is emphasized.

Table 1.5 gives an ingredient list of a modern whey-based soft drink. The milk serum forms 26% of the ingredients. Milk serum is refined whey produced through fermentation and extraction of fat and protein (Wagner *et al.*, 1975). Sparkling water, sweeteners and flavourings improve the product's acceptability for modern consumers. The preservatives enhance the shelf life of the drink, which is necessary in today's supply chain from producer to consumer, and reflects the demands of the market place. As opposed to the whey cure, where the whey itself was used, only a certain part of the whey is used in modern whey-based drinks. Furthermore, relatively small amounts are utilized in the soft drink industry. The remainder has to

Table 1.5 Ingredient list of a whey-based soft drink.

Sparkling water
Milk serum (26%)
Food acid (L + lactic acid)
Sweeteners (cyclamic, acesulfame K, saccharin)
Natural flavorings
Preservatives (potassium sorbate, sodium benzoate)

Source: Courtesy of Rivella and Vrumona B. V.

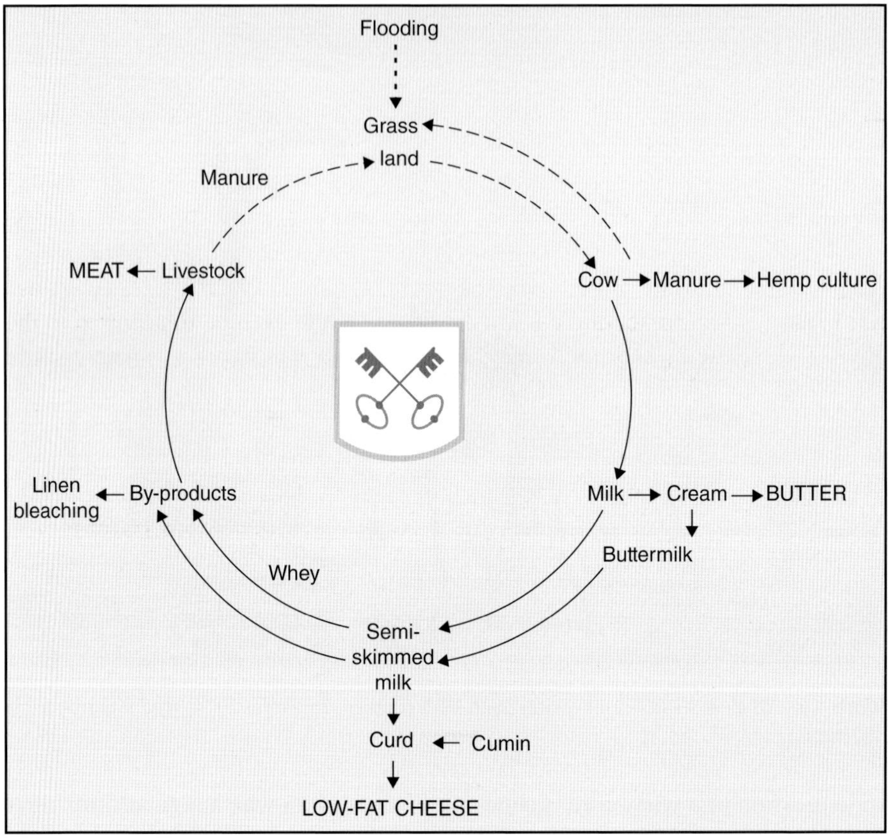

Figure 1.5 An example of a small cycle, depicting a farm area in the Middle Ages centred on the city of Leiden in The Netherlands. (*For colour details please see colour plate section.*)

be used elsewhere. It shows that much creativity is needed to fulfil both the demands of the market and the environment.

Developments in the dairy industry can also be described by distinguishing between the small and the big cycle. The small cycle can be found at farm level, where the production of milk and the processing into finished products takes place at the same location. In the big cycle milk production is separated from the processing part. Finished products and ingredients are manufactured in large-scale plants whose capacity can be several thousand times bigger than those on a farm.

Figure 1.5 is a diagram of a small cycle based on a situation dating back to the Middle Ages. In the centre is the coat of arms of the city of Leiden, which is located in the western Netherlands. In the past it was an important centre for trading dairy products. Around 1650 the average farm in that area had about 17 dairy cows and the milk production per cow per year was estimated at up to 2000 kg (Bieleman, 2010). Even in those days cleanliness was considered a vital aspect of processing milk (van Bavel & Gelderblom, 2009). The annual production of milk from such a farm could be processed within one hour in a modern plant. In those days the area around Leiden was noted for the production of butter and low-fat cheese. Cumin was added to the cheese, which resulted in a special flavour. Both products were made according to traditional methods; for instance, rennet – a coagulant essential for cheese making – was prepared on the farm itself. It was extracted from the abomasum of calves by means of brine or sour whey (van der Burg & Hartmans, 1952). One of the by-products of butter manufacture, sour buttermilk, had been used for cheese manufacture. It improved the action of the rennet and reduced cheese defects. Due to its firm texture the low-fat cheese could even be transported to the tropics. The remaining buttermilk and the whey released during cheese manufacture were by-products. They were used for feeding livestock and converted into meat. Another by-product, the manure produced by the cows and other livestock, was used partly in heavily manured gardens for hemp culture. Hemp was used for the production of basic fibres to make a wide range of products ranging from canvas and fishing nets to all kinds of ropes. The use of manure as a fertilizer of grassland was less known in those days (dashed lines in Figure 1.5). Flooding, which was common in that area during autumn until early spring, possibly had some fertilizing effect.

The remaining buttermilk and whey were probably among the first 'raw materials for industry'. Without the use of modern equipment the production of butter and cheese took several days. During this time the by-products had been spontaneously transformed into acid liquids. It is due to the action of lactic acid bacteria present on the surface of inadequately cleaned utensils. These bacteria convert part of the lactose in the milk into lactic acid. The acid by-products were applied in the bleaching process of spun and woven linen and cotton (Regtdoorzee Greup-Roldanus, 1936). This step in the bleaching process was called the 'lactic acid bath'. The bleaching procedure had several steps. In the first step the clothing was softened, defatted by a special caustic solution and rinsed. Then the garments were left out in the sun to destroy the natural colours. Subsequently the clothing was brought into a 'lactic acid bath', which part of the process was also called 'milking'. Here the calcium residues of the caustic treatment were made soluble and removed. It improved the colour of the clothing. The 'lactic acid bath' resulted also in a greater firmness of the garments. Sometimes this treatment was repeated. In a fourth step dressing took place and the bleaching process was finished. The 'lactic acid bath' was only used once and afterwards it was discharged into ditches together with the other effluents of the bleaching process. The ditches were called 'stinkers', so closing the cycle was certainly not the case. In spite of the fact that people were aware of the importance of cleanliness for the production of dairy products, knowledge about hygiene was very rudimentary. With our current understanding of hygiene, the use of cooling to low temperatures and the application of pasteurization mean that situations risking contamination can be reduced. Consequently, the period of a constant quality of the raw material is extended from hours to days, and the possibilities are expanded.

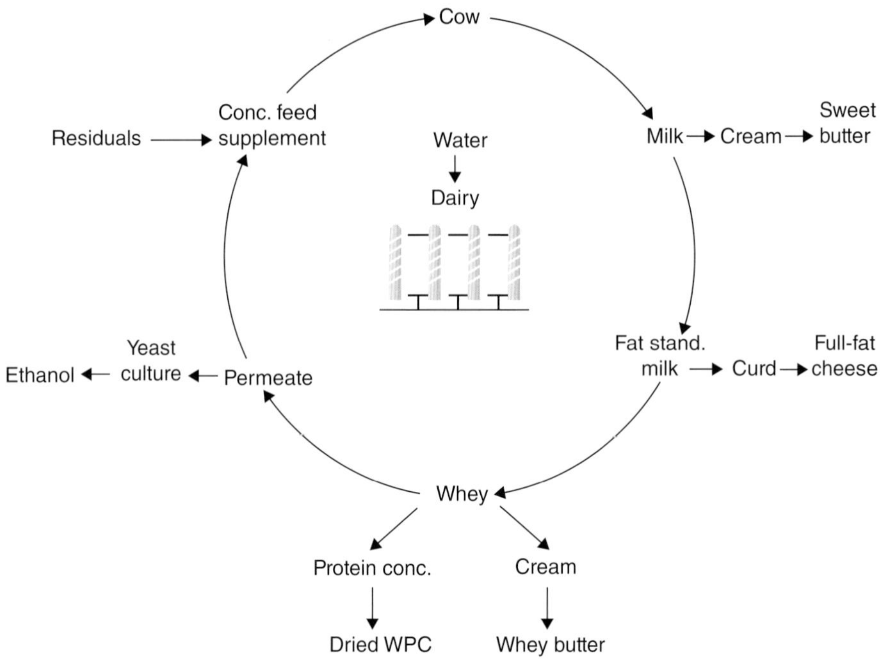

Figure 1.6 An example of a big cycle (industrial – environmental driven). WPC, whey protein concentrate.

An example of a big cycle is shown in Figure 1.6; it is a next step in closing the cycle. Plants based on this approach are known from Ireland, New Zealand and USA. A detailed description of the Corona plant in California was used to create the diagram of Figure 1.6 (Bush, 1991). That plant was in operation for about 20 years and was shut down recently. It could handle about 1.2 million kg of milk per day. In the centre of the diagram distillation towers are shown, which are quite unusual for a dairy plant. The most typical part of the plant is involved in the processing of milk into sweet butter and full-fat Cheddar cheese. In contrast to the process shown in Figure 1.5, cheese is made by means of commercially available starters and rennet. The fat of the released whey is separated and used in the production of another finished product called 'whey butter'. Only a modest quantity is produced as about one-tenth of the fat of the starting milk goes into the whey. The proteins are the most attractive constituents of the clarified whey. They can be separated by membrane processes to a large extent and dried to form an ingredient called whey protein concentrate (WPC).

The unusual part of the plant was involved in processing the remaining permeate. The driving force behind it was an environmental issue. The permeate contains mainly lactose and minerals. Its pollution value can be calculated (Scheltinga, 1972) to be about three-tenths that of full-cream milk, while its economic value is mostly a fraction of it. As direct waste water treatment of such enormous quantities was impossible, it was a challenging situation. The solution applied was to convert the lactose into ethanol according to the following reaction:

$$\underset{\text{Lactose}}{C_{12}H_{22}O_{11}} + H_2O \xrightarrow{\text{Yeast}} \underset{\text{Ethanol}}{4\,C_2H_5OH} + 4\,CO_2$$

In the batch process a special yeast strain was used for fermentation. The yeast is able to convert lactose into ethanol under exclusion of air/oxygen. However, the yield is modest: 0.46 g of ethanol per gram of lactose under industrial conditions (Mawson, 1987, 1990). Much of the ethanol production of the Corona plant was used as a fuel additive. Blending of ethanol with gasoline is advantageous as it significantly reduces vehicle emissions.

Another interesting feature is the way the Corona plant closes the cycle. All the remaining by-products – buttermilk from whey butter production, salty whey from Cheddar cheese manufacture, excess yeast and distillation residues from ethanol processing – are blended. After evaporation it is used as an animal feed supplement and returned to the surrounding farms. Here it is fed to the cows. The supplement is a miscellany of milk minerals/salt, nitrogen substances from whey/yeast and carbohydrates from yeast/lactose. Its composition is likely dominated by salt constituents. Also the water, which is released during evaporation, is used as process water in the plant. Overall there is quite a difference compared with the cycle in Figure 1.5. The destination of the remaining milk constituents is not the ditch but the stomach of the cow. The main reason that the Corona plant ceased to exist was its dependency on subsidies by the state of California.

Upgrading of the converted lactose to potable ethanol is an alternative. In New Zealand, for instance, its use is mentioned for the production of gin, vodka, cocktails, liqueurs and wine. However, for the production of ethanol, other raw materials are also available. Sugarcane, sugar beet, maize, potato waste and spoiled kiwi fruit possess potential for use as a feedstock in New Zealand (Acharya & Young, 2008). These less expensive raw materials counteract any widespread manufacture of ethanol from lactose in the dairy industry.

Figure 1.7 shows another example of a big cycle. It is characterized by market-driven activities, thus it can be found in many industrial settings all over the world. Sometimes the activities are divided over three plants with specialized units for butter manufacture, cheese production and whey processing. The different activities can also be found in one plant. For instance, this is the case in the USA, where distances play an important role in the decision-making process of production facilities. The cycle in Figure 1.7 is assembled in such a way that all activities are located in one plant. The plant produces finished products such as sour butter and fat-reduced cheese and the ingredients lactose, delactosed permeate and WPC. Besides WPC, fractions of whey proteins are also indicated. Detailed information can be found in the information sheets at the end of the book. In the centre of the diagram a spray drier is shown, which is a common piece of equipment for the production of

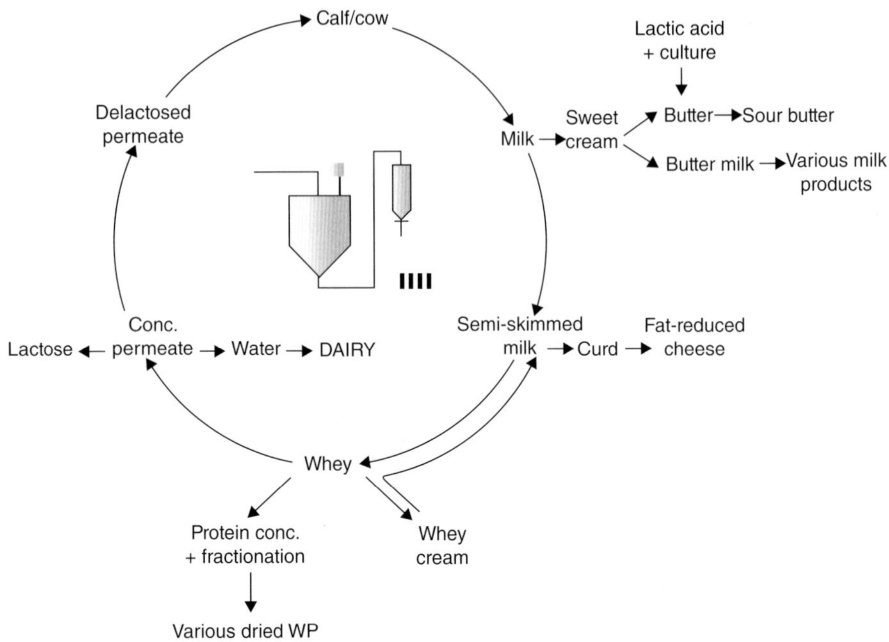

Figure 1.7 An example of a big cycle (industrial – market-driven). WP, whey protein.

ingredients. The main difference from Figure 1.6 is that the ethanol production has been replaced by the lactose manufacture. This was made possible by the tremendous growth of the lactose market in recent decades. In particular the applications in the food sector have increased substantially (Business Consulting 3A, 2008). This also has implications for the balance in the entire chain. The product remaining after crystallization of lactose from permeate is called delactosed permeate. It can, for instance, be used in the animal feed sector in liquid as well as dried form. A variety of whey ingredients can be used in feed for young animals, as indicated in a recent study (EWPA, 2007). In Figure 1.7 only the calf feeding application is indicated. The cattle feed supplement of Figure 1.6 is replaced by a better defined ingredient and shows that closing the cycle can be combined with further upgrading.

Other by-products of the big cycle in Figure 1.7 are buttermilk, whey cream and water. The first two by-products will be discussed in more detail in Chapter 4 ('Reuse'). Sour buttermilk is the by-product of lactic butter manufacture. This butter is particularly preferred in European countries. The by-product sour buttermilk is difficult to process and its applications are very limited. For these reasons the butter making process has been revised to yield sweet buttermilk. The use of sweet buttermilk within the dairy industry has been developed. The whey cream can be converted into whey butter as shown in Figure 1.6. However, it is a niche market and the keeping quality of the product is restricted. Whey cream is therefore typically reused in the cheese-making process itself. Water is an important by-product of dairy ingredient manufacture, as dried ingredients are by far the most commonly sold. Milk contains typically 87.1% water (Walstra *et al.*, 2006). The water is released during processing of milk or its derivatives and represents the by-product with the biggest volume. Most of it results from concentration by evaporators. The discharged vapour is, however, not pure water but may contain small amounts of volatile organic compounds (Verheyen *et al.*, 2009). This restricts its reuse in dairy operations. In Chapter 6 ('Vital Membrane Processes') we will discuss this topic in more detail.

1.3 Ingredient and adding value

From the information given earlier in this chapter it can be concluded that hundreds of ingredients from milk are available in the market. A driving force behind this diversification is the search for adding value. This means that an extra feature is added to an ingredient that goes beyond the standard version. The main objective of the ingredient manufacturer is to achieve a competitive advantage. The benefit for the customer is a more tailor-made ingredient for a specific application.

Table 1.6 gives an example for whey in which six characteristics with scope for diversification are given on the left-hand side. The first is the raw material itself: one can distinguish sweet and acid whey. Sweet whey is released during the manufacture of (semi)hard cheeses such as Gouda and Emmentaler. Acid whey is discharged, for instance, during the production of fresh cheese or quark. Both types of whey differ in composition as well as flavour, which has consequences for their use in end-products. Another aspect is the colour of the whey. During the manufacture of Cheddar cheese, for instance, colour agents are added to give this particular cheese the desired 'orange' appearance. Unfortunately not only the cheese is coloured but also the whey, giving the dried ingredient a light-brown colour. In the case of mozzarella cheese no colour agents are applied, which makes this type of whey more suitable for end-products with a white appearance. With regard to the characteristic 'Composition' in Table 1.6, the protein content of the starting material is increased. The WPC 34 version (protein expressed as percentage of dry matter) can be more easily produced than the WPC 80 counterpart. The latter is a more purified

Table 1.6 Diversification of ingredients – whey as an example (64 potential variants).

Characteristic	Variants
Raw material	Acid/sweet
Colour	White/light brown
Composition	High/medium protein
Functionality	Soluble/gel forming
Convenience	Instant/standard
Certificate	Kosher/halal

ingredient and will therefore be more expensive. The fourth feature, 'Functionality', is a modification of the raw material obtained, for instance, by applying a controlled heat treatment. In this way an ingredient can be obtained that more easily forms a gel in the final application. In contrast, by minimizing the heat treatment during the entire process an ingredient with a high solubility is the result. The latter will be better suited for applications in beverages. The fifth variant, 'Convenience', is a further extension of the possibilities. Here the ease of use, in which an ingredient can be applied in the process of the customer, is taken into consideration. For example, the 'instant' properties of a powder can be improved by agglomeration of the powder particles and/or coating them with an emulsifier. The sixth aspect, 'Certification', is related to the religion of certain groups of consumers. The Jewish population – in Israel but also, for instance, in the USA – require 'kosher' certified ingredients, whereas those with Islamic religious convictions need 'halal' approved ingredients. The most important markets for halal ingredients are located in the Far East and Middle East. Ingredients used for these purposes have to meet specific requirements. Manufacturers can only be certified if they fulfil certain demands regarding the use of excipients and the way the process is designed and controlled.

By considering all the possible combinations of the six characteristics, the number of different ingredients in this example could add up to 64. However, the potential of whey is greater still if we focus on its main constituents. In this context one can mention the various whey protein fractions, lactose in its different forms and the potential of milk salts. The versatility of ingredients from milk is further emphasized if one adds as raw materials milk itself, cream and cheese. This potential of value added ingredients will be utilized to its full extent as soon as customers show interest and the technology is made available.

Figure 1.8 again uses whey as an example, depicting the step-by-step improvement of three groups of constituents: proteins, lactose and minerals. As they are present in the same raw material the development of all three groups together is followed. The proteins, the most valuable constituents, are shown on the right-hand side of the diagram. Lactose, which is by far the most abundant element, is presented on the left-hand side. In the middle of the diagram the mineral-rich residue is depicted – this is generally the less valuable fraction of the whey. The cycle moves in an upward direction, showing the addition of value at each stage. Disposal of raw materials is avoided as much as possible, indicated by the closed curve. As the curve reaches a higher level this aspect becomes more important. In the curve you can also find examples of 'recycling', such as the reduction of use of potable water by application of so-called process water. From the point of view of the animal feed sector 'downgrading' takes place. Whey protein is increasingly being used in the more profitable food sector and less in whey powder. However, the overall picture is one of 'upgrading'. There is continuous development of new ingredients based on protein and lactose to supply, for example, various food and pharmaceutical markets. The mineral-rich residue, still containing some lactose and nitrogenous substances, is mostly made suitable for animal feed markets.

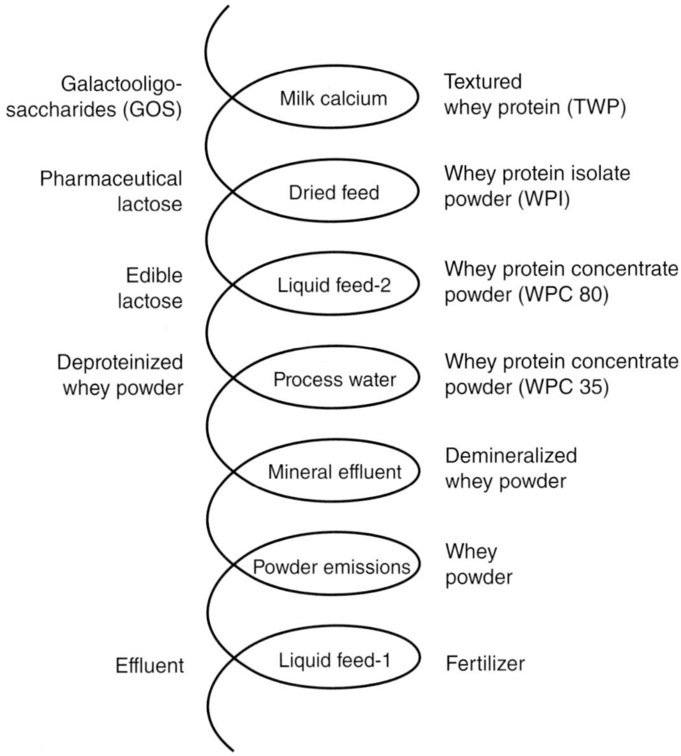

Figure 1.8 Upgrading the cycle for whey over the course of time. Lactose-based ingredients are shown on the left-hand side; protein-based ingredients on the right-hand side; minerals and other residues are in the middle of the scheme. TWP, textured whey protein; WPC, whey protein concentrate; WPI, whey protein isolate.

At the bottom of the curve of Figure 1.8 a situation is shown in which whey was considered less valuable. In the case of effluent whey was even considered a nuisance. It has a high pollution value, due mainly to its high lactose content, which caused problems in sewer systems. For instance, 100 kg of liquid whey has the pollution strength equivalent to sewage produced by 45 people (Webb & Whittier, 1970). The use of whey as a liquid feed-1[1] for livestock was an alternative as long as the dairy plants were small in size. Then the whey could be returned from the plant to the surrounding farms. But because of spontaneous fermentation the whey turned sour and also contained organic acids. After World War II this approach became less common in most countries. Another possibility was its application as a fertilizer. It was, for instance, suitable for maize (corn) crops, which demand nitrogen, potassium and phosphorus for growth. The nutrients were supplied by field spreading of whey (Peterson *et al.*, 1979). Due to stringent regulations, for instance

[1] Feed-1 is whey and feed-2 a by-product of processing whey into lactose and WPC.

to prohibit spreading close to streams, this application was restricted more and more in the course of time.

The drying of whey into powder was an important step in whey valorization. Its development in France started around 1950 (Vrignaud, 1983). The growth in the size of French cheese-making plants (average daily capacity was only around 10,000 L in 1937–38) was a key factor. A technical hurdle was the hygroscopic nature of the whey powder, which caused caking during storage. Not only the process but also the market had to be developed. A start was made with the successful use of whey powder in calf feeding. A supplementary value added step was demineralization, which made the whey suitable even for use in infant foods. Both whey powder and demineralized whey powder are still important outlets today. However, increasingly the applications developed in the three directions based on protein, lactose and minerals, as shown in Figure 1.8. They are the result of trends in the ingredient market and the availability of technologies such as membrane processes, crystallization of lactose and new drying procedures. Lactose is usually found in all dairy ingredients. In its pure form it occurs as edible and pharmaceutical lactose. The 'mother liquor' after crystallization of lactose from whey still contains lactose in solution, but also minerals and nitrogen substances. These mixtures of compounds are mostly used for animal feed (liquid feed-2 or dried feed). On the right-hand side of the curve in Figure 1.8 the ingredients are dominated by protein, but still contain some minerals and lactose. Advanced processing results in ingredients with a high protein content such as WPC 80 and whey protein isolate (WPI). Further processing of lactose, minerals and protein results respectively in galacto-oligosaccharides, milk calcium and textured whey protein (TWP). Some of these topics will be dealt with in the following chapters. At the top of the curve the processing becomes more complicated; for instance, fractions of protein are produced and the number of process streams increases. The addition of value is greatest in this part of the curve. The market price of a WPC 80 powder and the whey protein fraction lactoferrin might differ by a factor of 100. However, note that the higher product price is not only due to the addition of value but is also a reflection of the higher production costs.

The development of an ingredient might take a considerable amount of time. In the case of lactose (milk sugar) the time interval between discovery and industrial production was approaching 300 years (Funck, 1948; Visser *et al.*, 1988). Table 1.7

Table 1.7 History of milk sugar (lactose).

Year	Event
1633	Bertolettus described a sweetish solid substance of whey
1694	Testi introduced the name milk sugar
1900	Start of industrial milk sugar manufacture
1950	Yield improvement by the separation of denatured whey protein
1960	Introduction of continuous milk sugar separation
1980	Start of milk sugar manufacture by separation of soluble whey protein

Table 1.8 Excipients of a sleeping pill.

Lactose
Calcium hydrogen phosphate
Starch
Carboxymethylcellulose (carmellose) sodium
Magnesium stearate
Titanium dioxide
Hydroxylpropyl methylcellulose (hypromellose)

Source: Sandoz, 2011.

summarizes the history of this ingredient. In 1633 the Italian doctor Bartolettus informed the scientific world of the presence of 'salt of milk serum'. It was given the Latin names *manna* or *sal seri esentiale*. Bartolettus ascribed a laxative effect to this component of the milk, which was in line with the practice of farmers and shepherds at that time. In 1688 Etmüller purified the crude product obtained from whey by recrystallization. Shortly afterwards in 1694 the name *Saccharum lactis*, or milk sugar, was introduced by the Venetian doctor Testi. The 'secret agent' was dried in the sun and available as a powder. The product was probably a mixture of milk sugar and milk. It was used as a sort of panacea against various diseases, including rheumatism. Thus the application of lactose for medical purposes has a long history. Nowadays, it is replaced by more specific medicines. As far as composition is concerned it was not comparable to the present 'pharmaceutical' lactose. This specific lactose is the purest form of lactose and has the highest added value. It is widely used as a filler in tablets.

Table 1.8 lists the excipients of a sleeping pill. The drug itself is only present in a small amount. The bulk is formed by the excipients (carriers), which are the non-active compounds of the tablet. They act for instance as fillers, lubricants and whiteners. In the example chosen, lactose is the most abundant excipient on the list.

It was not until about 1900 that the industrial production of lactose was introduced. One of the oldest manufacturers was the N.V. Hollandsche Melksuikerfabriek (HMS) in Uitgeest, The Netherlands. It was founded by three chemists and processed the whey of the surrounding cheese plants. Figure 1.9 shows an artistic interpretation in stained glass of the situation in 1950. The picture illustrates a closed cycle in which the various compounds of the milk were utilized in an integrated approach. The compounds were processed into cheese, animal feed and, last but not least, lactose. Just above the letters HMS is the typical tomahawk shape of a lactose crystal. The staff of Asclepius illustrates the connection with the medical world, and emphasizes the application as pharmaceutical lactose – one of the first examples of upgrading the cycle. The top of the figure portrays the world as the market for this type of lactose.

Large-scale production of lactose started mainly after World War II. This was favoured by a new application, namely the use of 'crude lactose' in the culture

Figure 1.9 Artistic impression of the production and application of lactose dated 1950. Reproduced courtesy of Hollandsche Melksuikerfabriek-HMS. (*For colour details please see colour plate section.*)

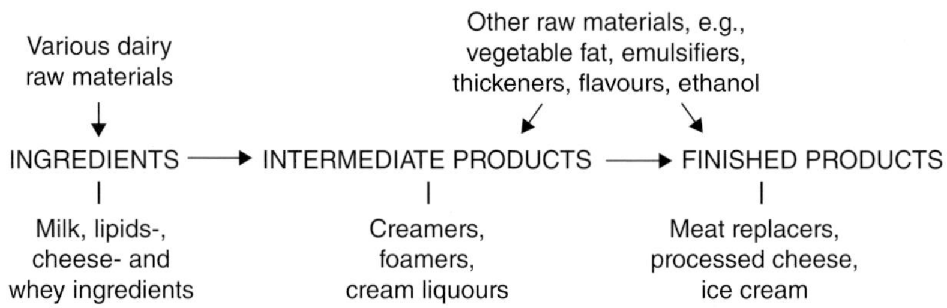

Figure 1.10 The value chain in food products: dairy as an example.

medium for penicillin-producing microorganisms. In the last few decades the development of the 'edible' lactose market has fuelled a tremendous expansion of lactose production and its use in all kinds of food products. Global lactose production was about 850,000 MT per year in 2006 (Business Consulting 3A, 2008). The increase in demand influenced the manufacturing process. As indicated in Table 1.7, improvements in lactose yield were achieved by separation of the denatured, insoluble whey protein ('lactalbumin'). This treatment reduced the viscosity of the lactose-containing liquid to some extent and therefore a higher degree of concentration could be achieved. As a consequence the yield of the lactose was enhanced. Further steps forward in the large scale manufacture of lactose were taken by making part of the production a continuous process, and using the decanting centrifuge for separating the crystals. The quality of the final product was improved by removing a high proportion of the residuals that remained after crystallization. Around 1980 the separation of WPCs through membrane processes commenced on an industrial scale. The resultant whey protein was soluble/undenatured. Compared with the inert 'lactalbumin' this marked a tremendous expansion in the potential applications of whey protein. Nowadays, separation of soluble whey protein is a usual step in making lactose from whey.

Figure 1.10 shows a value chain from the dairy industry as an example. It starts with raw materials ('ingredients') and via 'intermediate products' it continues to the product delivered to the consumer ('finished products'). The target is to add value at each stage of the chain. The various specific ingredients from milk will be discussed in the following chapters. Milk, lipids, cheese and whey-based ingredients will be distinguished (see information sheets).

'Intermediate products' are partially finished products, a field in which the dairy industry is quite active. Here the use of raw materials is not limited to dairy ingredients. Various examples of these intermediate products are given here. One of them is a cream liquor base. In this case the intermediate product is an alcohol-containing, stable, cream liquid that is shipped in bulk (containers) to the end-use stage. Here the product is presented in an attractive bottle and provided with a brand. Another example of this is a creamer or coffee whitener, where the less expensive vegetable fat replaces milk fat. The 'dairy' producer processes the various raw materials, converts it into a powder and packs it in bulk (big bags). Afterwards it is delivered it to the end user, where it is packaged in sachets suitable for the consumer market. Whipping powder is another example of an intermediate product, which can be used as a decoration cream for bakery products. This intermediate product, which includes for instance vegetable fat and emulsifiers, may replace the regular dairy cream. Advantages such as a longer shelf life, a higher overrun (product volume after whipping) and stability in an acid environment are claimed. It is for the end user to assess if these advantages outweigh the loss of the characteristic flavour of the fresh cream. The introduction of new technologies continues also in the field of 'intermediate products'. An example is the use of double emulsions (water in oil-in-water, or W/O/W) suitable for the encapsulation of active components in powders (Friesland Brands, 2008). A W/O/W emulsion consists of droplets of oil in a water-soluble matrix, in which the oil droplets are filled with small water droplets. The active substances can be incorporated in the inner water droplets. The goal is, for instance, to mask off flavours (bitter peptides), prevent chemical reactions (minerals) or to protect probiotics (viable bifidobacteria).

Probably the most widespread uses of intermediate products are as blends in a powder form. Costs, convenience and functionality play important roles. The blends may contain a variety of raw materials and are formulated for specific applications. These customized blends ease the task of the end user, for instance in cases of a long ingredient list. Successful blends can be found in a variety of applications such as ice cream/frozen desserts, cultured products, processed/imitation cheeses and nutritional supplements. The reduction or partial replacement of expensive milk protein by cheaper alternatives such as whey protein is a significant factor. Examples of such replacement can be found in frozen desserts and cultured products. In cheese products part of the milk protein may be replaced by vegetable protein or stabilizers. The end user decides which quality level is acceptable for the envisaged price level. An expanding area is sport nutrition, with the application of specific protein blends (Paul, 2009). Blends of whey protein, casein(ate) and soy protein isolate appear in commercial products. Besides a diversity in amino acid profile these proteins have different digestion rates. Whey protein is considered as 'fast', casein(ate) 'slow' and soy protein isolate 'intermediate'. The blend is claimed to prolong amino acid delivery to the tissue and produces superior gains in muscle mass compared with a single protein source.

The dairy industry is not only involved in ingredients and intermediate products but also in the more 'assembled end products'. Examples are ice cream and processed cheese. In these cases the whole value chain can be managed. Probably due to the large investments needed for a global market the dairy industry has gradually moved away from these type of activities and left them to more specialized companies. In Chapter 7, concerning end users, the application of ingredients from milk will be discussed.

One can argue that intermediate products need to be a part of this book, because they are a part of many dairy-oriented businesses. As was shown earlier, these intermediate products might contain minor amounts of dairy ingredients and a lot of non-dairy compounds. The consequence is that one has to deal with an enormous range of ingredients. This would affect the scope of the book and divert our attention. For these reasons we restrict ourselves to ingredients from milk only and highlight their special position.

References

Acharya V & Young BR (2008) A review of the potential of bio-ethanol in New Zealand. *Bulletin of Science Technology Society* **28**: 143–8.

Afoakwa EO, Paterson A & Fowler M (2007) Factors influencing rheological and textural qualities in chocolate – review. *Trends in Food Science & Technology* **18**: 290–8.

Anonymous (1975) Aus der Geschichte der Milchschokolade. *Schweizerische Milchzeitung* **101**: 557–8.

Bachmann HP (2001) Cheese analogues: a review. *International Dairy Journal* **11**: 505–15.

van Bavel B & Gelderblom O (2009) The economic origins of cleanliness in the Dutch Golden Age. *Past and Present* no. 205, November, pp. 41–69.

Bieleman J (2010) *Five Centuries of Farming – A Short History of Dutch Agriculture 1500 to 2000*. Wageningen Academic Publishers, pp. 51, 55.

Blomdahl L (1982) Glass – historia och tillverkning. *Vär Föda* **34**: 399–414.

van der Burg B & Hartmans S (1952) *Zuivelbereiding IV De Kaasbereiding in de fabriek*. FNZ, p. 44.

Bush P (1991) Gold to the last drop. *Dairy Foods* October, pp. 70–3.

Business Consulting 3A (2008) The world market for whey and lactose products 2006-2010. *European Dairy Magazine* **1**: 26–8.

EU regulation no. 1333/2008 on food additives. *Official Journal of the European Union* **31**.12.2008.

EU regulation no. 1169/2011 on provision of food information to consumers. *Official Journal of the European Union* **22**.11.2011.

EWPA (European Whey Products Association) (2007) Whey in animal nutrition – a valuable ingredient. EWPA, pp. 1–19.

Food Product Design (2007–2008) *Buyer's guide*. Available at: http://www.foodproductdesign.com

Friesland Brands (2008) Double emulsion and method to produce such. Patent Application WO2010039036.

Funck E (1948) Geschichtliches über Molke und Milchzucker. *Milchwissenschaft* **3**: 152–60.

Guinee TP & Kilcawley KN (2004) Cheese as an ingredient. In: Fox PF (ed.), *Cheese – Chemistry, Physics and Microbiology*. Vol. 2: *Major Cheese Groups*. Elsevier, pp. 395–428.

Haylock SJ & Dodds TM (2009) Ingredients from milk. In: Beckett ST (ed.), *Industrial Chocolate Manufacture and Use*. Blackwell Publishing, pp.76–100.

Hettinga K, van Valenberg H, Vries de S *et al.* (2011) The host defense proteome of human and bovine milk. *PLoS ONE* **6**(4), e19433.

Jana AH & Upadhyay KG (2003) Mozzarella cheese analogue – a review. *Journal of Food Science and Technology* **40**: 1–10.

Johnson B (2000) US whey products in snacks and seasonings. *Applications Monograph – Snacks*. US Dairy Export Council.

Johnson NM (1843) Artificial freezer. US 3,254 [patent].

Lucey JA, Maurer-Rothmann A & Kaliappan S (2011) Functionality of ingredients: emulsifying salts. In: Tamine AY (ed.), *Processed Cheese and Analogues*. Blackwell Publishing, pp. 110–32.

Mawson AJ (1987) Ethanol production from whey in New Zealand. *Australian Journal of Biotechnology* **1**: 64–73.

Mawson AJ (1990) Consideration of the fermentation of concentrated whey solutions for ethanol production. In: *Fermentation Technologies: Industrial Application*. Kluwer, pp. 264–70.

Marshall RT & Arbuckle WS (1995) *Ice Cream*. Chapman & Hall, pp. 178–85.

Midda M & Cooksey MW (1986) Clinical uses of an enzyme-containing dentifrice. *Journal of Clinical Periodontology* **13**: 950–6.

Nestec (2008) Ice-structuring peptide from lactic origin. Patent Application EP1917865.

O'Rourke R (2005) *European Food Law*. Sweet & Maxwell, p. 60.

Paul GL (2009) The rationale for consuming protein blends in sports nutrition – review. *Journal of the American College of Nutrition* **28**: 464S–472S.

Peterson AE, Walker WG & Watson KS (1979) The effect of whey applications on chemical properties of soils and crops. *Journal of Agricultural Food Chemists* **27**: 654–8.

Raffald E (1769) *The Experienced English Housekeeper*. Manchester: J Harrep, p. 228.

Regtdoorzee Greup-Roldanus SC (1936) Geschiedenis der Haarlemmer blekerijen. PhD thesis, University of Amsterdam.

Sancisi-Weerdenburg H (1995) Over koken en kookboeken in de Oudheid (1). *Hermeneus Tijdschrift voor antieke cultuur* **67**: 23–7.

Scheltinga HMJ (1972) Measures taken against water pollution in dairies and milk processing industries. *Pure and Applied Chemistry* **9**: 101–12.

Thompkinson DK & Suman Kharb (2007) Aspects of infant food formulation. *Comprehensive Reviews in Food Science and Food Technology* **6**: 79–102.

Tripathi V & Vashishtha B (2006) Bioactive compounds of colostrum and its application. *Food Reviews International* **22**: 225–44.

Udabage P & Augustin MA (2003) Dairy ingredients in ice-cream. *Australian Journal of Dairy Technology* **58**: 21–5.

Verheyen V, Cruickshank A, Wild K *et al.* (2009) Soluble, semi-volatile phenol and nitrogen compounds in milk-processing wastewaters. *Journal of Dairy Science* **92**: 3484–93.

Visser RA, van den Bos MJ & Ferguson WP (1988) *Lactose and its chemical derivatives*. Bulletin of the International Dairy Federation no. 233, pp. 33–44.

Vogt CW (1930) Apparatus and process for treating materials. US Patent 1,783,864. Machine for processing materials. US Patent 1,783,865. Process and apparatus for treating materials. US Patent 1,783,866. Machine for processing materials. US Patent 1,783,867.

Vrignaud Y (1983) Valorisation du lactosérum – une longue histoire. *Revue Laitière Française* no. **422**, 41–8.

Wagner KH, Albig K, Ali A & Maiwald KG (1975) Die ernährungsphysiologische Bedeutung von Getränken auf Molkenbasis für die Osmolarität. *Milchwissenschaft* **30**: 724–9.

Walstra P, Wouters JTM & Geurts TJ (2006) *Dairy Science and Technology*. CRC Press/Taylor & Francis, p. 4.

Webb BH & Whittier EO (1970) *Byproducts from Milk*. Westport, AVI Publishing Co. International, pp. 106, 124, 405.

Whittier EO & Webb BH (1950) *Byproducts from milk*. New York: Reinhold.

Zollikofer E (1974) Aus der Geschichte der Molkenverwertung. *Schweizerische Landwirtschaftliche Forschung* **13**: 413–22.

2
Milk Specifics

More than 2000 years ago the compounds butter, cheese and serum were already distinguished in milk (Funck, 1948). It is clear that in this connection, the term 'serum' has to be interpreted as cheese whey. However, the distinction is useful even today, and we will make use of it to discuss the multiple ingredients obtained from milk. The term 'milk lipids' rather than butter will be used due to its broader meaning. In this chapter, the various ingredients from milk will be introduced. The information sheets mentioned will relate to the composition of milk and its major components – milk proteins, milk salts, milk sugar and milk lipids.

2.1 Palette of ingredients

Figure 2.1 depicts a palette of ingredients of milk. Four categories of ingredients are distinguished and indicated by different coloured lines. The blue line tracks the milk ingredients, which are most commonly derived from skimmed milk with its bluish-white tint. The lipids ingredients are designated by the yellow line, indicating the colour of fat. The various cheese ingredients are indicated by the orange line. Cheese is in many cases slightly coloured by the use of the natural colour agent annatto. For this reason, some cheeses such as Cheddar cheese have a yellow-orange colour. The whey left after cheese manufacture has a green colour. The ingredients derived from whey are marked by the green line.

The palette of ingredients of Figure 2.1 shows a wide diversity. In each coloured line, various commodities are present. Examples of commodities along the blue line (milk) are, for instance, whole milk powder (WMP), skimmed milk powder (SMP), caseinates and milk protein concentrates (MPC). Butter and anhydrous milk fat (AMF) are the commodities found on the yellow line (lipids). Unmatured cheeses and processed cheese (PC) are marked on the orange line (cheese). Whey powder (WP), edible lactose and whey protein concentrates (WPCs) are examples of commodity ingredients derived from whey (green line). Besides these commodities, specialized

From Milk By-Products to Milk Ingredients: Upgrading the Cycle, First Edition. Ruud de Boer.
© 2014 John Wiley & Sons, Ltd. Published 2014 by John Wiley & Sons, Ltd.

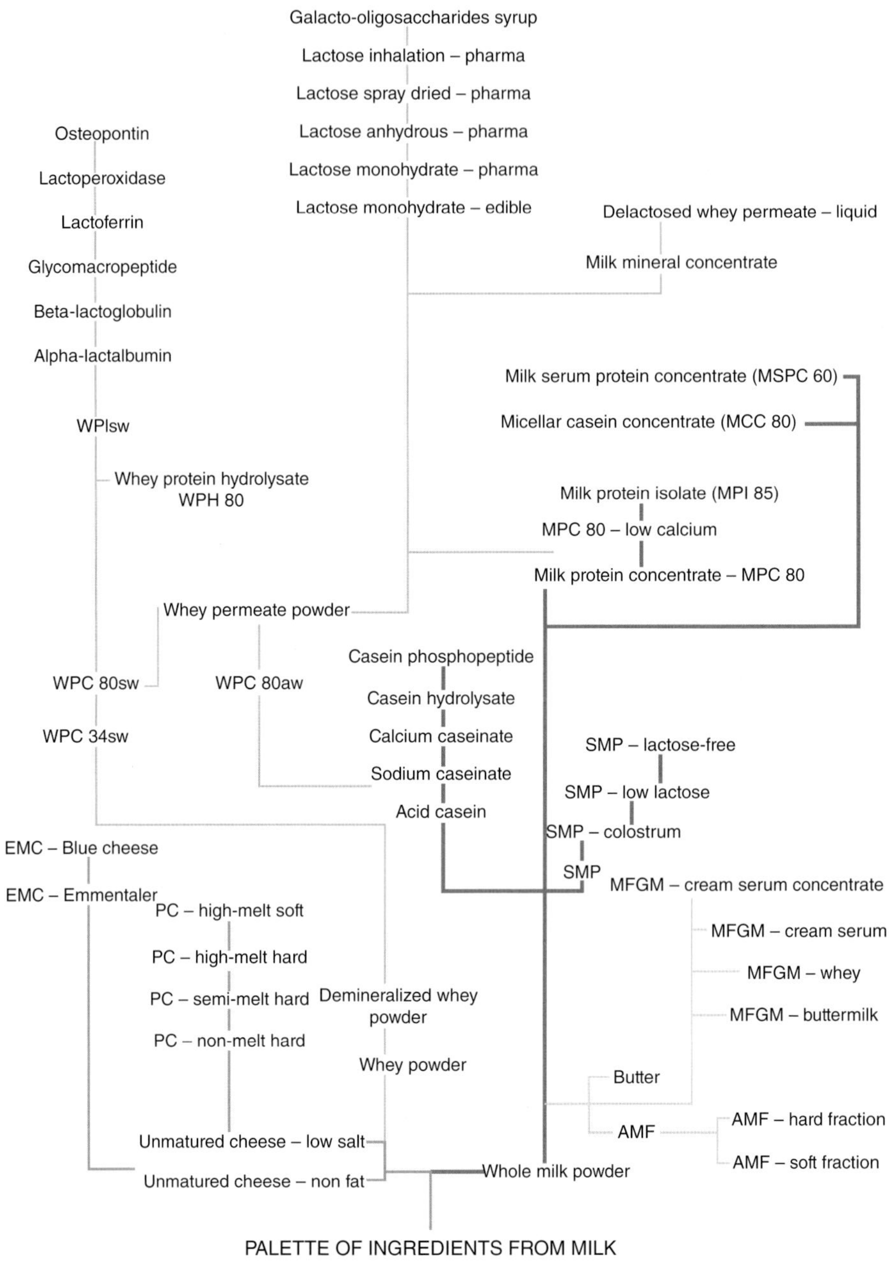

Figure 2.1 A palette of ingredients obtained from milk. AMF, anhydrous milk fat; EMC, enzyme modified cheese; MFGM, milk fat globule membrane; MPC, milk protein concentrate; PC, processed cheese; SMP, skimmed milk powder; WPC, whey protein concentrate; WPI, whey protein isolate. (*For colour details please see colour plate section.*)

ingredients can also be distinguished. Specialized ingredients indicated with the green line include the various whey protein fractions (lactoferrin, osteopontin, etc.), pharma lactose inhalation (for inhalers) and milk mineral concentrate. Enzyme modified cheeses (EMC) are examples of specialized flavour ingredients; they are marked with the orange line. Some milk fat globule membrane (MFGM) compounds are expected to be high value-added ingredients of the yellow line. With respect to the blue line, representatives of the speciality category are casein hydrolysates, milk casein concentrates (MCC) and milk serum protein concentrate (MSPC).

Figure 2.1 does not provide information regarding the quantity of ingredients obtained in the above mentioned four categories. This will be illustrated by the amount of milk produced by a highly productive Holstein-Friesian cow during 10 lactation periods (Melkcontrole Nijland, 2010). The lifetime production was the impressive quantity of 100,000 kg of milk at 3.93% fat and 3.25% protein. The data are summarized in Figure 2.2. On the milking machine jar, which records the quantity of milk released during milking, an estimate of the amount of finished products and ingredients is indicated. Two options are provided in Figure 2.2. The first has cheese production as a starting point, and the second the manufacture of SMP. The protein figure of the highly productive cow is not sufficient to calculate the amount of cheese. For such calculations the casein content is needed, and in particular the paracasein level (van den Berg *et al.*, 1996). Also, other factors such as moisture

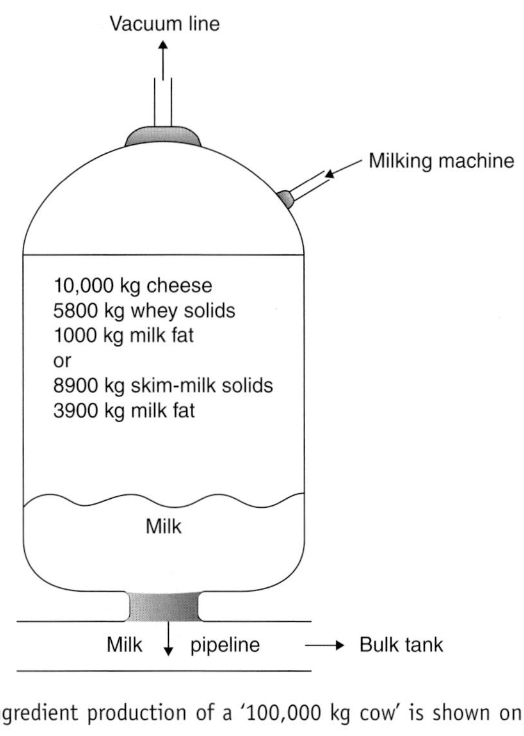

Figure 2.2 The ingredient production of a '100,000 kg cow' is shown on the jar of a milking installation.

content, fat and additional dry matter are important to make an accurate estimate of cheese yield (Walstra *et al.*, 2006, pp. 631–9). For this purpose, a rough estimate is sufficient, which results in 10,000 kg of young, full-fat Gouda cheese. In cheese whey, an average 5800 kg of solids are left (Olling, 1963). The quantity of remaining milk fat is estimated to be 1000 kg. In the second scenario, 8900 kg skimmed milk solids are produced, and the quantity of milk fat is increased to 3900 kg. The quantity of products/ingredients obtained is dependent on the production direction chosen, and the type of solids available in the starting milk. Apart from the whey solids, the remaining product streams are divided into various finished products and ingredients, depending on the added value, that can be achieved. Within the whey segment, additional choices can also be made, for instance the production of whey powder or an alternative such as the production of WPC and lactose. Economic factors play a decisive role, and the calculations become more complex (Peters, 2005). Large investments in the whey stream with only modest levels of solids (approximately 6%) need to be justified. In the case of WPC/lactose, a solution also has to be found for the remaining delactosed whey permeate.

The information sheets at the end of this book cover 53 ingredients from milk. The selection presents the numerous possibilities of ingredients from milk. All ingredients are offered by manufacturers. They have a long shelf life, ranging from approximately 6 months to 2 years. Three-quarters of them are available in dried form. The clearly structured information sheets each contain a short description of how the ingredient is manufactured. The heading ‘Product analysis’ includes the chemical analysis and the microbiological data. If relevant, nutritional and physical information is presented too. The information usually comes from the producers of ingredients; their names are referenced in a footnote. The data can be used as an illustration; producers provide the analytical techniques on which the information is based. The analytical data may relate to a (guaranteed) ingredient specification or a typical analysis, which is dependent on the available source. Information regarding applications and properties are collected from manufacturers’ brochures and general literature. Sources of further information related to description, applications and properties can be found under the heading ‘Literature’.

In the following sections, the specific aspects of milk composition, milk protein, milk salts, milk sugar and milk lipids will be discussed.

2.2 Milk composition

The palette of ingredients is a reflection of the composition of milk, which has a myriad of constituents. Table 2.1 gives examples of several constituents. Most of them are of interest for the manufacture of ingredients. The average composition of the milk of the Holstein-Friesian breed is shown (major source: Walstra *et al.*, 2006, pp. 3–16, 26, 73). However, the content of the major constituents may vary considerably, by up to 30%. At present the milk protein content of the above mentioned breed is 3.5% in The Netherlands. Minor contents may show an even

Table 2.1 Several constituents of bovine milk and their approximate concentration (mg/kg)[a].

Constituent	Concentration
Water	871,000
Lactose	46,000
Protein	33,000
Mineral substances	7000
Miscellaneous/organic acids	3000
Total solids-not-fat	89,000
Lipids	40,000
Triglycerides	39,000
Milk fat globule membrane[b]	1000
Phospholipids	320
Gangliosides	4
Oligosaccharides[c]	50
Casein	26,000
α_{s1}-Casein	10,700
α_{s2}-Casein	2800
β-Casein	8600
κ-Casein	3100
γ-Casein	800
Serum proteins	6300
β-Lactoglobulin	3200
α-Lactalbumin	1200
Serum albumin	400
Proteose peptone	800
Immunoglobulins[d]	300–600
Lactoferrin[d]	50–70
Lactoperoxidase[d]	8–20
Growth factors[d]	<0.06
Osteopontin[e]	18
Lysozyme[f,g]	0.008–0.6
Non-protein nitrogen	680
Minerals (fraction in serum)	
Sodium	480 (0.95)
Potassium	1430 (0.94)
Calcium	1170 (0.32)
Magnesium	110 (0.66)
Chloride	1100 (1.0)
Sulphate (SO_4)	100 (1.0)
Inorganic phosphate	2030 (0.53)
Citrate	1750 (0.92)
Miscellaneous	
Vitamin B2	1.8

[a] Typical data from lowland breeds.
[b] Includes membrane lipids and proteins.
[c] From Mehra & Kelly (2006).
[d] From Smithers (2008).
[e] From Schack *et al.* (2009).
[f] From Król *et al.* (2010).
[g] From Tripathi & Vashishtha (2006).

bigger fluctuation. The variation depends on the phase of lactation, the feed provided and the genetic predisposition of the individual cow.

The word 'serum' is mentioned in Table 2.1. Serum is the liquid left after the removal of casein and fat from the milk. It is a rich source of all kinds of constituents and contains lactose, serum proteins, non-protein nitrogen substances, minerals and a range of vitamins. The term 'milk serum' is used here to distinguish this liquid from other serums containing streams such as cream serum, acid whey and cheese whey. The composition of milk serum or 'ideal whey' is slightly different from a very dominant source of ingredients, called the 'cheese whey'. Besides the serum constituents, cheese whey contains glycomacropeptide (GMP), also called casein macropeptide (CMP), which is released from casein during cheese-making. GMP represents about 5% of the casein part of the milk. Other minor residuals can also be found in the whey after the manufacture of cheese. They are, for instance, fat, metabolites, starter cultures, coagulant, calcium chloride, colouring agent and curd fines.

2.2.1 Milk composition: diversity

The diversity of constituents in milk is phenomenal, and the number increases rapidly due to the availability of new analytical tools (Hettinga *et al.*, 2011). The range, from macro constituents to minor constituents, runs easily into the thousands. It is estimated that about 500 separate fatty acids have been detected in milk lipids (Hettinga, 2005). The macro constituents with a high nutritional value, such as various proteins and calcium, are generally known to the public. Vitamin B2 is a minor constituent and its concentration in milk is 1.8 mg/kg. The vitamin is a mainstay in supporting the consumption of milk and milk products. In the case of toddlers, it provides 52% of the daily allowance of vitamin B2 (Oude Elferink, 2011). Some other minor constituents shown in Table 2.1, however, have a less clear physiological function, such as gangliosides and various oligosaccharides.

Colostrums are well known for their great diversity of major bioactive proteins. Colostrum is the milk obtained during the first days of milking after parturition. It is essential for the development and the immune status of the newborn calf. It should be noted that the composition is dependent on the cow and the number of times they have been milked. The composition of bovine colostrum changes rapidly after 72–96 hours into that of mature milk. Several interesting bioactive proteins and their concentrations in colostrum and mature milk are indicated in Table 2.2 (Scammell, 2001). The concentrations of immunoglobulins (IgA, IgG1, IgG2, IgM) in colostrums are almost a hundred-fold higher than in milk. Of the immunoglobulins, the IgGs are by far the largest fraction, and are currently the primary marker of colostrum quality. Also, lactoferrin and various growth factors are present at greatly elevated concentrations. Well-known growth factors are insulin-like growth factor (IGF), transforming growth factor (TGF) and epidermal growth

Table 2.2 Some major bioactive protein compounds in colostrum and mature cow's milk expressed in terms of weight per litre.

Constituent	Colostrum	Milk
IgA	3.2–6.2 g	0.1 g
IgG1	48–87 g	0.4 g
IgG2	1.6–2.9 g	0.05 g
IgM	3.7–6.1 g	0.05 g
IGF-1	0.1–2 mg	25 μg
IGF-11	0.1–2 mg	2 μg
TGF-β	20–40 mg	1–2 μg
EGF	4–8 mg	2 μg
Lactoferrin	1.5–2 g	0.1 g
Lysozyme	0.1–0.7 mg	0.1–0.3 mg
Lactoperoxidase	30 mg	20 mg

EGF, epidermal growth factor; Ig, immunoglobulin; IGF, insulin-like growth factor; TGF, transforming growth factor.
Source: Scammell, 2001. Reproduced with permission from Dairy Industry Association of Australia.

factor (EGF). They have specific biological functions and play a role in stimulating growth and differentiation of mammalian cells (Tripathi & Vashishtha, 2006).

Advances in research and the availability of new isolation procedures will very likely increase the number of ingredients from milk in the future.

2.2.2 Milk composition: concentration

The differences in concentration between the various constituents in milk can be immense. The most abundant are lactose, fat and protein. They are called macro constituents. Of the minor constituents such as minerals, vitamins, lysozyme and growth factors, some of them are present in microgram amounts per kilogram of milk, while the total protein content in milk is expressed in tens of grams. This variation has an impact on the production cost of the different ingredients. Minor constituents can be produced economically only in large-scale operations. For instance, 25,000 kg of milk or milk derivative is needed for the production of 1 kg of active lactoferrin. Given this background, it is understandable that attempts were made to increase its concentration in milk. One of these was the creation of transgenic cows expressing recombinant human lactoferrin in their milk (van Berkel *et al.*, 2002). Compared to bovine milk, two factors were changed. The bovine lactoferrin (bLF) was replaced by recombinant human lactoferrin (rhLF) and the concentration increased considerably. In the most successful transgenic cattle line the concentration in the milk increased to 2800 mg/kg, which remained consistent throughout a lactation period of 280 days. This concentration is approximately 50 times higher than in normal bovine milk. Furthermore, it was demonstrated that the LF variants produced in most expression systems are similar or often indistinguishable from those of native LF. Aspects such as molecular weight, degree of glycosylation, antimicrobial and anti-inflammatory activity, thermostability and ability to bind Fe^{3+} were considered (González-Chávez *et al.*, 2009). In spite of the fact that the experiment with cows was successful, the adverse comment in the media could not be overlooked. The hurdle for applications in food as there remains consumer resistance to genetically modified food in some parts of the world, especially in Europe. The offspring of the famous bull 'Herman' were restricted to experimental purposes. 'Herman' ended his life in the museum (now the Naturalis Biodiversity Centre, in Leiden, The Netherlands) that exhibits the development of life on Earth in all its great diversity. The availability of technology, in this case for rhLF derived from cow's milk, does not always mean that its use is warranted.

2.2.3 Milk composition: size

In addition to their immense diversity and large differences in concentration, the constituents in milk also show a wide variation in size. The size of the constituents varies by a factor of approximately 10,000, as shown in Table 2.3. Membrane processes (see Chapter 6, 'Vital membrane processes') make use of the differences in size to separate and purify various constituents. The 'cow water' and ions, such as chloride and calcium, represent the smallest category, while the fat globules are on average by far the largest constituents. As opposed to the ions, the fat globules and their surrounding membrane are, however, composed of many different constituents. The size of the fat globules varies by a factor of approximately 100. The difference in size in the various protein compounds is also considerable. The serum proteins are approximately 50 times smaller than the milk caseins; the latter are called casein micelles. As opposed to the globular serum proteins, the casein micelles form a spherical shape, and vary in size substantially. This indicates that milk exists in a dynamic state, as evidenced by the phenomena that some important minerals such as calcium and phosphate are only partially present in the serum phase (Table 2.1). The other portions exist as undissolved salts. The dynamic equilibrium can easily be influenced by, for instance, souring, enzymatic action or heating. There is a close connection, in particular, between the protein and the mineral part of the milk.

Table 2.3 Size of some milk constituents.

Constituent	Size (nm)
Water	0.3
Chloride, calcium ions	0.4
Lactose	0.8
Serum proteins	3–6
Casein micelles	20–400
Fat globules	100–10,000

2.3 Proteins

Milk protein is the most valuable constituent of milk. It determines to a great extent the price of the milk, its nutritional value and the yield of cheese. In this section some aspects relevant for ingredients from milk will be discussed. These include: conversion factor, amino acid profile, isoelectric point of various milk proteins, spherical casein micelles and globular whey proteins.

2.3.1 Proteins: conversion factor

An accurate determination of the protein content is crucial for the dairy industry. In the standard method for estimation of the protein level, the total nitrogen content is estimated by the Kjeldahl method. The 'crude protein' content is calculated by multiplying the nitrogen content with a conversion factor, also called the Kjeldahl factor. The 'crude protein' also includes the modest fraction of non-protein nitrogenous substances. Different food proteins contain different amounts of nitrogen thus the conversion factor will vary accordingly. It is not surprising that this has sometimes been a subject of intense debate. For unmodified cow's milk protein, a nitrogen conversion factor of 6.38 was determined in the 19th century, and is widely used in, for instance, Codex Standards and European Council Regulations on milk proteins, both for casein and whey (Bulletin IDF, 2006). Recent technical developments make additional diversification of milk proteins possible. In the information sheets, various proteins have been selected such as glycomacropeptide and osteopontin. The conversion factors were calculated for the different milk protein fractions (van Boekel & Ribadeau-Dumas, 1987). As a basis for their calculation, the amino acid composition and sequence were used. This was considered to be more specific than the Kjeldahl method. A survey of the various conversion factors of milk proteins is shown in Table 2.4. According to the authors cited in the table, the conversion factor for milk protein (including carbohydrates and phosphates bound to proteins) is 6.35, which is slightly lower than the traditional factor of 6.38. As some of the proteins are glycoproteins, a distinction is made between 'without carbohydrate' and 'with carbohydrate' in Table 2.4. In cases where carbohydrates are available, the nitrogen content of the molecule is lower and as a consequence the conversion factor is higher. An example is κ-casein. Two-thirds of this molecule contains a carbohydrate group. The conversion factor increases from 6.12 for pure protein to 6.35 for the glycoprotein. Another example is the enzymatic hydrolysis of casein by rennet, which results in paracasein and CMP. As is demonstrated by Karman & van Boekel, 1986 the released CMP or GMP results in an elevated conversion factor of 7.37 ('with carbohydrate'/genetic variant A), however, the factor for paracasein is slightly reduced compared to casein. The difference in conversion factors between (acid) whey and (cheese) rennet whey can also be explained by the presence of GMP in the latter (Table 2.4). The conversion factor of milk fat globule membrane (MFGM) is surrounded by more uncertainties. These proteins

Table 2.4 Conversion factors of milk and milk protein fractions.

Protein	Without carbohydrate	With carbohydrate
Milk	6.31	6.35
α_{s1}-Casein	6.36	–
α_{s2}-Casein	6.29	–
β-Casein	6.37	–
κ-Casein	6.12	6.35
γ-Casein	6.34	–
β-Lactoglobulin	6.29	–
α-Lactalbumin	6.25	–
Serum albumin	6.07	–
Immunoglobulin	6.00	6.20
Lactoferrin	5.72	6.14
MFGM	6.60	7.08
(Acid) casein	6.33	6.36
Paracasein	6.31	–
(Acid) whey	6.21	6.28
Rennet whey	6.28	6.41

MFGM, milk fat globule membrane.
Source: van Boekel & Ribadeau-Dumas, 1987. Reproduced with permission from Elsevier.

are difficult to characterize, therefore only a rough estimate of the conversion factor (7.08) was given. For osteopontin Arla Foods Ingredients (2009) uses a factor of 7.17. This value is also based on the protein's amino acid composition and sequence. It is applied to calculate the true content of osteopontin in order to avoid an underestimation of a highly value-added ingredient. The conversion factors for other minor glycoproteins, for instance immunoglobulin (a heterogeneous protein fraction) and lactoferrin, are assessed to be slightly below the average value for milk protein of 6.35 (Table 2.4).

2.3.2 Proteins: amino acid profile

The building blocks of proteins are amino acids. The amino acid patterns of various milk protein fractions are shown in Table 2.5. The contents of these fractions in milk vary as well as the amino acid profiles. It is important to note that the data provided in Table 2.5 are based on industrial sources (see information sheets). To some extent they are dependent on the nature of the raw material used, as shown by comparing WPC 80_{aw} and WPC 80_{sw}. WPC 80_{aw} is produced from acid whey and WPC 80_{sw} from cheese whey. The latter contains GMP, which has a significant effect on the amino acid composition: that is, the content of lysine, tryptophan and S-containing amino acids are reduced, while the threonine level is enhanced (Mallee & Steijns, 2007). In order to compare the concentrations, the way the amino acid content is expressed must be considered. The total of amino acids can exceed 100

Table 2.5 Amino acid composition of various milk proteins and reference protein (g/100 g protein).

	Milk powder	Calcium caseinate	WPC 80_{aw}	WPC 80_{sw}	α-Lactalbumin	Reference protein	GMP	Osteopontin
Histidine[a]	2.8	3.2	2.4	2.1	3.0	**1.6**	0.3	3.4
Isoleucine[a,b]	5.2	5.8	6.0	7.2	6.0	**3.0**	10.1	2.6
Leucine[a,b]	9.7	10.1	12.8	11.4	10.8	**6.1**	2.6	5.7
Lysine[a]	7.9	8.3	10.9	9.8	10.9	**4.8**	5.9	8.2
SAA[a]						**2.3**		
Methionine	2.7	3.0	2.5	2.3	1.0		1.8	0.2
Cysteine/cystine	0.8	0.4	3.2	2.5	4.8		0.1	0.1
AAA[a]						**4.1**		
Phenylalanine	4.8	5.4	4.0	3.8	4.1		0.5	4.1
Tyrosine	4.5	5.8	3.6	3.2	4.4		0.1	1.8
Threonine[a]	4.4	4.6	5.4	7.9	4.3	**2.5**	15.8	7.2
Tryptophan[a]	1.3	1.4	2.6	2.2	4.8	**0.66**	0.0	0.9
Valine[a,b]	6.3	7.4	5.3	6.9	4.3	**4.0**	8.0	4.3
Alanine	3.3	3.1	5.1	5.2	2.0		5.5	3.3
Arginine	3.5	3.8	2.9	2.7	1.3		0.5	2.2
Aspartic acid	8.0	7.3	12.1	11.3	16.4		8.6	23.4
Glutamic acid	20.8	22.3	18.6	18.2	12.6		20.5	15.8
Glycine	2.0	1.9	1.9	2.0	2.7		1.1	1.5
Proline	9.8	10.5	5.5	6.4	3.3		12.5	7.4
Serine	5.6	6.3	4.8	6.1	3.4		6.1	10.4
Source	CVB[c]	DMV	DMV	DMV	Davisco	FAO[d]	Davisco	Arla

[a] Essential amino acid.
[b] Branched-chain amino acid.
[c] CVB Feed tables (2007).
[d] Reference protein for older child, adolescent, adult (FAO, 2011).
AAA, aromatic amino acids; aw, acid whey; GMP, glycomacropeptide; SAA, sulphur-containing amino acids; sw, sweet whey; WPC, whey protein concentrate.
Total amino acids >100 due to water-containing molecules.

due to the water-containing molecules. In this example the total amino acid content of caseinate, WPC 80_{aw} and WPC 80_{sw} is about 110.

In Table 2.5 the amino acid compositions of various proteins and milk protein fractions are indicated. These show great variations. For instance, calcium caseinate is rich in glutamic acid whereas osteopontin is very high in aspartic acid. On the other hand, there are relatively few sulphur amino acids (SAA) in osteopontin. Glycomacropeptide(GMP) contains no tryptophan at all and is also low in aromatic amino acids (AAA). Milk powder, calcium caseinate, WPCs and α-lactalbumin have a more balanced amino acid profile. They are mostly used as nutritional agents. For nutritional applications, the amino acid profile is crucial, in particular the 'indispensable' or essential amino acids (EDA, 1997). Essential amino acids are not produced by the human body and have to be provided in food. Branched-chain amino acids (BCCA) are among the nine essential amino acids (Table 2.5 indicates both the essential amino acids and the branched-chain amino acids). It is reported that BCCAs in whey proteins might, for instance, create an additional advantage for muscle mass conservation (Graf *et al.*, 2011). Table 2.5 also gives data for a reference protein (FAO, 2011). Milk powder, calcium caseinate, WPCs and α-lactalbumin all have a surplus of essential amino acids, which makes them a favoured source of protein.

Glycomacropeptide has a unique amino acid composition, as shown in Table 2.5. It deviates considerably from the regular milk proteins such as milk powder, caseinates and WPCs. GMP is a natural, palatable protein source that is nearly free of phenylalanine. The technology applied – membrane processes or chromatography – has a slight effect on the phenylalanine content. In Table 2.5 a low level of 0.5 g/100 g of protein is shown. The low level makes it useful to treat individuals with phenylketonuria (PKU). Such patients often tolerate dietary levels of less than 500 mg phenylalanine per day. But they must follow a lifelong low-phenylalanine diet to prevent neurological impairment. The GMP-based formula is adapted in order to provide three essential amino acids – histidine, leucine and tryptophan – as well as tyrosine. If GMP is supplemented with these amino acids, it offers a nutritionally adequate source of protein for PKU patients (Ney *et al.*, 2009). Besides its nutritional application, GMP is also a bioactive protein with several promising possibilities. For instance, GMP interacts with toxins, viruses and bacteria and exerts immunomodulating activities (Thomä-Worringer *et al.*, 2006).

Osteopontin is the other minor milk protein shown in Table 2.5. It is also involved in various biological processes such as the repair of tissue (e.g. in wounds or cancers) and 'biomineralization' (Mazzali *et al.*, 2002). Osteopontin is not a balanced protein source compared to calcium caseinate and the major whey proteins. It has, for instance, very low values of S-containing amino acids, while the aspartic acid, serine and threonine levels are high. Calcium binding is related to the elevated level of the latter three amino acids. Bovine osteopontin is used, for instance, to mimic the composition of human milk. Osteopontin in human breast milk comprises on average about 2% of the total protein (Schack *et al.*, 2009). Addition of approximately

100 mg/kg osteopontin to a common infant formula is needed to imitate the level in human milk.

2.3.3 Proteins: electric charge

Due to their various promising biological properties, there is an increasing need to produce individual milk proteins on a large scale without damaging protein structures and their functional properties. Table 2.6 shows the enormous variation in molecular weight (MW) of various milk proteins, which range from 7×10^3 to 1.3×10^9 Da. Molecular weight is closely related to the size of the molecule and can be utilized to separate some of the major protein fractions (see Chapter 6, 'Vital Membrane Processes'). The isoelectric point (IP) of the various fractions shows large variation too. Differences in IP within immunoglobulins are remarkable. This is due to the diversity of this fraction (Table 2.2). The IPs of the other proteins in Table 2.6 are average values that are used in the separation of various milk proteins. The net charge of the protein molecule is zero at its IP. In case of the long-established production of casein, this property is exploited. The casein is precipitated and separated from the acid whey at pH 4.5 and purified immediately afterwards. Protein fractions can also be isolated by keeping the protein in its native state. Differences in IP, which depend on the amino acid residues of the proteins, may be used for isolation. Amino acids are 'zwitterions', as shown here:

$$\underset{\text{Positive ion}}{\text{R-CH(NH}_3{}^+\text{)-COOH}} \xleftarrow{\text{H}^+} \underset{\text{Zwitterion}}{\text{R-CH(NH}_3{}^+\text{)-COO}^-} \xrightarrow{\text{OH}^-} \underset{\text{Negative ion}}{\text{R-CH(NH}_2\text{)-COO}^-}$$

The terminal carboxyl and amino groups in the polypeptide backbone as well as the side chains (R) determine the diversity in IP. Side chains can contain a carboxyl group (aspartic acid, glutamic acid) or an amino group (lysine, arginine). At a neutral pH value they ionize to yield $-COO^-$ and $-NH_3{}^+$. For milk proteins this results in

Table 2.6 Some physicochemical properties of various milk proteins.

Protein	Isoelectric point	Molecular weight (Da)	Proportion (%) total protein
Casein	4.5	1.3×10^9 (hydrated)	79
GMP (part of casein)	2.6	7000	4
β-Lactoglobulin	5.4	18,300	10
α-Lactalbumin	4.4	14,000	4
Bovine serum albumin	5.1	69,000	1
Immunoglobulins	5.5–8.3	15×10^3–1×10^6	1
Lactoferrin	7.9	77,000	0.2
Lactoperoxidase	9.5	77,500	<0.1
Osteopontin	3.5	41×10^3–75×10^3	<0.1

Source: Data from Goodall *et al.*, 2008, Strohmaier 2004 and Azuma *et al.*, 2006.

an overall negative charge at the pH of milk, which is 6.7. As is demonstrated in Table 2.6 the IPs of lactoferrin and lactoperoxidase have the highest IP values, 7.9 and 9.5, respectively. The elevated arginine levels of both protein fractions (Cals *et al.*, 1991; Pierce *et al.*, 1991) probably contribute to these high IPs.

Protein fractions can be derived from dairy liquids by selecting the pH value and the starting material. In the case of sweet cheese whey (pH ~ 6.5), lactoferrin and lactoperoxidase are charged positively while the major whey proteins have a negative charge. Under these circumstances both minor proteins can be absorbed by using cation exchange chromatography (negatively charged resin). Subsequently they are removed from the resin. In contrast to the earlier mentioned basic proteins, the phosphorylated osteopontin and GMP are highly acidic protein fractions with IP values of 3.5 and 2.6 respectively. The isolation of osteopontin is preferably carried out from a GMP-free starting material. Instead of cheese whey the by-product of casein production, acid whey, can be used advantageously. This acid whey has a pH of about 4.5. Osteopontin is the only protein fraction that is negatively charged under these circumstances. For its production, anion exchange chromatography is used. The latter uses a positively charged resin. Chromatography is particularly useful for the production of fractions occurring in low concentrations. The proportion of total protein represented by the various component proteins is set out in Table 2.6.

2.3.4 Proteins: casein micelles

Casein micelles are one of the three biomolecular aggregate structures in milk. The others are the fat globules and lactose, which will be discussed in the following sections. The main function of the micelle is to fluidize the casein molecules and solubilize relatively high levels of calcium and phosphate. Casein is the most important protein of milk. Both protein and calcium are essential for the fast growth of the neonate.

Casein micelles, or colloidal calcium-protein complexes, are important constituents for several ingredients produced from milk (see information sheets). In WMP, SMPs, MPCs and milk protein isolate (MPI), casein constitutes approximately 80% of the total protein, which is similar to that of the original milk (Table 2.1). In MCC the casein level is increased to about 92% of total protein by membrane processes. As opposed to these ingredients, the casein component in caseinates and cheese is significantly modified.

In addition to their nutritional properties, casein micelles have a large impact on the functional properties of the derived products. An understanding of the behaviour of casein micelles under different process conditions is vital for the dairy industry. The heat stability of concentrated milk has been a subject of extensive research for many years. Insight into the micelle structure is also important for the coagulation of milk for cheese-making and aggregation and solubilization of casein for the manufacture of caseinates. In the course of time various models have been developed in an effort to understand their behaviour (Qi, 2007; Dalgleish, 2011; Dalgleish &

Corredig, 2012). Some details of the structure of casein micelles are a subject of ongoing debate. An example is the intriguing question: 'Casein sub-micelles: do they exist?' (Walstra, 1999). The most straightforward model, in which the casein micelle is considered as a whole, is chosen for our purposes (de Kruif & Holt, 2003). The model will be used to provide the background for some interesting technological properties of ingredients. It is important to keep in mind that a model is an oversimplification of reality. Casein micelles are not fixed structures and they change under influence of temperature, pH and ionic strength.

A simplified survey of casein micelles and its modifications is shown in Figure 2.3. The casein micelles vary in magnitude from 20 to 400 nm, and follow a Gaussian distribution. The average size and size distribution are influenced by the individual cow and the lactation period. Casein micelles exist in a very heterogeneous system. Within the casein family four members can be distinguished. They are referred to by a Greek letter (sometimeswith a subscript): α_{s1}-casein, α_{s2}-casein, β-casein and κ-casein. In bovine milk they represent, respectively, approximately 38%, 10%, 36% and 12% of the whole casein on a weight basis. The calcium-sensitive caseins, α_{s1}-, α_{s2}- and β- casein, are sequestered mainly into the micelle structure through strong linkages to calcium phosphate. The extent of the calcium binding is directly related to the number of phosphoserine residues in the casein molecules. The calcium phosphate nanoclusters, also known as colloidal calcium phosphate (CCP), are shown by red dots in Figure 2.3. They form the centrepiece of the casein micelle structure and are distributed fairly randomly. About 800 of these amorphous calcium phosphate nanoclusters are present in an average sized casein micelle of 100 nm diameter. The α_{s1}-, α_{s2}- and β-caseins are depicted as black curved lines and are predominantly located in the centre of the micelle. In contrast to the other caseins, κ-casein is calcium-insensitive as it does not contain phosphoserine residues. κ-Casein has a 'hairy layer' structure and a hydrophilic character. It is thought to be predominantly located at the outside of the casein micelle. The hairy layer or brush can stabilize up to 10 times the weight of the calcium-sensitive α_{s1}-, α_{s2}- and β-caseins via the formation of micelles. In Figure 2.3 κ-casein is shown as blue curved lines.

Casein micelles can be destabilized by the use of rennet. By adding the enzyme, the hairs of the casein micelles, containing GMP, are cut off at random. The remaining paracasein micelles start to aggregate when about 70% of the κ-casein hairs are split off. A gel is formed, which is converted into a cheese curd during processing. In Figure 2.3 the paracasein micelles and the released GMP are schematically drawn. The GMP molecules have a length of approximately 10 nm. GMP is indicated by short blue lines that surround the micelle. The cheese curd is depicted with red dots of calcium phosphate nanoclusters inside the paracasein molecule. On the outside part of the κ-casein layer remains. The latter is shown as short blue lines attached to the paracasein micelles.

During the manufacture of casein the calcium phosphate nanoclusters are released from the casein micelles by lowering the pH value. At the isoelectric point of casein

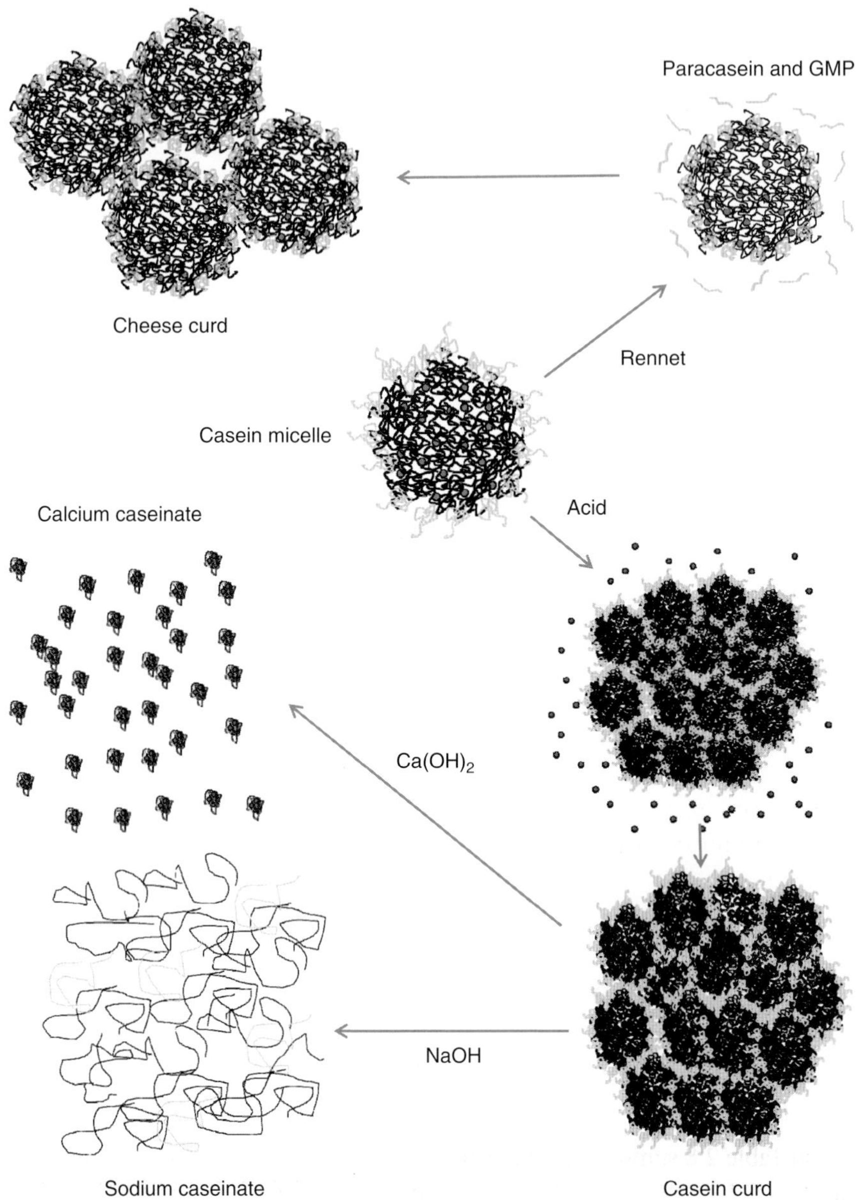

Figure 2.3 Casein micelles and modification. *Source:* Courtesy of FrieslandCampina DMV. (*For colour details please see colour plate section.*)

(pH 4.5), the κ-casein hairs collapse. The casein micelles destabilize and form aggregates if the temperature is increased. A casein curd is also shown in Figure 2.3. The collapsed κ-casein hairs are depicted as thick blue lines and the remaining casein as a compact black structure. The casein curd can be solubilized in water by the use of alkali. The pH is usually adjusted to 6.7. Sodium hydroxide is most commonly

Table 2.7 The typical composition of two types of whey (%).

Constituent	Casein whey	Gouda whey
Total solids	6.5	6.5
Proteins (N × 6.38)	0.8	0.86
Lactose	4.7	4.7
Lipids	0.03	0.05
Minerals (ash)	0.79	0.53
Calcium	0.16	0.06
Phosphorus	0.1	0.07
pH	4.7	6.4

Source: de Wit, 2001. Reproduced from the EWPA.

used, and the casein is converted into sodium caseinate. This caseinate has lost its micellar structure and can be characterized by a random coil configuration. The only connection between the amino acids is the peptide bond. If $Ca(OH)_2$ is applied as an alkali, a calcium caseinate is formed. It consists of small particles with a diameter of approximately 10 nm (Qi, 2007).

The changes in the casein micelles, caused by the use of acid or rennet, are also reflected in the composition of the released whey. In Table 2.7 the composition of casein whey and Gouda whey (as an example of cheese whey) is shown (de Wit, 2001). The protein content of cheese whey is higher than that of acid whey, which is the result of the removal of the GMP part of the casein micelle. In case of acid whey, the mineral content is higher. This is caused by the release of the calcium phosphate component of the micelle during casein manufacture. Both mineral constituents are therefore increased in the casein whey. The lower protein and the higher mineral content have a negative effect on the value of the casein whey.

2.3.5 Proteins: whey

Examples of commodity whey protein ingredients are WPC and WPIs. Their composition and properties are listed on the information sheets. The overall properties of these whey protein ingredients can be distinguished from casein-containing ingredients, and Table 2.8 summarizes the main differences. Whey proteins, in their native form, are soluble over a wide pH range in contrast to casein ingredients. The latter

Table 2.8 Some differences between caseinates and whey proteins.

	Casein	Whey proteins
Solubility	–	+
Heat stability	+	–
Gel forming	–	+

undergo sedimentation close to the IP of 4.5. Another striking difference is that whey proteins can be transformed into a gel, while casein-containing ingredients are generally very heat stable. These differences in properties have consequences for applications in food products.

Whey proteins are compact globular proteins that differ both in structure and in properties from each other.Their sensitivity to heat is particularly important. β-Lactoglobulin is the most abundant whey protein. At room temperature and between pH 5 and 7 it occurs as a dimer consisting of two identical subunits, each with a molecular weight of 18,400. Above 40°C the dimers start to dissociate into monomers. These monomers have a higher reactivity and play an important role in heat sensitivity. The second whey protein is α-lactalbumin. It is the smallest and most heat-resistant whey protein. Unlike α-lactalbumin, the immunoglobulins are very heat-labile. Commodity whey protein ingredients are a mixture of various heat-sensitive protein fractions. Their properties are not only dependent on the composition of the mixture, but even more on the temperatures applied during processing. High temperatures may have a negative effect on solubility. The properties of whey proteins can also be modified in a desired direction, for instance, to induce gelling or to improve heat stability. Novel non-thermal processes such as pressurization and ultrasound are mentioned in this context, but they are currently largely centred on food preservation (Abd El-Salam *et al.*, 2009). We will focus on the mainstream processes and modifications in which heat treatments play a crucial role.

One of the first steps in the manufacture of whey proteins was the production of an insoluble, denatured form, as is mentioned in Table 1.7. The unconcentrated cheese whey was heated to approximately 90°C/15 min and $CaCl_2$ was added to precipitate the protein. The inert, dried ingredient was introduced as 'lactalbumin' around 1950 and used for nutritional purposes. Advances in protein technology have broadened the applications of whey proteins considerably. The opposite of 'lactalbumin' is the native whey protein. The latter is mentioned in the information sheet 'Milk Serum Protein Concentrate' (MSCP). This protein is prepared directly from milk prior to cheese-making. Compared to the usual spray-dried WPC and WPI ingredients, the apparent viscosity of a freeze-dried MSPC, both without and with a very mild heat treatment of 65°C/2′, is considerably lower (Marcelo & Rizvi, 2008). This result suggests that the amount of native whey protein, which is obtained under optimal conditions, is greater than those in commercial products. The commercial WPC products usually contain 8–10% denatured and aggregated proteins (Havea *et al.*, 2009). In this respect the 'temperature history' is important. Temperature history means the cumulative effect of the various heat treatments resulting in changes in the native conformation of whey protein molecules. The changes are caused by the usual applied whey processes such as pretreatment, pasteurization, evaporation and drying. The effects of two extreme heating rates on β-lactoglobulin were calculated and are shown in Figure 2.4 (de Wit, 1998). A heating rate of 1°C/s may occur in the heating section of a pasteurizer, and a heating rate of 1°C/min in the

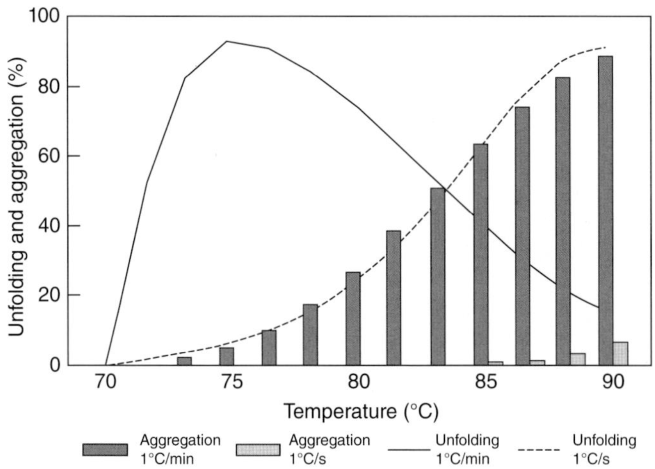

Figure 2.4 Percentage of unfolding and aggregation of β-lactoglobulin in whey during heating processing between 70 and 90°C at various heating rates. *Source:* de Wit, 1998.

regeneration section of a heat exchanger. The lines indicate that slow heating/cooling is most detrimental, as is reflected in the percentage of unfolding and aggregation. Aggregation is irreversible denaturation, and appears to start when 60% unfolding has occurred. The heating rate of 1°C/s does not show aggregation below 83°C, but at a heating rate of 1°C/min aggregation has started already at 73°C. This example shows that careful selection of processing is important to obtain consistent functionality of whey protein ingredients.

The properties of the various whey protein ingredients can be exploited to give desired modifications. In this case use is made of variations in temperature, pH, time and ionic strength. In a review, de Wit (2009) differentiated as many as seven stages in the thermal behavior of β-lactoglobulin at pH exceeding 6.8 in the temperature range 20–150°C. He also included dissociation, molten globule and partial unfolding. We will restrict ourselves to considering temperatures below 100°C. In order to explain the various functional properties we use a less complex distinction – between the native, denatured, aggregated and gel status. Figure 2.5 gives an overview of these four stages of modification. In the native state of whey proteins, the hydrophobic residues are hidden inside the globular structure. Upon denaturation, the hydrophobic residues are exposed to the surface leading to an improvement in protein functionality. An increase in hydrophobic residues enhances the emulsifying properties of the protein. This occurs under moderate heat treatment, such as 84°C/30 s at neutral pH (Moon & Mangino, 2004). Controlling the extent of denaturation is of crucial importance in tailoring the desired properties. Good foaming properties are also correlated with the hydrophobicity of the protein molecules (Sullivan *et al.*, 2008). Gelation is usually the result of a heat treatment at higher temperatures than the denaturation temperature. For the WPI ingredient this temperature is 77–88°C

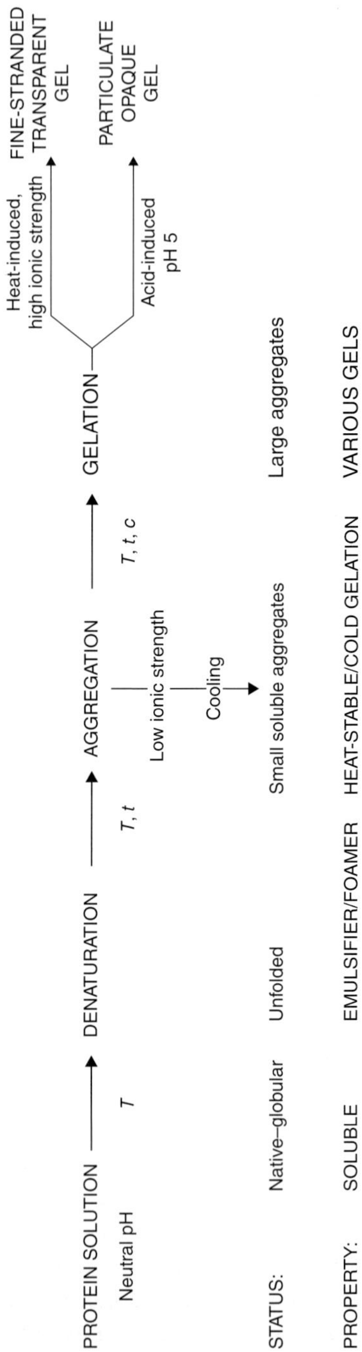

Figure 2.5 Whey proteins and their modifications (T, temperature; t, residence time; c, protein concentration).

(Aguilera, 1995). Furthermore, a minimum protein concentration is required, for instance more than 5%. At a pH close to the IP of major whey proteins, pH 4–5, the net surface charge is nearly zero, which reduces the repulsive interactions between protein molecules. The result is the formation of large aggregates that form opaque particulate gels. The gel consists of weak, non-covalent interactions (Havea *et al.*, 2004). At a pH far removed from the IP, the surface charge on the proteins is large and the repulsive forces prevent the random formation of large aggregates. When salt, for instance $CaCl_2$ or NaCl, is added the ionic strength increases. These ions shield the surface charge of the proteins, which results in a reduced repulsion and facilitates aggregation. Under these circumstances a transparent, fine-stranded gel with a high water-holding capacity is formed. In this gel, disulphide bonding dominates (Havea *et al.*, 2004). Moreover, modified whey protein concentrates have been developed with the ability to form gels simply by the addition of salt without heating (Elofson *et al.*, 1997). For this purpose whey protein solutions (6–10%) of a very low ionic strength and pH 7 were heated at 70–90°C and subsequently quickly cooled. Instead of gels, small soluble aggregates were formed. In this scenario the protein molecules were highly charged and as a result the repulsion forces were too high. In its final application, gelation can be induced by adding salt or by changing the pH toward the IP of the proteins (Alting *et al.*, 2004). Similar soluble whey protein aggregates can also be formed to survive thermal processing without detrimental changes. For this purpose a WPI solution (7% protein) at a pH of 6.8 was heated at 90°C/10 min and subsequently cooled rapidly (Ryan *et al.*, 2012). The treated protein could survive the usual heat treatment of meal replacement beverages at pH exceeding 6.5.

In short, whey proteins can be modified to fulfil functions such as emulsifiers, foamers and gels, to withstand heat treatments and to induce gelation in their final application.

2.4 Salts

The milk salt fraction is complex and dynamic and interacts strongly with the protein part of the milk (Gaucheron, 2005). The structure of casein micelles is a good example of this interaction. Milk salts are not bound in significant amounts to fat globules and lactose. The salt fraction forms only a modest part of the milk (see Table 2.1). The information sheets in this book refer to the ash content instead of the salt content; this is a standard procedure and is a measure of the total amount of minerals in an ingredient. During the ash determination, organic salts, of which citrate is the principal constituent, are lost. For this reason, the ash content is somewhat lower than the milk salt content. In contrast to the inorganic phosphorus, the organic part – present in the casein micelles and milk fat globule membrane – is not considered part of the milk salts. In this section the focus will be on calcium

Table 2.9 The composition (g/100 g) of skimmed milk and ultrafiltration milk permeate.

Constituent	Skimmed milk	Milk permeate
Total solids	9.0	5.5
Protein (N × 6.38)	3.3	0.16
Lactose	4.7	4.7
Citrate	0.15	0.15
Calcium	0.125	0.040
Magnesium	0.010	0.006
Phosphorus	0.10	0.04
Sodium	0.05	0.05
Potassium	0.15	0.15
Chloride	0.10	0.10

Source: Brule *et al.*, 1978.

phosphates as they play an important role in the properties of ingredients from milk. Its effects will be discussed in the absence and presence of protein.

2.4.1 Salts: permeate

The composition of milk without fat and proteins is provided in the second column of Table 2.9. This clear yellow-greenish liquid, milk permeate, is obtained directly from raw milk. In this way the effect of heat processes on its composition is omitted (Brule *et al.*, 1978). Casein micelles and globular whey proteins are both absent in permeate. Besides milk minerals and lactose, milk permeate contains some non-protein nitrogen substances, which were calculated using a conversion factor of N × 6.38. The milk salt fraction is similar to the serum part of milk, which is shown in Table 2.1.

An important application of milk permeate is the standardization of protein, for instance for the manufacture of skimmed milk powder. According to recent regulations either lactose or milk-derived permeate are permitted. The protein on a solids-not-fat basis should be at least 34%. For milk with a high protein content, the milk permeate portion can be as much as 20% of the total. This addition also changes the milk salt composition and increases the soluble fractions of calcium and phosphate. In turn it could slightly influence the heat classification of skimmed milk powders expressed in the whey protein nitrogen index (see 'Skimmed Milk Powder' information sheet). This index is used as a criterion in the selection of food applications. Sikand *et al.* (2008) has proposed to adapt the calculation of the index.

The stability of protein-free permeates is different from that of the original milk. This is caused by the absence of casein micelles in the permeate. Calcium phosphate is a complex range of materials containing calcium and phosphate as key components. At the pH of milk a basic and an acidic type are present (Gao, 2010). The basic type resembles hydroxyapatite (HAP) and the acidic type is dicalcium phosphate (DCP). The latter has a higher solubility and is the more common form at the pH of milk. The milk permeate is susceptible to changes in heat treatment and pH. Heating to 90°C reduces solubility and results in the precipitation of 60% of the calcium and 30% of the phosphorus. However, precipitation can be avoided if the pH is reduced from 6.7 to 6.0. Both phenomena are applied in the processing of milk permeate or its less pure counterpart, whey permeate. In the manufacture of milk mineral concentrates, as set out in the information sheet (see 'Milk mineral concentrate'), the permeate is heated to high temperatures at a pH of approximately 7. Under these conditions a calcium phosphate deposit is formed. Anema (2009) characterizes the deposit isolated from (casein) whey permeate as a poorly crystallized HAP or amorphous calcium phosphate (ACP). Both deposits have an extremely low solubility product. Pouliot *et al.* (1991) obtained a crystallized calcium phosphate from (cheese) whey permeate by applying a combination of heating, alkalinization and seeding with dicalcium phosphate (DCP). DCP can be converted into a more soluble form by reducing the pH value, which is indicated as monocalcium phosphate (MCP). This process step is applied in the concentration of permeate to avoid undesirable deposits on the equipment. A schematic overview of certain variants of calcium phosphate relevant for the processing of permeate is shown here:

$$\underset{\substack{\text{pH} \approx 6 \\ \text{MCP}}}{Ca(H_2PO_4)_2} \xleftrightarrow[]{\substack{\text{Low temperature} \\ H^+}} \underset{\substack{\text{pH} \approx 6.7 \\ \text{DCP}}}{Ca(HPO_4)} \xleftrightarrow[]{\substack{\text{High temperature} \\ OH^-}} \underset{\substack{\text{pH} \approx 7 \\ \text{ACP} \qquad \text{HAP}}}{Ca_3(PO_4)_2/Ca_5(PO_4)_3OH}$$

2.4.2 Salts: in milk

Precipitation of calcium phosphate can be reduced considerably if the permeate is mixed with proteins. Very few deposits are detected if milk – or whey proteins – are added before the heat treatment. This protective effect was observed when the protein content was higher than 8 g/kg (Brule *et al.*, 1978). Particularly, caseins are effective stabilizers due to the presence of CCP in the micelles. In Figure 2.6 the solubilization or association of CCP and casein are shown after the most common treatments during the production of ingredients. The solubility of the calcium phosphate decreases after heating. The heat-induced CCP associates with the micelles, which is indicated by the direction of the arrow. A similar effect occurs when the milk is concentrated and some additional CCP is formed. During the cooling of milk the solubility of the calcium phosphate increases and the arrow points to the serum phase. In the case of ultrafiltration the soluble calcium and phosphate ions are partially removed. The ions pass through the membrane and are lost in the permeate. Some casein dissociation occurs in the protein-rich concentrate. Emulsifiers or calcium sequestrants remove CCP from the paracasein micelles during the preparation of processed cheese. It affects the cheese structure and a smooth texture is created. All the CCP become soluble during the acidification of milk to pH 4.6 resulting in disintegration of the micelles (Figure 2.3). In the following section, attention will be focused on the way

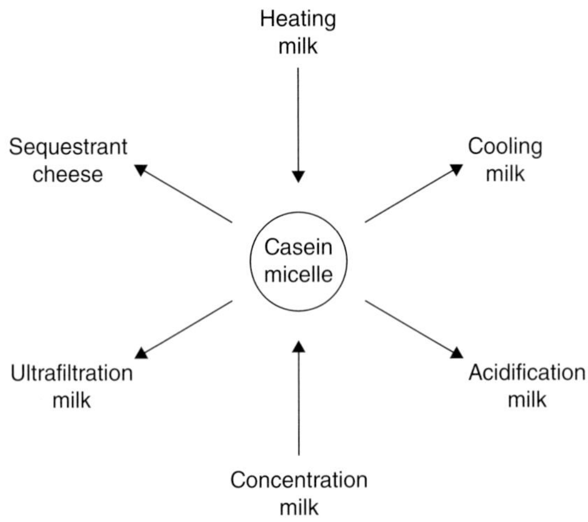

Figure 2.6 Colloidal calcium phosphate: solubilization (arrow towards periphery) or association (arrow towards centre).

in which the combination of acidification and heat affects the structure of cheese ingredients.

An example of a successful ingredient cheese is mozzarella. Some variants of this cheese are available, such as mozzarella low moisture part-skim, cagliata (see information sheet 'Unmatured Cheese Low Salt') and imitation mozzarella (Table 1.4). Here we discuss 'mozzarella' cheese, which is usually made from cow's milk. Mozzarella is a fresh cheese that can be characterized by its stretch properties. Stretch is the ability of cheese to form fibrous strands during heating. Due to this property, mozzarella is preferred for use on pizzas. The important role of insoluble calcium phosphate (i.e. CPP) in modulating cheese functionality is reviewed in detail by Lucey (Lucey *et al.*, 2003; Lucey, 2008; Lucey & Horne, 2009). In the usual procedure for making mozzarella, the pH of the cheese is lowered by means of lactic acid cultures to 5.2–5.3. The calcium is partially removed from the para-casein micelles, which become more flexible due to the reduction of the cementing agent calcium phosphate. If acidification during the initial stage of cheese-making is avoided the cheese structure will become more rigid because of excessive calcium phosphate cross-linking (nanoclusters). After the fermentation process the stretch properties have been given to the curd. The stretching process is typically carried out at a temperature of 58°C by using a brine solution of about 80°C. The relative low curd pH and the high temperature favour a limited aggregation of the paracasein and the formation of fibres of high tensile strength (Guinee, 2002). During the stretching process the bacterial numbers and the residual enzymes, including rennet, are lowered too. The result is a low level of proteolysis, which restricts the development of shorter fibres and extends the storage time of the mozzarella cheese. Stabilization is further improved when the cheese is frozen after a short ripening period (individual quick frozen – IQF cheese). In this case any further decrease in insoluble calcium phosphate, which is a chemical process, is brought to an end and further changes in texture are avoided. Besides the factors mentioned earlier, the composition of the cheese is also vital for its texture (Kindsted, 1993). Mozzarella has a moisture content of approximately 46%, which is considerably higher than that of Cheddar at around 37%. The result is that the mozzarella cheese structure is less rigid. In mozzarella low moisture part-skim the content of the fat is lowered to control 'oiling off' when the pizza is being baked. However, it results in a firmer texture of the cheese. The properties of 'pizza cheese' are optimized by controlling various parameters:

- pH after fermentation;
- insoluble calcium phosphate content of the curd;
- temperature treatment and moisture content;
- fat content of the cheese.

2.5 Milk sugar

Milk sugar, or lactose, is abundantly present in the milk of nearly all mammals. In cow's milk its average level is higher than that of the other constituents (Table 2.1). Lactose is a rather homogeneous substance. Only trace amounts of monosaccharides and other carbohydrates, for instance sialyl oligosaccharides, are present (McJarrow & Amelsfort van Schoonbeek, 2004). As shown in the information sheets, milk sugar is present in an almost pure form in pharmaceutical and edible lactose and it is a major constituent in whey powder, permeate powder and skimmed milk powder. In other ingredients from milk, its content is minimized for reasons of lactose intolerance (lactose-free or low-lactose milk powder) or purity of the protein source (caseinates, MPCs).

In the centre of Figure 2.7 the structure of a molecule of lactose is shown. It is a disaccharide consisting of galactose and glucose linked by a β1-4 glycosidic bond. On the right-hand side is the glucose part, a reducing carbohydrate. The configuration around the C1 of glucose is not stable. The glucose moiety may exist in two isomers, the α and the β forms. In the structural formula of α-lactose, the hydroxyl group on the C1 of glucose is orientated downwards. In case of β-lactose, it is positioned upwards. The change in solution from one form into the other is called 'mutarotation'. It is a consequence of the fact that the hemiacetal group (ring) is in equilibrium with the open-chain aldehyde form.

In Section 2.5.1 the physicochemical properties of lactose will be reviewed. The focus of the physical aspects will be on its crystallization behaviour, which is among others important for the manufacture of various lactose ingredients. In Section 2.5.2, dealing with chemical aspects, the various modifications will be discussed.

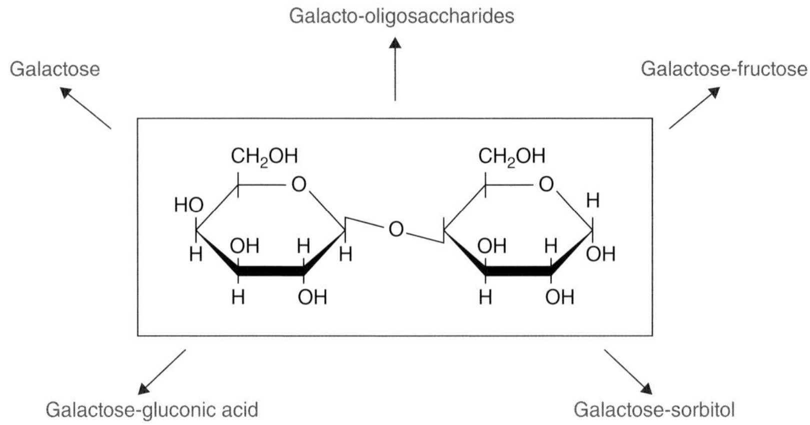

Figure 2.7 Variety in milk sugar derivatives.

2.5.1 Milk sugar: physical aspects

The most common solid forms of lactose are amorphous lactose, α-lactose monohydrate and β-lactose anhydride. Amorphous lactose is the non-crystallized form and results from rapid evaporation/drying. Amorphous lactose contains the same ratio of α- and β-lactose that was initially present in the solution. The presence of amorphous lactose, for instance in whey powder, often leads to caking due to its highly hygroscopic nature (Mol & Roetman, 1974). The 'pure' lactose variants are present in crystallized form. α-Lactose monohydrate ($C_{12}H_{22}O_{11}\cdot H_2O$) is by far the most common form, while the use of β-lactose anhydride ($C_{12}H_{22}O_{11}$) is limited to a few applications. Depending on the processing method, each crystal variant includes small amounts of the other isomer. For instance β-lactose contains about 20% of anhydrous α-lactose (see information sheet 'Lactose anhydrous pharma') and α- lactose monohydrate includes about 5% β-lactose anhydride.

The solubility of lactose plays an important role in the manufacture of edible as well as pharmaceutical lactose. Figure 2.8 shows the solubility of lactose at various temperatures. The figure is based on the data of Whittier (1944). Curves for the initial solubility of both α- and β-lactose are drawn, showing that the solubility of β-lactose is considerably greater than that of α-lactose. The curve for α-lactose is particularly steep at higher temperatures and intersects the β-lactose line at 93.5°C. The solution is saturated with the α-form below 93.5°C and with the β-form above 93.5°C. The values of the final solubility are the same regardless of the form initially dissolved and refer to solutions in α/β equilibrium. Furthermore, a super solubility curve is drawn. The practical significance of super solubility lies in the fact that solutions below this line and above the final solubility line, will not crystallize unless lactose seed crystals are introduced. This region is called the metastable area. Above the super solubility line, the labile area, spontaneous crystallization takes place. Control

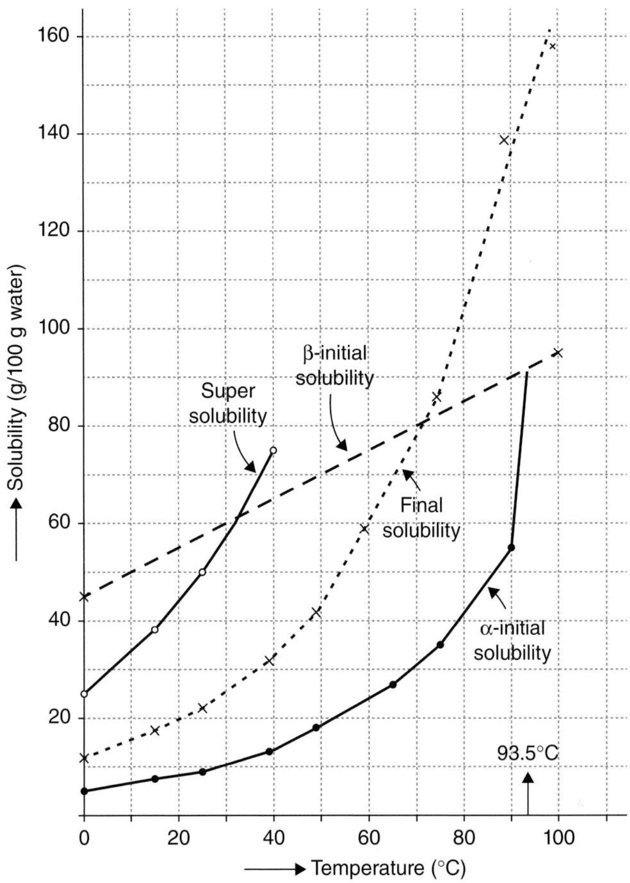

Figure 2.8 Solubility of lactose at various temperatures. *Source:* Data from Whittier, 1944.

of crystallization is vital in order to maximize the yield and achieve a desired quality of lactose.

Nowadays, the starting material for the manufacture of lactose is mostly whey permeate. The rate of concentration is the basis for a good yield. Usually, the permeate is concentrated to 60% total solids and contains about 120 g lactose per 100 g water. The final solubility will be about 20 g lactose per 100 g water, if the temperature is brought to 20°C (Figure 2.8). In this example the theoretical yield is slightly above 80%. The mentioned yield can be further improved by decreasing the temperature, for instance to 5°C, or increasing the lactose content of the permeate to 200 g per 100 g water (Paterson, 2009). In the latter case the start of crystallization – nucleation – is allowed during concentration. It can be the beginning of a continuous operation; however, such a process is more difficult to stabilize (Vu *et al.*, 2006). Batch processing is therefore the common practice in lactose factories. The cooling curve applied is often considered confidential. Temperatures experienced industrially range from 70 to 10°C and the batch time is approximately 24 h. The target is to control crystal size and final particle size distribution for convenient downstream processing. Models have been developed to unravel the various factors involved in the major crystallization mechanisms (Butler, 1998). The process has great complexity as the different reaction steps – mutarotation, nucleation, crystal growth – occur simultaneously (Mimouni, 2009). Crystallization of lactose is a slow process, which can be explained by mutarotation. During the crystallization, as the α-form crystallizes from the supersaturated solution, the concentration of α-lactose present in the solution decreases. The molecules of the β-anomer then mutarotate into the α-form to restore equilibrium. The rate of mutarotation is strongly affected by temperature. For instance at 20°C it takes many hours before the mutarotation equilibrium has been reached. At 70°C it only takes a few minutes. The β/α equilibrium itself also changes to some extent from 62/38 at 5°C to 59/41 at 70°C (Roetman, 1982). The nucleation rate of α-lactose is particularly a function of absolute supersaturation (McLeod *et al.*, 2011). For the production of large particles with a narrow size distribution, the nucleation rate has to be carefully controlled. Seeding in the metastable region is often applied and hence the growth of large crystals promoted. Minimum spontaneous (secondary) nucleation is required to produce large crystals with a narrow size distribution. The agitation rate in the crystallizer is kept at a low level, one that is just sufficient to maintain the particles in solution (Wong *et al.*, 2011). After induction of nucleation the solution is gradually cooled to promote further growth of the crystals and to optimize the yield. The α-lactose monohydrate crystals can easily be separated from the liquid. However, the theoretical yield mentioned earlier cannot be achieved due to losses of lactose during processing (fines, recycling after washing). The yield achieved under industrial circumstances is typically 65%.

In practice various shapes of α-lactose monohydrate crystals can be found such as diamonds, prisms, pyramids and tomahawks depending on the conditions of crystallization. All the forms found are shown to be crystallographically equivalent to the tomahawk form (van Kreveld & Michaels, 1964). The tomahawk shape, being the most common morphology, is demonstrated in Figure 2.9. The Miller indices of the various lactose faces are included in the illustration (Michaels & van Kreveld, 1966). The crystal shown has a sharp top ($0\bar{1}0$) and is formed at a low supersaturation level. Crystal growth has started at the top of the picture (apex) and extends downwards while forming its characteristic faces.

Lactose crystals can incorporate or absorb substances such as calcium phosphate deposits, proteins, lactose phosphate and riboflavin (Wang *et al.*, 2001). Calcium phosphates in permeate interfere with the lactose crystallization process. They cause undesirable nucleation and the deposits are incorporated in the crystals, which increases the ash content of the edible lactose. Deposit formation needs to be prevented at the start of the lactose process. Usually the pH of the permeate is adjusted to 7 and the liquid is heated (see Section 2.4). Subsequently the deposits can be separated. An alternative pretreatment is an ion-exchange treatment in which the calcium ions are exchanged for sodium ions. In this way deposits of calcium salts can be avoided. The yellowish colour of the crystal in Figure 2.9 indicates

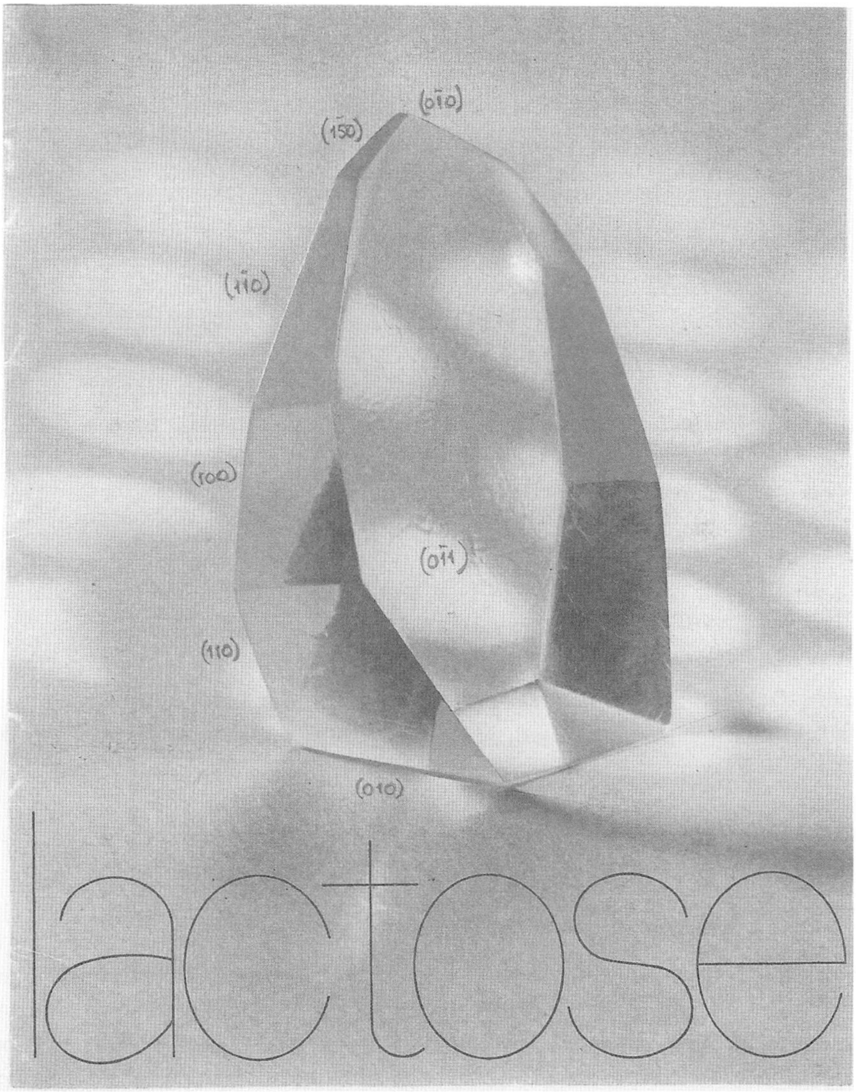

Figure 2.9 Tomahawk shape crystal of α-lactose monohydrate. The crystal faces are indicated by Miller indices (adapted from CCF brochure). (*For colour details please see colour plate section.*)

the presence of riboflavin. Enhanced levels of riboflavin (50 mg/kg) result in an elongated shape and affect the growth on the (100), (1$\bar{1}$0) and (010) faces. Riboflavin is mainly present at the surface. It can be partially removed by washing the crystals during the centrifugation process. In this way the quality of the edible lactose is improved. Lactose phosphate is difficult to remove, even in the case of further purification into pharmaceutical lactose. It is integrated in the crystal lattice. Lifran *et al.* (2006) suggest that lactose phosphate results from the metabolism of lactic acid cultures that are added during cheese-making. This constituent from cheese whey has growth-retarding properties, particularly for the faces (010) and (110) (Visser, 1984, 1988; Lifran *et al.*, 2007). Lactose phosphate is present in trace amounts of 300–400 mg/kg lactose, which means on average one molecule of lactose phosphate for every 3500 molecules of lactose.

2.5.2 Milk sugar: chemical aspects

Figure 2.7 contains an overview of various derivatives of lactose (de Boer & Dijksterhuis, 1998; Timmermans, 1998; Gänzle *et al.*, 2008; Lifran *et al.*, 2009). The galactose part is characteristic for the various milk sugar derivatives. The glucose component of the molecule can be converted by isomerization from an alkaline solution into fructose (lactulose). Sorbitol (lactitol) can be produced from glucose by catalytic hydrogenation. Gluconic acid (lactobionic acid) can be obtained by oxidation from glucose. The galactose part can be released by acid hydrolysis at high temperature. The basis of these chemical derivatives is a highly refined lactose. Its purity limits possible poisoning of catalysts by nitrogen substances and makes downstream processing less complicated. Besides the chemical process, enzymatic hydrolysis can also be applied to modify lactose. The bond between glucose and galactose is split by β-galactosidase, for instance, in the production of low-lactose or lactose-free milk powders. From highly purified lactose, galacto-oligosaccharides (GOS) can also be manufactured. The same type of enzyme is used under specific conditions to convert lactose into a oligosaccharides. GOS consists of a chain of galactose molecules with a glucose molecule at the reducing terminus. The number of 'mers' is two to eight. Another, less known modification is lactosucrose. It is a trisaccharide consisting of glucose, galactose and fructose. Lactosucrose is synthesized from lactose and sucrose using the enzyme fructosyltransferase.

In most of the value-added derivatives discussed above the dairy industry is involved only as a supplier of raw materials. The background is that the production of these compounds requires equipment and procedures not in use elsewhere in the dairy industry. As opposed to chemical processes, enzymatic modification can be considered a core activity. Cheese production is an example of complicated enzymatic processes. This familiarity with enzymatic conversions was useful for the application of enzymatic hydrolysis as a tool for modification. The technology is commonly based on the use of the soluble enzyme β-galactosidase, or lactase enzyme. The target of the enzymatic hydrolysis is alleviating lactose intolerance and facilitating global use of dairy products. These considerations were a starting point for the selection of the information sheets provided at the back of the book.

2.6 Lipids

Lipids are the second most abundant constituent of milk, as shown in Table 2.1. Compared to milk sugar their composition is very complex. Besides milk fat, for instance, fat-soluble vitamins, flavour compounds and colour agents are included in the lipid fraction. Milk lipids are primarily present in milk in the form of fat globules. These globules have a diameter in the range of 1 to 10 μm and are considerably larger than casein micelles (Table 2.3). Fat globules are surrounded by a very thin membrane with a thickness of about 10 to 20 nm. By encapsulation of the fat core, the membrane allows the fat to stay in suspension and this prevents coalescence. The core contains approximately 98% of the milk fat (Jensen, 2002). This membrane is a dynamic, unstable structure that can easily be destroyed by processing. In its natural state the membrane contains not only a variety of lipid components but also various proteins. The proteins account for 1–4% of the total milk proteins (Vanderghem *et al.*, 2010).

In this section, various aspects of milk lipids will be discussed. A distinction will be drawn between the real milk-fat component – triglycerides, or triacylglycerols (TAG) – and the milk-fat globule membrane (MFGM). Amongst the milk-fat ingredients in the information sheets, butter is also listed; for the most part butter is used as a consumer product. Therefore reference is made to an excellent overview in which the process and properties of butter are described (see Walstra *et al.*, 2006, pp. 467–88).

2.6.1 Lipids: triacylgycerols

The TAG component of milk lipids contains numerous fatty acids and unique short-chain fatty acids. About 500 fatty acids are currently known (Hettinga, 2005). The mean fatty acid composition of Dutch bovine milk was investigated in 2005 (Heck, 2009). It was found that 34 fatty acids comprise about 96% of the total fat. The other 4% is composed of numerous different fatty acids (>100). Table 2.10 gives an overview of the 12 major fatty acids. If we consider only this number of fatty acids, the possible number of tracylglycerol structures is 12^3 (1728) using a purely random model. The fatty acid spectrum is dominated by acids containing an even number of carbon atoms. The principal fatty acids in bovine milk are palmitic, oleic, stearic and myristic acid. The short-chain fatty acids, which are a vital element of butter flavour, show lower numbers, at least when expressed on a weight basis. In the information sheet 'Butter' both weight percent (wt.%) and molecular weight percent (mol%) are provided. The mol% – about twice the value of the wt.% – demonstrates more clearly the important contribution of the short-chain fatty acids to the total fatty acid profile. The range of values of all fatty acids shows a large variation as indicated in the first column of Table 2.10. The fatty acids arise from two sources: synthesis *de novo* in the mammary gland and plasma lipids originating from the feed. The fatty acids produced *de novo* are those with a carbon chain of 4–14 atoms, whereas the remainder of the C_{16} fatty acids and almost all of the C_{18} acids arise from the cow's blood (Huppertz *et al.*, 2009). The influence of the feed, via the blood of the cow, is illustrated in the columns headed 'Winter fat', 'Summer fat' and 'Special feeding program'. The data in the columns 2, 3 and 5 are provided by

Table 2.10 The fatty acid composition of various types of bovine milk fat (wt.%).

Fatty acids	Range of values	Winter fat	Summer fat	Summer fat olein-13°C	Special feeding programme
4:0 butyric acid	2–5	3.8	3.7	4.5	3.6
6:0 caproic acid	1–5	2.3	2.2	2.7	2.3
8:0 caprylic acid	1–3	1.4	1.3	1.7	1.2
10:0 capric acid	2–4	3.1	2.8	3.7	2.8
12:0 lauric acid	2–5	4.1	3.8	4.1	3.2
14:0 myristic acid	8–14	11.6	10.9	10.4	11.0
16:0 palmitic acid	22–35	32.2	29.3	22.3	26.1
16:1 palmitoleic acid	1–3	1.7	1.7	2.3	1.7
18:0 stearic acid	9–14	9.2	10.4	3.1	13.0
18:1 oleic acid	20–30	19.8	22.7	28.0	30.0
18:2 linoleic acid	1–3	1.5	1.8	2.3	3.6
18:3 linolenic acid	0.5–2	0.4	0.5	0.6	0.9
Saturated	71	73.0	69.6	59.8	62.5
Unsaturated	29	27.0	30.4	40.2	37.5

Source: Data from Jensen, 2002 and FrieslandCampina, 2006.

FrieslandCampina and relate to lowland cattle breeds. The various feed sources have an impact on the unsaturated fat levels, which are 27.0, 30.4 and 40.2 in columns 2, 3 and 4, respectively. The amount of unsaturated fatty acids is increased by feeding green fodder during summertime. A special feeding programme, which makes use of an extruded linseed supplement, further increases the unsaturated fatty acid level (VanVolsem, 2000). In these cases average values are used as it is not possible to say that an exact quantity of any single component is representative of these feeding schedules.

As indicated in the information sheet 'Anhydrous Milk Fat (AMF)', the increased shelf life at ambient temperature is an important advantage of AMF over butter. A block chart of the manufacture of AMF is provided by Tetra Pak Processing (2003). AMF can be produced from butter or cream as a starting material. Cream is mostly preferred as a raw material for reasons of a consistent AMF quality. A subsequent step in upgrading AMF is the production of various fractions ranging from soft to hard consistency. The differences are expressed by the melting point indicating the temperature (range) at which the fat changes from solid to liquid state.

Fractionation is based on the ability of fats and oils to form crystals. Unlike milk sugar, the crystallization of milk fat is a less precise separation process. The solid fraction always carries substantial amounts of the liquid fraction with it (Fouad *et al.*, 1990). Underlying this phenomenon is the diverse composition of the triacylglycerols. Generally speaking the shorter the chain length and the greater the number of double bonds, the lower the melting point. The melting points range from −8°C to 70°C. The fatty acid distribution on the glycerol backbone also plays a

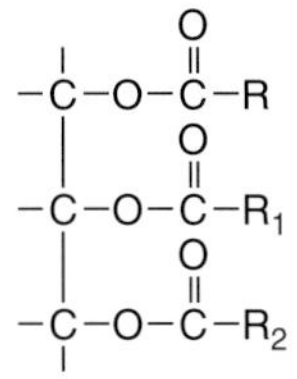

	4:0	16:0	18:1	
$-C-O-C(=O)-R$	2	46	43	*Sn*-1 position
$-C-O-C(=O)-R_1$	1	42	25	*Sn*-2 position
$-C-O-C(=O)-R_2$	97	12	32	*Sn*-3 position

Figure 2.10 Average arrangement of three fatty acid residues in designated positions on glycerol backbone, expressed as mol%. *Source:* Data from Walstra & Jenness, 1984. Reproduced with permission from Wiley-Blackwell.

role. An example of the arrangement of three fatty acid residues is given in Figure 2.10. Nearly all the butyric acid is attached to the *sn*-3 position, while the palmitic acid is mainly located in *sn*-1 and *sn*-2 positions, and the oleic acid is more evenly distributed over the three positions. For this reason milk fat is a highly asymmetric fat (Rajah & Burgess, 1991). Beyond compositional variation, triacylglycerols also show temperature-dependent polymorphic behaviour and formation of mixed crystals, complicating an efficient crystallization process (Walstra *et al.*, 2006, pp. 37–63). The consequence thereof is that multiple steps are needed for obtaining the desired tailor-made fat fractions.

The production of the various milk fat fractions is carried out by a 'dry fractionation' process. In this process the use of solvents is avoided, thereby maintaining the 'natural status' of milk fat fractions, which is important for food applications. In dry fractionation long residence times are required for nucleation and crystal growth (Vanhoutte *et al.*, 2002, 2003). The melted oil is cooled at a low rate to a selected temperature and induces a dense crystal network that results in a low level of oil entrapment. Agitation is required to keep the crystals in suspension and enhance the heat transfer. Crystals of a size 200–350 μm are preferable for separation by filtration in order to keep the inclusion of liquid fat as low as possible (Bulletin IDF, 1991).

In Figure 2.11 an example is shown of a triple fractionation process (Deffense, 1993). The hard fraction is called 'stearin' and the soft fraction is named 'olein'. Starting with AMF of a melting point 33°C (AMF-33°C), an olein-8°C fraction

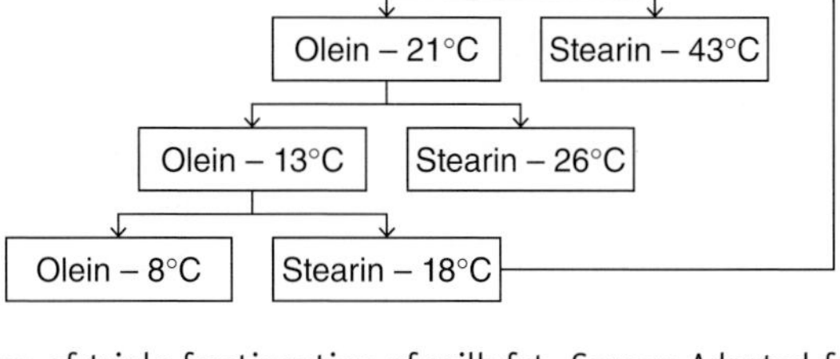

Figure 2.11 Scheme of triple fractionation of milk fat. *Source:* Adapted from Deffense, 1993. AMF, anhydrous milk fat.

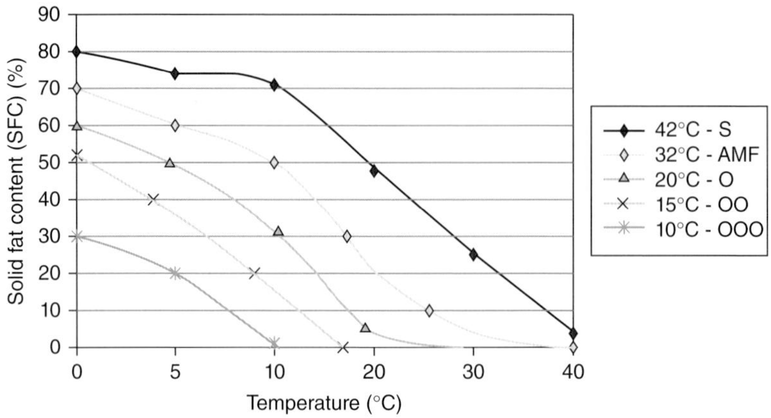

Figure 2.12 Melting curves of various milk fat fractions. *Source:* Boer, 2008. Reproduced with permission from the author, F. Boer. AMF, anhydrous milk fat; S, stearin; O, OO, OOO, different olein fractions. (*For colour details please see colour plate section.*)

can be obtained in a three-step crystallization. The high-melting-point fractions will be removed in the first steps. The minimum overall yield of olein-8°C is 20%, but can be increased by recycling the intermediate soft stearin fraction (stearin-18°C) into the feed material. In Table 2.10 column 4 the fatty acid composition of an olein-13°C fraction, produced from summer fat, is shown based on data received from FrieslandCampina. The fatty acid composition does not change dramatically, but the unsaturated fatty acid percentage in this fraction is still the highest of all indicated. Besides the triacylglycerols, this fraction also contains enhanced levels of flavour compounds, colour agents (carotenoids) and cholesterol.

The melting curves of various milk fat fractions are shown in Figure 2.12 (Boer, 2008). The fractions are produced following a multi-step fractionation process. The selected crystallization temperatures are indicated and range from 42 to 10°C. The melting profiles of the AMF, stearin and the three olein fractions (O, OO, OOO) show distinctive differences in the melting behaviour. They clearly demonstrate the various options available for application.

2.6.2 Lipids: milk fat globule membrane (MFGM)

The MFGM component of milk represents only a minor proportion of the total composition (Table 2.1). So it is amazing that such a small quantity of material is able to enclose such large numbers of fat globules. Assuming a fat content of the milk of 4% and an average fat globule diameter of 3 μm, the surface area of all globules in 1 kg of milk is approximately 800,000 cm^2. The composition of the MFGM is completely different from either the milk fat or the milk serum. It resembles the natural membranes occurring in all living organisms. For this reason it is particularly interesting for health-related topics. Our main focus will be the

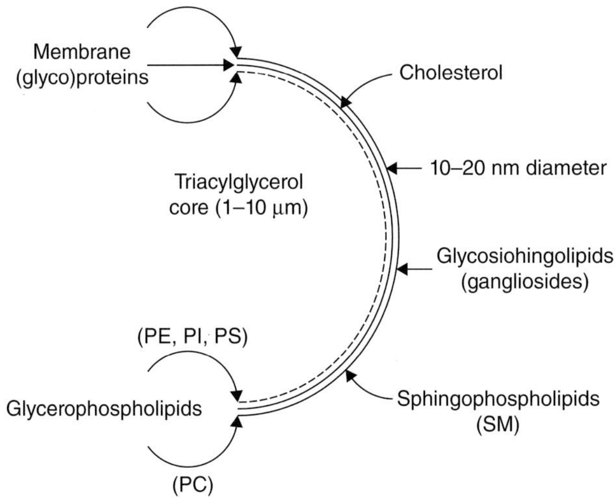

Figure 2.13 Schematic overview of the milk fat globule membrane from unprocessed milk with some components of interest for ingredients. *Source:* Data from Lopez *et al.*, 2008. PC, phosphatidylcholine; PE, phosphatidylethanolamine; PI, phosphatidylinositol; PS, phosphatidylserine; SM, sphingomyelin (SM).

connection with ingredients from milk. Additional background information can be obtained from various recent reviews (Singh, 2006; Dewettinck *et al.*, 2008; Lopez *et al.*, 2008; Huppertz *et al.*, 2009; Vanderghem *et al.*, 2010; El-Loly, 2011).

A schematic overview of the milk fat globule membrane from unprocessed milk is given in Figure 2.13. Only those compounds that are relevant for the MFGM information sheets are mentioned. In its natural state the membrane consists of a complex mixture of proteins, phospholipids and minor constituents. The lipids and proteins are asymmetrically arranged in the membrane. The membrane is a trilayer with a proteinaceous lining on the inner face bounded by an outer bilayer of phospholipids. Some of the proteins act as anchor points, thereby forming a complex that interconnects the inner and outer layers. The bilayer consists of five types of phospholipids of which three are located on the inside and two on the outside. The five phospholipids are: phosphatidylethanolamine (PE), phosphatidylcholine (PC), phosphatidylserine (PS), phosphatidylinositol (PI) and sphingomyelin (SM). The molecular structures of these phospholipids are set out in Figure 2.14.

Data on the levels of the various phospholipids are included in the information sheets. The principal phospholipids are PE, PC and SM. Each represents approximately 30% of the total phospholipids. The share of the minor phospholipids, PI and PS, is about 10% each (see also Gallier *et al.*, 2010). SM and PC are both a source of choline, which is an essential nutrient for brain development. It has a beneficial influence on brain development in neonates and on memory function in adults (Zeisel, 2006; Hellhammer *et al.*, 2010). However, more research in humans is required to confirm these results. The remaining constituents of the MFGM are

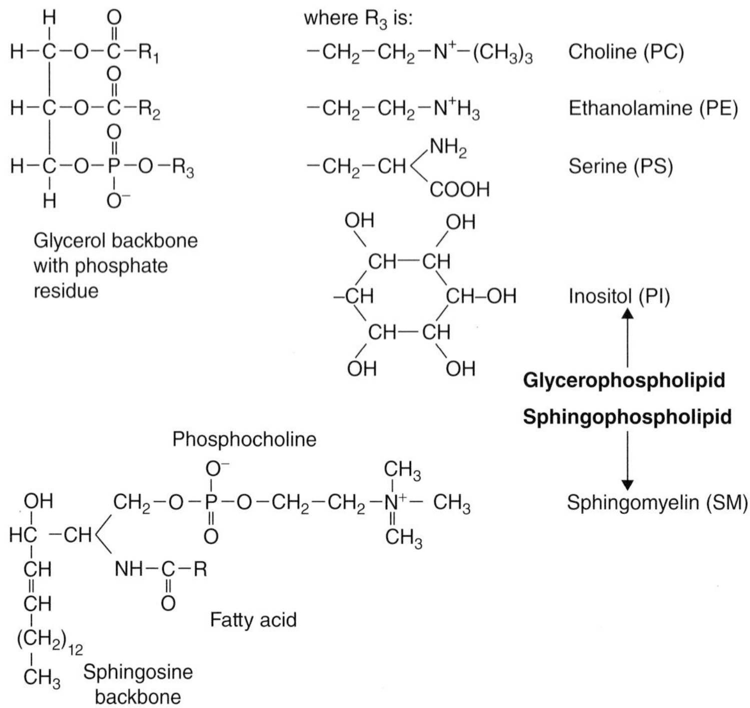

Figure 2.14 Molecular structure of various phospholipids. PC, phosphatidylcholine; PE, phosphatidylethanolamine; PI, phosphatidylinositol; PS, phosphatidylserine; SM, sphingomyelin (SM).

cholesterol and minor amounts of (sialo)glycosphingolipids. The latter are referred to in the MFGM information sheets as gangliosides. They are complex molecules containing oligosaccharides and sialic acid. The sialic acid is understood to be an important 'brain nutrient' (McJarrow *et al.*, 2009). This kind of compound can be of interest for infant food.

Four MFGM information sheets are included in this book. The ingredients were distinguished by the starting material used for their production (buttermilk, cream serum and whey). They are side streams from the manufacture of regular dairy products and their valorization into special components can be feasible. The processes are focused on the most common products in which cooling and heating is part of the treatment. Cooling can induce migration of membrane material to milk serum, while heating causes association of some whey proteins with the outside membrane layer. These process steps change the composition of the original membrane material to some extent. In industrial practice the MFGM is released from the triacylglycerol core by churning (MFGM-buttermilk). During processing the fat globule and air bubbles come into contact with each other, and the MFGM is ruptured and migrates into the buttermilk, whereupon the air bubbles collapse or coalescence takes place.

In case of MFGM-cream serum, the membrane is released during the phase inversion from an oil-in-water emulsion (cream) into a water-in-oil emulsion (AMF). MFGM-whey is derived from cheese-making. The fat is concentrated in the cheese curd and the MFGM is damaged during processing and fragments migrate into the whey.

As is shown in the four MFGM information sheets the phospholipid level is rather modest and ranges from about 1.6% to 16% on dry basis. For value-added ingredients even higher levels of active components are desirable. However, some problems in the further purification of MFGM ingredients must be overcome. Cream serum will be used as an example as this raw material has the most favourable phospholipid content. The cream serum contains not only phospholipids but also other milk proteins, lactose and minerals. The lactose content can be considerably reduced by applying membrane processes. Through this method the phospholipid content can be enhanced (cream serum concentrate). Selective removal of casein is an additional option to elevate the phospholipid content. However, the separation of casein by membrane processes is hardly possible as the sizes of the casein micelles and MFGM components are similar. Another option to reduce the casein content of the cream serum is an additional washing step, which can be carried out before the cream is converted into AMF (Britten *et al.*, 2008). Besides treatments relating to the aqueous phase, the reduction of the neutral lipids can be considered for further purification. According to the information sheet on cream serum the neutral lipids form about 50% of the total lipid content. These neutral lipids are contaminants that attach to the MFGM and are the residue of the phase conversion in the manufacture of AMF. Separation of the neutral lipids and phospholipids is virtually impossible without the use of solvents for extraction and fractionation (Rombaut & Dewettinck, 2006). Although solvents such as ethanol or acetone are not desirable for food usage, such a method of extraction might be acceptable in cosmetics manufacture. MFGM shows exciting possibilities. But more research and development work is needed before it yields well-established ingredients from milk.

References

Abd El-Salam MH, Safinaz El-Shibiny & Aida Salem (2009) Factors affecting the functional properties of whey protein products: a review. *Food Reviews International* **25**: 251–70.

Aguilera JM (1995) Gelation of whey proteins. *Food Technology* October, 83–9.

Alting AC, Weijers M, de Hoog EHA *et al.* (2004) Acid-induced cold gelation of globular proteins: Effects of protein aggregate characteristics and disulfide bonding on rheological properties. *Journal of Agricultural and Food Chemistry* **52**: 623–31.

Anema SG (2009) Stability of milk-derived calcium phosphate suspensions. *Dairy Science and Technology* **89**: 269–82.

Arla Foods Ingredients (2009) Lacprodan OPN-10 Revision of the Kjeldahl factor. *Technical Bulletin BANU/2009-1.*

Azuma N, Maeta A, Fukuchi K & Kanno C (2006) A rapid method for purifying osteopontin from bovine milk and interaction between osteopontin and other milk proteins. *International Dairy Journal* **16**: 370–8.

van den Berg MG, van den Berg G & van Boekel MAJS (1996) Mass transfer processes involved in Gouda cheese manufacture, in relation to casein and yield. *Netherlands Milk & Dairy Journal* **50**: 501–40.

van Berkel PHC, Welling MM, Geerts M *et al.* (2002) Large scale production of recombinant human lactoferrin in the milk of transgenic cows. *Nature Biotechnology* **20**: 484–7.

van Boekel MAJS & Ribadeau-Dumas B (1987) Addendum to the evaluation of the Kjeldahl conversion. *Netherlands Milk and Dairy Journal* **41**: 281–4.

Boer F (2008) Fractioneren maakt melkvet meer dan een fractie beter. *Genootschap ter bevordering van Melkkunde*, November Seminar.

de Boer R & Dijksterhuis J (1998) Permeate, milk sugar, and derivatives – an overview. In: *Whey Conference Proceedings*. International Dairy Federation, pp. 220–6.

Britten M, Lamothe S & Robitaille G (2008) Effect of cream treatment on phospholipids and protein recovery in butter-making process. *International Journal of Food Science and Technology* **43**: 651–7.

Brule G, Real Del Sol E, Fauquant J & Fiaud C (1978) Mineral salts stability in aqueous phase of milk: influence of heat treatments. *Journal of Dairy Science* **61**: 1225–32.

Bulletin IDF (1991) Utilizations of milk fat. *Bulletin of the International Dairy Federation* no. 260.

Bulletin IDF (2006) Comprehensive review of scientific literature pertaining to nitrogen protein conversion factors. *Bulletin of the International Dairy Federation* no. 405.

Butler B (1998) Modelling industrial lactose crystallization. PhD thesis, University of Queensland.

Cals MM, Mailliart P, Brignon C, Anglade P & Ribadeau-Dumas B (1991) The primary structure of bovine lactoperoxidase, a fourth member of a mammalian heme peroxidase family. *European Journal of Biochemistry* **198**: 733–9.

CVB Feed Tables (2007) *Chemical Compositions and Nutritional Values of Feed Ingredients*. The Hague: Product Board Animal Feed.

Dalgleish DG (2011) On the structural models of bovine casein micelles – review and possible improvements. *Soft Matter* **7**: 2265–72.

Dalgleish DG & Corredig M (2012) The structure of the casein micelle of milk and its changes during processing. *Annual Reviews of Food Science and Technology* **3**: 449–67.

Deffense E (1993) Milk fat fractionation today: a review. *Journal of the American Oil Chemists' Society* **70**: 1193–201.

Dewettinck K, Rombaut R, Thienpont N, Le TT, Messens K & Van Camp J (2008) Nutritional and technological aspects of milk fat globule membrane material. *International Dairy Journal* **18**: 436–57.

EDA (1997) *Nutritional Quality of Proteins*. European Dairy Association.

El-Loly MM (2011) Composition, properties and nutritional aspects of milk fat globule membrane – a review. *Polish Journal of Food and Nutrition Sciences* **61**: 7–32.

Elofson C, Dejmek P, Paulsson M & Burling H (1997) Characterization of a cold-gelling whey protein concentrate. *International Dairy Journal* **7**: 601–8.

FAO (2011) Dietary protein quality evaluation in human nutrition. Report of an FAO Expert Consultation. *Food and Nutrition Paper 92*. UN Food & Agriculture Organization.

Fouad FM, van de Voort FR, Marshall WD & Farrell PG (1990) A critical evaluation of thermal fractionation of butter oil. *Journal of the American Oil Chemists' Society* **67**: 981–8.

Fox PF (2001) Milk proteins as food ingredients. *International Journal of Dairy Technology* **54**: 41–55.

Funck E (1948) Geschichtliches über Molke und Milchzucker. *Milchwissenschaft* **3**: 152–60.

Gallier S, Gragson D, Cabral C, Jiménez-Flores R & Everett DW (2010) Composition and fatty acid distribution of bovine milk phospholipids from processed milk products. *Journal of Agricultural and Food Chemistry* **58**: 10503–11.

Gänzle MG, Haase G & Jelen P (2008) Review lactose: crystallization, hydrolysis and value-added derivatives. *International Dairy Journal* **18**: 685–94.

Gao R (2010) Ion speciation in milk-like systems. PhD thesis, Wageningen University.

Gaucheron F (2005) The minerals of milk. *Reproduction Nutritional Development* **45**: 473–83.

González-Chávez SA, Arévalo-Gallegos S & Rascón-Cruz Q (2009) Review Lactoferrin: structure, function and applications. *International Journal of Antimicrobial Agents* **33**: 301.e1–301.e8.

Goodall S, Grandison AS & Jauregi PJ (2008) Selective removal of the major whey proteins using ion exchange membranes. *Journal of Dairy Science* **91**: 1–10.

Graf S, Egert S & Heer M (2011) Review: Effects of whey protein supplements on metabolism: evidence from human intervention studies. *Current Opinion in Clinical Nutrition and Metabolic Care* **14**: 569–80.

Guinee TP (2002) The functionality of cheese as an ingredient: a review. *Australian Journal of Dairy Technology* **57**: 79–91.

Havea P, Carr AJ & Creamer LK (2004) The roles of disulphide and non-covalent bonding in the functional properties of heat-induced whey protein gels. *Journal of Dairy Research* **71**: 330–9.

Havea P, Watkinson P & Kuhn-Sherlock B (2009) Heat-induced whey protein gels: protein–protein interactions and functional properties. *Journal of Agricultural and Food Chemistry* **57**: 1506–12.

Heck JML (2009) Milk genomics – Opportunities to improve the protein and fatty acid composition in raw milk. PhD thesis, Wageningen University, pp. 88–94.

Hellhammer J, Waladkhani AR, Hero T & Buss C (2010) Effect of milk phospholipid on memory and psychological stress response. *British Food Journal.* **112**: 1124–37.

Hettinga D (2005) Butter. In: Shahadi F (ed.), *Bailey's Industrial Oil and Fat Products.* John Wiley & Sons, Ltd, pp. 1–59.

Hettinga K, van Valenberg H, de Vries S *et al.* (2011) The host defense proteome of human and bovine milk. *PLoS ONE* **6**(4): e19433.

Huppertz T, Kelly AL & Fox PF (2009) Milk lipids – composition, origin and properties. In: Tamine AY (ed.), *Dairy Fats and Related Products*. Wiley-Blackwell, pp. 1–27.

Jensen RG (2002) The composition of bovine milk lipids: January 1995 to December 2000. *Journal of Dairy Science* **85**: 295–350.

Karman AH & van Boekel MAJS (1986) Evaluation of the Kjeldahl factor for conversion of the nitrogen content of milk and milk products to protein content. *Netherlands Milk and Dairy Journal* **40**: 315–36.

Kindsted PS (1993) Effect of manufacturing factors, composition and proteolysis on the functional characteristics of Mozzarella cheese. *Critical Reviews in Food Science and Nutrition* **33**: 167–87.

van Kreveld A & Michaels AS (1964) Measurement of crystal growth of α-lactose. *Journal of Dairy Science* **48**: 259–65.

Król J, Liwińczuk Z, Brodziak A & Barlowska J (2010) Lactoferrin, lysozyme and immunoglobulin G content in milk of four breeds of cows managed under intensive production system. *Polish Journal of Veterinary Sciences* **13**: pp. 357–61.

de Kruif CG & Holt C (2003) Casein micelle structure, functions and interactions. In: Fox PF & McSweeney PLH (eds). *Advanced Dairy Chemistry*. Kluwer Academic/Plenum Press, pp. 233–76.

Lifran EV, Vu TTL, Durham RJ, Hourigan JA & Sleigh RW (2006) Lactose phosphate and its origin in cheese making: a case study. *Australian Journal of Dairy Technology* **61**: 163–6.

Lifran EV, Vu TTL, Durham RJ, Hourigan JA & Sleigh RW (2007) Crystallization kinetics of lactose in the presence of lactose phosphate. *Powder Technology* **179**: 43–54.

Lifran EV, Hourigan JA & Sleigh RW (2009) Lactose derivatives: turning waste into functional foods. *Australian Journal of Dairy Technology* **64**: 89–93.

Lopez C, Briard-Bion V, Menard O, Rousseau F, Pradel P & Besle JM (2008) Phospholipid, sphingolipid, and fatty compositions of the milk fat globule membrane are modified by diet. *Journal of Agricultural and Food Chemistry* **56**: 5226–36.

Lucey JA (2008) Some perspectives on the use of cheese as a food ingredient – review. *Dairy Science and Technology* **88**: 573–94.

Lucey JA & Horne DS (2009) Milk salts: technological significance. In: Fox PF & McSweeney P (eds), *Advanced Dairy Chemistry: Vol. 3 Lactose, Water and Minor Constituents.* Springer, pp. 351–89.

Lucey JA, Johnson ME & Horne DS (2003) Perspectives on the basis of the rheology and texture properties of cheese. *Journal of Dairy Science* **86**: 2725–43.

Mallee L & Steijns J (2007) Whey protein concentrates from acidic whey: benefits for use in infant formulas. *Agrofood Industry Hi-tech* **18**: 14–16.

Marcelo PA & Rizvi SSH (2008) Physiochemical properties of liquid virgin whey protein isolate. *International Dairy Journal* **18**: 236–46.

Mazzali M, Kipari T, Pohascharoensuk V, Wesson JA, Johnson R & Hughes J (2002) Review: Osteopontin – a molecule for all seasons. *Quarterly Journal of Medicine* **95**: 3–13.

McJarrow P & Amelsfort van-Schoonbeek J (2004) Bovine sialyl oligosaccharides: seasonal variations in their concentrations in milk, and comparison of the colostrums of Jersey and Friesian cows. *International Dairy Journal* **14**: 571–9.

McJarrow P, Schnell N, Jumpsen J & Clandinin T (2009) Influence of dietary gangliosides on neonatal brain development. *Nutrition Reviews* **67**: 451–63.

McLeod J, Paterson AHJ, Jones JR & Bronlund JE (2011) Primary nucleation of alpha-lactose monohydrate: The effect of supersaturation and temperature. *International Dairy Journal* **21**: 455–61.

Mehra R & Kelly P (2006) Review – milk oligosaccharides: Structural and technological aspects. *International Dairy Journal* **16**: 1334–40.

Melkcontrole Nijland (2010) 100,000 kg milk production 2010, cow Sissi 432. See: http://melkcontrolenijland.nl

Michaels AS & van Kreveld A (1966) Influences of additives on growth rates in lactose crystals. *Netherlands Milk Dairy Journal* **20**: 163–81.

Mimouni A, Schuck P & Bouhallab S (2009) Isothermal batch crystallization of alpha-lactose: a kinetic model combining mutarotation, nucleation and growth steps. *International Dairy Journal* **19**: 129–36.

Mol JJ & Roetman K (1974) Manufacture of non-caking whey powder. *Officieel Orgaan FNZ* **66**: 458.

Moon B & Mangino ME (2004) The effect of preheating on solubility and emulsion stability of whey protein concentrates. *Milchwissenschaft* **59**: 165–9.

Ney DM, Gleason ST, Calcar van SC *et al.* (2009) Nutritional management of PKU with glycomacropeptide from cheese whey. *Journal of Inherited Metabolic Disease* **32**: 32–9.

Olling CCJ (1963) Composition of Friesian whey. *Netherlands Milk & Dairy Journal* **17**: 176–84.

Oude Elferink S (2011) Het goede van melk. *Genootschap ter bevordering van Melkkunde*, April seminar.

Paterson AHJ (2009) Production and uses of lactose. In: McSweeney & Fox PF (eds), *Advanced Dairy Chemistry, Vol. 3: Lactose, Water, Salts and Minor Constituents.* Springer, pp. 105–20.

Peters RH (2005) Review: Economic aspects of cheese making as influenced by whey processing options. *International Dairy Journal* **15**: 537–45.

Pierce A, Colavizza D, Benaissa M *et al.* (1991) Molecular cloning and sequence analysis of bovine lactoferrin. *European Journal of Bichemistry* **196**: 177–84.

Pouliot Y, Landry J & Giasson J (1991) Induction de la précipitation phosphocalcique dans le perméate de lactosérum de fromagerie. *Lait* **71**: 313–20.

Qi PX (2007) Review – Studies of casein micelle structure: the past and present. *Lait* **87**: 361–83.

Rajah KK & Burgess KJ (1991) *Milk Fat: Production, Technology and Utilization*. Society of Dairy Technology.

Roetman K (1982) Quantitative aspects of crystalline lactose in milk products. PhD thesis, Wageningen University.

Rombaut R & Dewettinck K (2006) Review: Properties, analysis and purification of milk polar lipids. *International Dairy Journal* **16**: 1362–73.

Ryan KN, Vardhanabhuti B, Jaramillo DP, van Zanten JH, Coupland JN & Foegeding EA (2012) Stability and mechanism of whey protein soluble aggregates thermally treated with salts. *Food Hydrocolloids* **27**: 411–20.

Scammel AW (2001) Production and uses of colostrum. *Australian Journal of Dairy Technology* **56**: 74–82.

Schack L, Lange A, Kelsen J, Agnholt J, Christensen B, Petersen TE & Sørensen SE (2009) Considerable variation in the concentration of osteopontin in human milk, bovine milk and infant formulas. *Journal of Dairy Science* **92**: 5378–85.

Sikand V, Tong PS & Walker J (2008) Impact of protein standardization of milk powder with lactose or permeate on whey protein nitrogen index and heat classification. *Dairy Science and Technology* **88**: 105–20.

Singh H (2006) The milk fat globule membrane – a biophysical system for food applications. *Current Opinion in Colloid & Interface Science* **11**: 154–63.

Smithers GW (2004) Isolation of growth factors from whey and their application in the food and biotechnology industries – a brief review. *Bulletin International Dairy Federation* no. 389, pp. 16–19.

Smithers GW (2008) Review: Whey and whey proteins – from gutter-to-gold. *International Dairy Journal* **18**: 695–704.

Strohmaier W (2004) Chromatographic fractionation of whey proteins. In: *Bulletin IDF 389: Advances in fractionation and separation: process for novel dairy applications*. International Dairy Federation, pp. 29–35.

Sullivan ST, Khan SA & Eissa AS (2008) Whey proteins: Functionality and foaming under acidic conditions. In: Onwulata CI & Huth PJ (eds), *Whey Processing, Functionality and Health Benefits*. Wiley-Blackwell & IFT Press, pp. 114–26.

Tetra Pak Processing (2003) *Dairy Processing Handbook*. Tetra Pak, pp. 294–9.

Thomä-Worringer C, Sørensen J & López-Fandiño R (2006) Review: Health effects and technological features of caseinomacropeptide. *International Dairy Journal* **16**: 1324–33.

Timmermans E (1998) Lactose derivatives: functions and applications. In: *Whey* [conference proceedings]. International Dairy Federation, pp. 233–50.

Tripathi V & Vashishtha B (2006) Bioactive compounds of colostrum and its application. *Food Reviews International* **22**: 225–44.

Vanderghem C, Bodson P, Danthine S, Paquot M, Deroanne C & Blecker C (2010) Milk fat globule membrane and buttermilks: from composition to valorization. *Biotechnology, Agronomy, Society and Environment* **14**: 485–500.

Vanhoutte B, DeWettinck K, VanIerberghe B & Huyghebaert A (2002) Monitoring milk fat fractionation: effect of agitation, temperature, and residence time on physical properties. *Journal of the American Oil Chemists' Society* **79**: 1169–76.

VanHoutte B, DeWettinck K, VanIerberghe B & Huyghebaert A (2003) Monitoring milk fat fractionation: filtration properties and crystallization kinetics. *Journal of the American Oil Chemists' Society* **80**: 213–18.

VanVolsem T (2000) Lait de vache et produit d'alimentation d'une vache pour obtenir un tel lait. EP 1106077B_2 [patent].

Visser RA (1984) Experiments for tracing growth retarders in lactose. *Netherlands Milk Dairy Journal* **38**: 107–33.

Visser RA (1988) Crystal growth retarding of α-lactose hydrate by sugar phosphates; a continued story. *Netherlands Milk Dairy Journal* **42**: 449–68.

Vu TTL, Durham RJ, Hourigan JA & Sleigh RW (2006) Dynamic modeling optimization and control of lactose crystallizations: Comparison of process alternatives. *Separation and Purification Technology* **48**: 159–66.

Walstra P (1999) Casein sub-micelles: do they exist? *International Dairy Journal* **9**: 189–92.

Walstra P & Jenness R (1984) *Dairy Chemistry and Physics*. John Wiley & Sons, Inc., p. 67

Walstra P, Wouters JTM & Geurts TJ (2006) *Dairy Science and Technology*. CRC Press/Taylor & Francis.

Wang HC, Kurimoto M, Kahr B & Chmielewski J (2001) α-Lactose monohydrate single crystals as hosts for matrix isolation of guest biopolymers. *Bioorganic & Medicinal Chemistry* **9**: 2279–83.

Whittier EO (1944) Lactose and its utilization: a review. *Journal of Dairy Science* **27**: 505–37.

de Wit JN (1998) Nutritional and functional characteristics of whey proteins in food products. *Journal of Dairy Science* **81**: 597–608.

de Wit JN (2001) *Whey – Lecturer's Handbook*. European Whey Products Association, pp. 14–15.

de Wit JN (2009) Thermal behaviour of bovine β-lactoglobulin at temperatures up to 150°C: a review. *Trends in Food Science & Technology* **20**: 27–34.

Wong SY, Bund RK, Connelly RK & Hartel RW (2011) Determination of the dynamic metastable limit for α-lactose monohydrate crystallization. *International Dairy Journal* **21**: 839–47.

Zeisel H (2006) The fetal origins of memory: the role of dietary choline in optimal brain development. *Journal of Pediatrics* November, S131–S136.

3 Patents

Science plays a dominant role in the progress achieved in the field of 'Milk Specifics'. In this chapter the focus will be on the development and protection of 'know-how' in the dairy industry. In science, development of a theory, or 'know-why', is crucial in order to make progress. The driving force in science is sharing of knowledge. Scientific knowledge is easily accessible and the publication of scientific results is common practice. The objective of industry is to acquire exclusivity on knowledge. The application of 'know-how' and its usefulness is emphasized. Access to company information is limited to get an edge on the competition.

An example of extreme exclusivity is the secret formula of the soft drink Coca-Cola (Allen, 1994). The formula, which consists of a variety of ingredients, dates back to 1886. Its composition is deposited in a safe and treated as a kind of state secret. This approach is commonly characterized as a 'trade secret'. Only a few confidants are informed, and Allen (1994) reported that these employees are not even allowed to travel together. The composition of the syrup concentrate is further protected by using code names for the different ingredients. In 1975 a competitor was able to imitate and improve the secret formula. Excellent marketing of Coca-Cola based around the mysterious recipe had already been effective, and the formula remained untouched even after more than a century. Nowadays, the end result is still sold under the name Coca-Cola Classic. A variation on this theme was practised in the manufacture of soup in powder form. Maggi presented the product in 1883. Good social conditions, increased employee loyalty to the company and helped to keep the recipe secret (Seifert, 2008). A more recent strategy to safeguard a company's secrets is by construction of a 'concrete wall' around the production facilities. Outsiders are not permitted to enter the plant and even equipment manufacturers have to deliver their apparatus at the gate. Assembly of the production line is carried out by the technical staff of the company. DMV International (nowadays FrieslandCampina) used this approach extensively in order to protect their sophisticated technology in

From Milk By-Products to Milk Ingredients: Upgrading the Cycle, First Edition. Ruud de Boer.
© 2014 John Wiley & Sons, Ltd. Published 2014 by John Wiley & Sons, Ltd.

the field of caseinates and pharmaceutical lactose. A disadvantage of this approach is that it may be more expensive to maintain a well-equipped technical staff than hiring contractors on a temporary basis. A more open attitude of the food industry with respect to the outside world makes the wall more porous. We live in a time of turbulent change, as clearly described in *Wikinomics* (Tapscott & Williams, 2008). A new balance has to be found between the closed and the open side of a company. In-house innovation alone will not be enough to survive in a fast-changing and intensely competitive economy. An example of a different mindset is 'Ideagoras', which refers to places on the internet where large numbers of people or businesses gather to exchange ideas and solutions. This tool could make the development of new products and services faster and more efficient. Also under the changed circumstances it is wise to keep some applications of knowledge proprietary in order to justify high investments in company R&D.

Besides restricted access to production facilities and confidentiality agreements, applying patents is an additional tool to protect valuable knowledge. In this way companies try to build a 'virtual wall' around their facilities. However, patents form a kind of window in the wall, providing the opportunity to trace developments in the industrial world from a distance. Because of this book's technical nature, consideration of patents is relevant to the overall content. In the following sections, the general aspects of patents and the criteria applied will be discussed (Section 3.1). Various infant food patents will be given as illustrations – infant food was one of the first proprietary foods to emerge in the dairy industry. Some examples of the early developments in infant food from the 19th century (Section 3.2) will be compared to the approach of ingredient suppliers in infant food in the 21st century (Section 3.4). Over the course of time a transition from national to more international patent procedures has taken place, which will be described in Section 3.3, which is not specifically related to infant food.

3.1 General

Aspects of intellectual property that are relevant to industry include trademarks, designs and patents (O'Donnell *et al.*, 2008; Couzens, 2009). A trademark is a sign that distinguishes the goods or services of one company from another. An example of such practice is the previously mentioned Coca-Cola brand. Trademarks, as well as other intellectual property rights, need to be filed in individual countries. In case of trademarks, there is no maximum term for maintaining these rights. Designs describe the shape or appearance of an object. Technical issues need to be avoided here in order to avoid an overlap between designs and patents. In the food industry designs are commonly used to protect special packages of consumer products. As opposed to trademarks, the maximum time span of designs is 25 years. A patent is related to a technical improvement; it provides protection against infringements over a time period typically of 20 years. As patents are an important source of technical information, they will be discussed in more detail. The legal aspects of the patent procedure will be touched upon but are not the main objective of this chapter.

The United States Patent and Trademark Office (USPTO) defines a patent as 'the right to exclude others from making, using, offering for sale, or selling in the United States'. Once the patent is issued, the patentee must enforce the patent without aid of the USPTO (www.uspto.gov). A patent confers an exclusive right for a limited period of time in return for the full disclosure of an invention. The idea behind a patent system is the encouragement of the exchange of information instead of keeping it confidential. In this way the development of science and technology is advanced and the innovation process is stimulated.

A patent is a solution of a technical nature to a technical problem. In order to be patentable an invention must fulfil four criteria:

- new/novel;
- inventive/non-obvious;
- useful/industrially applicable;
- sufficiently disclosed.

New requires that the invention is not yet available in the public domain, in any form. In this context also the terms 'prior art' or 'state of the art' are used. Prior to filing a patent, a novelty search in the general literature and already published patents is the usual procedure. In this way the present state of the art is described as accurately as possible and the 'newness' is established. The *filing date* – the date when a patent application is first filed at a patent office – is an essential element in the patent procedure. 'First-to-file' is a common element in judging the criterion 'new' in most countries. In the USA the approach is different; here the earliest invention date can be used. 'First-to-invent' is a more complicated criterion as it is has to be proven, for instance based on detailed laboratory notebooks. The application must also be filed to secure patent rights. The patent application is then dated before the actual filing date. This effective filing date is called the 'priority date'. The difference between 'first-to-invent' and 'first-to-file' will be soon be obsolete through further harmonization of patent legislation. Recently it was announced that the USA switched to a 'first-to-file' system from 16 March 2013 (USPTO, 2011). Both terms, filing date and priority date, will be used interchangeably in this chapter.

Inventive is a more arbitrary criterion and starts where 'new' ends. 'Novelty' needs to be established over all prior art available, where 'inventive' is assessed in relation to prior art in that and related fields only. The issue here is whether an invention can be anticipated in the industry (Abhinav, 2011). The word 'surprisingly' is frequently mentioned in the patent literature and indicates that the invention cannot be anticipated. The usual approach is that the invention is not obvious to someone skilled in the art. This hypothetical person does not have to be an expert, but a person with average skill (Frischling & Orso, 2005). Of course the skill relates to the field of technology concerned.

Useful means that the invention is applicable in the industry. A scientific theory can be breaking news but is not patentable itself. However, its translation into products and processes, for instance, can be a rich source for patents. It boils down to whether the invention involves technical steps or renders a technical effect.

Sufficiently disclosed is perhaps the most difficult criterion. It is a direct consequence of the principle underlying the patent monopoly. A patentee should share his or her knowledge with others to establish technological progress throughout society. To that end, the patent should provide examples of how a third party could reproduce the invention once the monopoly period ends.

Based on the data of the European Patent Office (www.espacenet.com) approximately 70 million documents are available in the worldwide patent literature including the less accessible Japanese database (www.jpo.go.jp). They can be divided into various fields of technology. The 'food chemistry' field is a modest player according to the international patent filings recorded by the World Intellectual Property Organization (www.wipo.org). It represents approximately 1% of the total filings during the period from 2007 to 2011. Another field related to 'ingredients from milk' is 'pharmaceuticals'. During the same period these filings constituted approximately 5% of the total. The list of the top 100 applicants is dominated by companies like Panasonic, Toyota, Philips, Siemens and Samsung, while a big food company like Nestlé ranked a modest 77th place in 2011. From these figures it can be concluded that food, and as a consequence also the 'milk ingredient' topics, are of minor importance in the enormous mass of recorded patents today. However, it does not mean that their role within the dairy industry can be neglected. Thousands of dairy-related documents are involved, as will be indicated below.

In Figure 3.1 three groups of ingredients from milk are recorded based on the worldwide database of Espacenet (advanced search). The investigated period begins in 1890, which is approximately the starting time of the dairy industry. The period from 1890 to 2009 is divided into intervals of 20 years. The groups of ingredients are 'milk powder', lactose and 'whey protein'. The quotation marks indicate that the two words are used in relation to each other. Whole milk powder will be included in 'milk powder' and whey protein concentrates are part of the 'whey protein' category. If a constituent is not referred to as 'whey protein' in the title or abstract of the patent, it will not be included in the numbers. An example of such a 'whey protein' is osteopontin. The trends in the three groups of ingredients are clearly demonstrated in Figure 3.1. The first commercial drying of milk and manufacture of lactose dates back to approximately 1900. Both show a gradual increase in numbers since that time. A sudden increase in the number of documents occurs in the most recent period from 1990 to 2009. The development of the European Patent Organization might have played a role in this increase too. We will limit ourselves here to the technical issues. The manufacture of undenatured whey protein

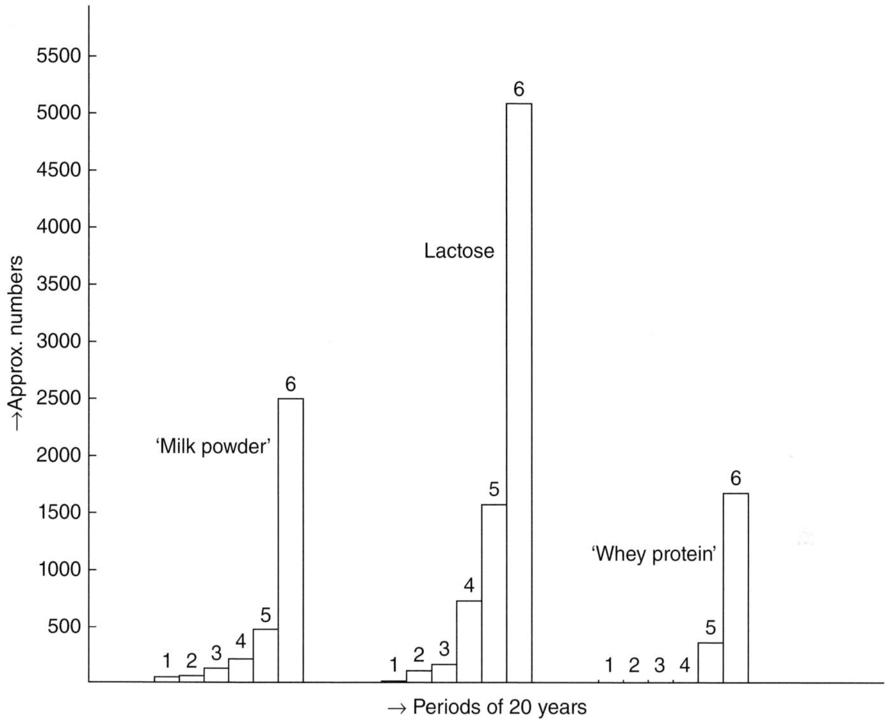

Figure 3.1 Records of three ingredients in the espacenet patent database during 1890–2009 (1: 1890–1909, 2: 1910–29, 3: 1930–49, 4: 1950–69, 5: 1970–89, 6: 1990–2009).

started around 1970; in this case the development was rapid, with a relatively steep increase in the last 20 years. Lactose shows the highest numbers in Figure 3.1. The uses of lactose outnumber by far the production-related patents, which is also the case for 'milk powder' and 'whey protein'. Lactose is used in many ways. The most common patent applications are for use in infant food, pharmaceutical formulations, ice cream, confectionery and meat products, but they are also filed for unusual purposes such as cosmetics, vaccines and hair care. Relatively new players in this field are located in Japan (JP) since 1970 and in China (CN) during the most recent years (1990–2009). Major contributions to the lactose patent portfolio are performed by pharmaceutical companies. 'Milk powder' shows an enormous range of uses such as chocolate, mousses, milk/soya combinations, fish products, seasonings, etc. Sometimes unexpected uses are referred to, for instance bath agents and beauty enhancers. Their recent numbers are also swollen by companies located in China and Japan. 'Whey protein' shows a similar picture and a big variety of uses too. Nutritional applications play an interesting role – examples are allergy reduction, brain development, cardiovascular health, obesity/weight control and sleep assistance. The more common uses are desserts, drinks, cheese products and bakery products, for instance.

Table 3.1 Information regarding some competitors in 'white dairy' consumer products dated 2000.

	General information				White dairy information		
	R&D % total turnover	Total turnover (bn €)	Dairy turnover (bn €)	EU milk processed (bn kg)[a]	EU white dairy turnover (bn €)	Published patents '95–'00[b]	Patents per year per bn € turnover
Nestlé	1.4	55.8	15.1	2.8	2.7	62	3.7
Danone	0.9	14.3	6.9	2.2	3.9	34	1.5
FCDF	0.4	4.4	4.4	5.9	1.7[c]	14	1.4
Campina	0.6	3.9	3.9	4.7	2.5	10	0.7
Müller	0.3[d]	1.7	1.7	1.8	1.2[e]	2	0.3
Ehrmann	0.1[d]	0.42	0.39	0.25	0.35	2	1.0

Source: Courtesy of FrieslandCampina DMV.
[a]Rabobank International *Dairy companies in the European Union* (2001).
[b]Patent search based on International Patent Classification A23C comprising dairy products; excluded were cheese, butter, ingredients, as well as ice cream (A23G) and babyfood (A23L); searched until March 2001.
[c]The white dairy turnover includes the divisions Friesche Vlag, Riedel, Friesland Madibic and Friesland consumer products.
[d]Müller 48 FTE, Ehrmann 5 FTE; number of FTEs were multiplied by €104,000 (€67 per h, 1550 h/year).
[e]Best guess of white dairy turnover of Müller based on 3 bn cups per year and information from Zuivelzicht May 2, 2002.

Obtaining exclusivity of know-how is in many cases a major objective of the industry, as mentioned earlier. Competition plays an important role in achieving such a favourable position. Competitor analysis can be an important tool in the long-term strategy of a company, in which patents can be a significant factor. An example of such an analysis, which was carried out in 2000, is shown in Table 3.1. The footnotes to this table indicate how the information was collected. Six important companies in the so-called 'white dairy' sector were selected. The comparison was carried out with a focus on dairy consumer products with an indulgence character – sweet desserts and fermented desserts – and dairy-based drinks; their processing and packaging were included too. As indicated in Table 3.1, various sources of information were combined to obtain a comprehensive picture. Information was collected regarding R&D efforts, general turnover, 'white dairy' turnover, amount of milk processed and patents and patent applications during a period of 5 years. The 'white dairy' turnover was combined with the number of 'patents' to obtain the amount of 'patents' per year per billion euro turnover. This figure makes possible a comparison of patent activity between the companies. Nestlé scores high, with 3.7 'patents' per year per € bn turnover. The patent portfolio of Nestlé is well balanced and reflects a long term R&D effort in a sophisticated international network. R&D expenditures were estimated at 1.4% of the total turnover. Danone is placed second, with 1.5 'patents' per year per 'white dairy' turnover, while the German based participants close the list. Danone distinguished itself in a concentrated effort in the field of fermentation/functional foods; their R&D expenses were estimated to

Table 3.2 Criteria for filing a patent in industry.

Patent chances
Market prospects
Core technology
Release of critical information
Enforceability
Costs

Source: Courtesy of FrieslandCampina DMV.

be 0.9% of the turnover. The FCDF (Friesland Coberco Dairy Foods) patent portfolio was particularly related to packaging. The attention of Müller and Ehrmann to their patent portfolios was modest and restricted to Germany; trademarks and advertising expenditures are probably more important for their success. Also the R&D efforts of these two private companies were modest in the time span studied. These companies most likely rely on intensive collaboration with suppliers of ingredients. A well-established patent portfolio of the ingredient manufacturer is in these cases also a valuable tool and protects the buyers of ingredients in a competitive world.

Filing a patent is an activity that needs careful consideration. Six criteria are given in Table 3.2. They all play a role in the decision-making process in a company. The 'patent chances' of the potential invention is investigated first. A preliminary search of the literature is carried out based on the criterion 'new'; subsequently the criterion 'inventive' is judged, as mentioned earlier. The 'market prospects' are considered if the conclusion regarding patentability is favourable. As a patent might play a decisive role in the future of a company, it is obvious that the business department of a company is involved. Indeed, a unique market position can be worth a substantial amount of money. The prospects of commercial success of a potential patent should be estimated as well as possible. In a subsequent step it will be determined if the invention belongs to the 'core technology' of the company or its competitors. Examples of core technology of a company's ingredient activities can be, for instance, protein technology, emulsion and foam, and nutrition and physiology. A fourth criterion indicated in Table 3.2 is the possible release of critical information. If the expected invention can be kept proprietary, a company is less likely to start a patent filing procedure. The release of know-how is acceptable if its value is less than the exclusivity of a patent. The fifth criterion is 'enforceability'. This aspect relates to the likelihood of imitation and infringement as well as the willingness to enforce. It is important that the invention can be detected, for example, in the composition of the ingredient. For this reason a product patent is often preferred over a process patent, as will be explained later. At the end of this procedure a go/no go decision will be made. In the decision-making process the costs of a patent will be considered carefully too. In general it is difficult to predict the final costs. The initial costs are modest; they include the cost of a patent attorney for writing the patent application and the filing costs. The fees for a patent attorney will increase as the contents of the application become more complex. The costs mount up significantly with time and depend, for instance, on the patent route chosen (national or international) and the number of countries involved (Couzens, 2009). In a later stage costs associated with defending a patent might be incurred. Usually the supervision of an intellectual property portfolio is one of the duties of general management. A company faces certain risks when applying for a patent. First there is the risk that the patent application will not be granted finally. A Japanese study revealed that the ratio of granted patents to patent applications in the field of chip technology varied from 0.12 to 0.32, dependent on the company involved (Kimura & Tanaka, 2010). Several factors might influence this figure, for instance an unfavourable search report of the patent office or a change in strategy of the company involved. Furthermore, inventions show a highly skewed distribution of returns in which the majority are of relatively low value and a fairly small minority are of extremely high commercial value (Bosworth & Jobome, 1999). In an EU study (Giuri *et al.*, 2007) inventors were asked to rank the present value of their patents. Only 7.2% of the patents in this study were estimated to be worth more than 10 million euros. Smart manoeuvring is required to obtain the desired exclusivity.

3.2 Infant foods in the 19th century

Many great inventions happened in the 19th century. A hundred of them were described by van Dulken and Thompson (2001) as they appeared in the patent literature in Britain and the USA. Most patents were of a technical nature, such as automobiles, bicycles, aircraft, electric bulbs, radios and telephones. An exception in this list was the food-related patent entitled 'Flaked cereals and process of preparing the same', nowadays called a breakfast cereal (Kellogg, 1896). van Dulken and Thompson (2001) did not refer to developments in the dairy industry. However, this sector was also a part of this revolutionary phase, with the invention of evaporators, cream separators, milk driers and pasteurizers, for instance. This took place during the second part of the 19th century. It enabled the eventual transition of milk processing from being a farm-based cottage industry to one involving modern dairy plants (Tetra Pak, 2003).

Infant food was also in its first stages (1865) and efforts were made to reduce high infant mortality. In many cities in the 1890s more than one-third of all children died before their fifth birthday (Apple, 1987). The first patents were filed at that time; no less than 27 brands of patented baby foods were already available by 1883 (Bracken, 1956). Infant food has a rich history, which is described in various references (Meigs, 1896; Radbill, 1981; Barness, 1987; Fomon, 2001; Castilho & de Azevedo Barros Filho, 2010). Nowadays, it is still recognized that the ideal food for newborn infants is mother's milk. However, alternatives are sometimes needed. Over time wet nurses (a woman who breastfeeds another child), animal milks, porridges and other exotic foods have been used. Due to its availability cow's milk was an important source of milk. When an analysis of cow's milk and human milk could be further refined, it was concluded that cow's milk had to be modified to improve its results. This

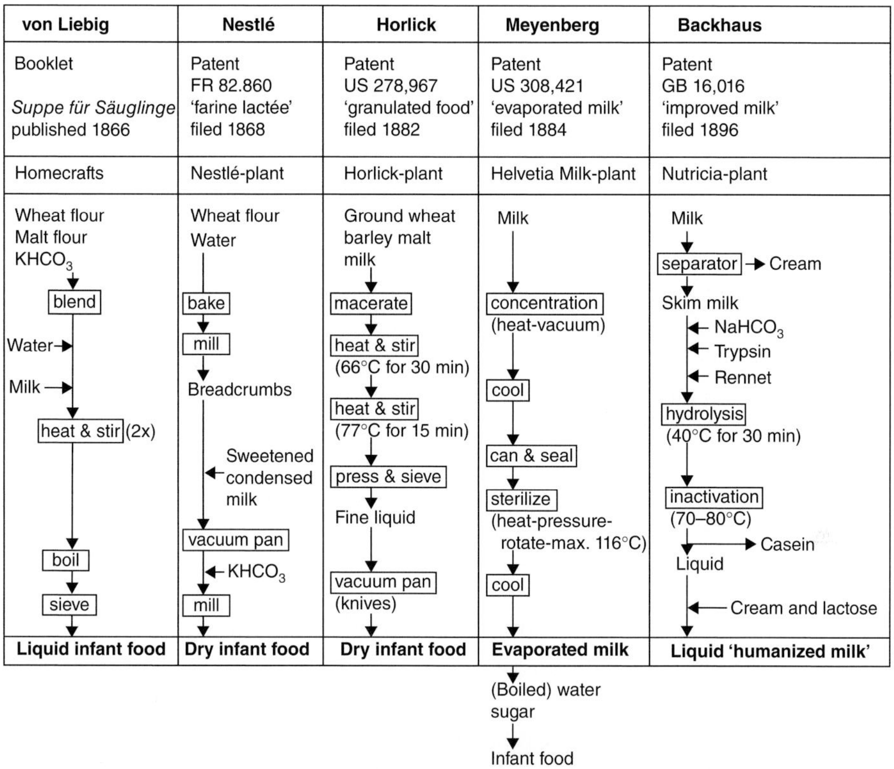

Figure 3.2 Some interesting developments in infant food preparation during the second half of the 19th century.

approach is still valid and serves as a basis for modern infant feeding. Gradually, it also became clear that hygiene is crucial to reduce the mortality rates of infants. Both progress in analysis and awareness of hygiene play important roles in judging the state of the art of proprietary infant food formulas, which were introduced in the second part of the 19th century.

Figure 3.2 gives an overview of the main developments in infant food preparation, in which milk is involved. The process schemes of each of the five types of food are depicted so they can be compared more easily. The unit operations are indicated in the rectangles. Four of the five developments were patented, with patents filed in France (FR), the United States (US) and Great Britain (GB), and registered in different years. Each patent was used on an industrial scale. The end-products obtained could be in a liquid or in a dry form. One of them is a liquid in a sterilized form (evaporated milk). This intermediate product needs further handling in the household, with (boiled) water and sugar needing to be added before it is ready for use.

The publication of von Liebig's (1866) *Suppe für Säuglinge,* with its home-made formula, served as a precursor to industrial development. The recipe and procedure

Table 3.3 Infant food: recipe and preparation[a].

Measure one tablespoon (tbsp) of wheat flour, 1 tbsp of malt flour and add a thimble of a potassium bicarbonate solution (11%)
Mix thoroughly and add 2 tbsp of water and 10 tbsp of a low-fat milk (2.5%)
Heat the liquid[b] in a feeding-bottle under continuous stirring until it becomes viscous
Stop heating and stir for 5 min
Repeat heating until the liquid turns viscous again
Start boiling when the liquid is sweet and the viscosity is low
The food is ready for use after sieving[c]

Source: von Liebig, 1866.
[a]Double the concentration of human milk.
[b]Maximal 66°C.
[c]Shelf life maximal 24 hours.

are explained in more detail in Table 3.3. The aim was to obtain the same nutritional value as human milk. von Liebig's product consisted of milk, wheat flour and malt flour with potassium bicarbonate to reduce the acidity of the flours. Flour or its baked form was used a great deal by the poor and country folk at that time (Radbill, 1981). It was an inexpensive raw material. Malt is a germinated flour that is able to convert starch into simpler carbohydrates. von Liebig distinguished two main functions in human milk. The casein ('Käsestoff') was used for 'blood building' and was an important nutrient for meat. The milk fat ('butter') and milk sugar were used for many other purposes and ultimately converted into heat. The ratio of 'blood building' and 'heat generating' substances in human milk was estimated to be 1 to 3.8. In von Liebig's infant formula, starch largely took the place of milk fat and milk sugar. During the two-step heating process (maximal 66°C and stirring) the insoluble starch of the wheat flour was converted into dextrin and glucose by the malt flour. It was stated that these carbohydrates were more easily digested as milk sugar. Subsequently the mixture was boiled, and the shelf life was estimated to be a maximum of 24 hours at room temperature. A disadvantage was that the procedure was difficult to follow.

At approximately the same time as von Liebig's publication, Nestlé considered the industrial production of infant food. Henri Nestlé developed a new product and called it *farine lactée* (milk-bread). Being a resident of Switzerland, he was free to manufacture the product and did not have to consider any patent legislation in Switzerland at that time (Ritter, 2004). Nestlé applied for a French patent in 1868 as he probably anticipated a profitable business in France too (Nestlé, 1868). But here he had to take into account French patent law. The first page of the document is shown in a handwritten form in Figure 3.3. In the text of the patent application, the state of the art is not mentioned. It is very likely that Nestlé built on the work of von Liebig, who was a famous pioneer in the field of applied chemistry. Both processes are described in Figure 3.2. Nestlé made the preparation of infant food more convenient and extended its shelf life considerably. Only water, previously boiled, needed to be added to the end-product. The main differences with

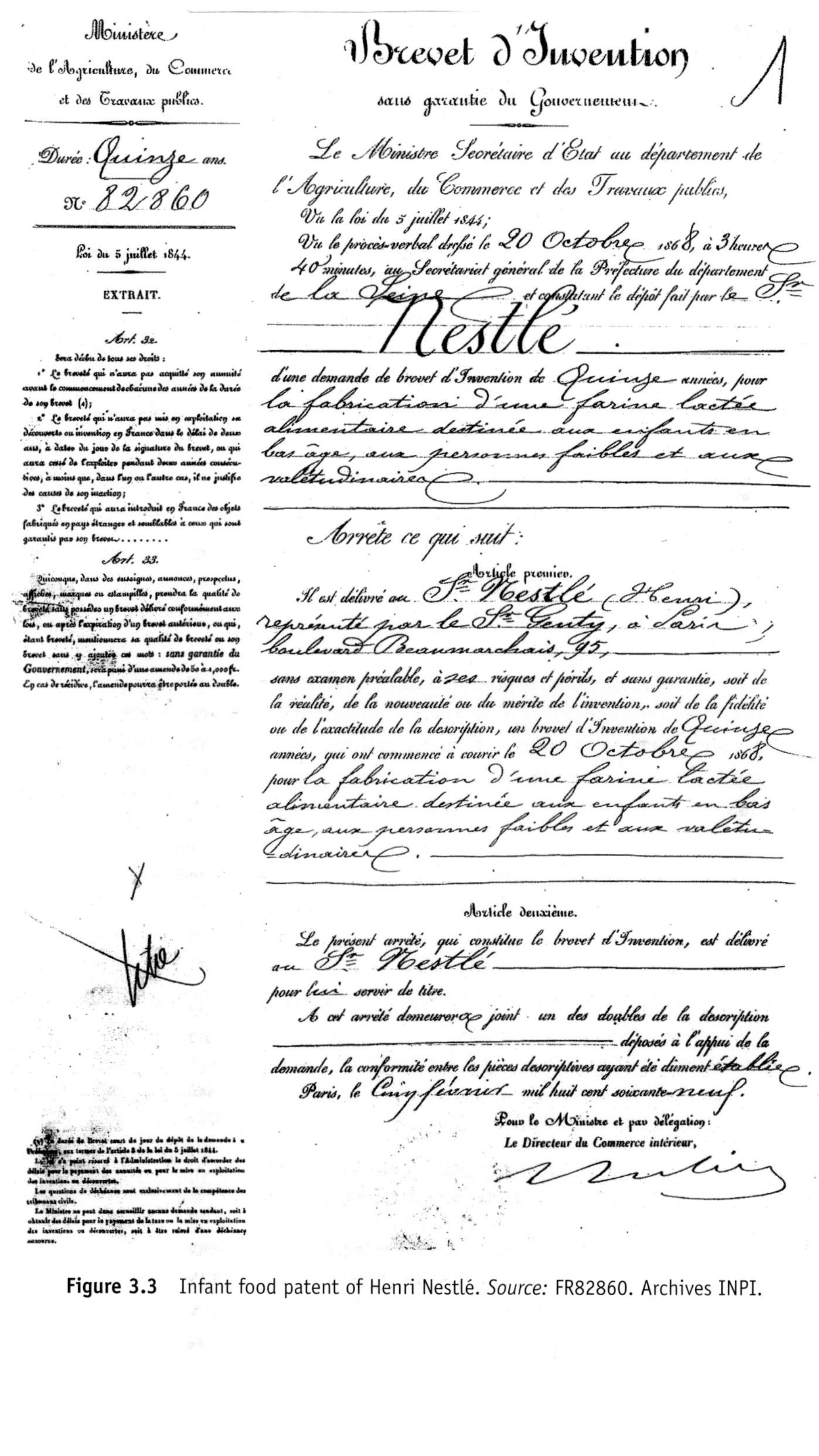

Ministère de l'Agriculture, du Commerce et des Travaux publics.

Durée : Quinze ans.

N° 82860

Loi du 5 juillet 1844.

EXTRAIT.

Art. 32.

Sera déchu de tous ses droits :

1° Le breveté qui n'aura pas acquitté son annuité avant le commencement de chacune des années de la durée de son brevet (1);

2° Le breveté qui n'aura pas mis en exploitation sa découverte ou invention en France dans le délai de deux ans, à dater du jour de la signature du brevet, ou qui aura cessé de l'exploiter pendant deux années consécutives, à moins que, dans l'un ou l'autre cas, il ne justifie des causes de son inaction;

3° Le breveté qui aura introduit en France des objets fabriqués en pays étranger et semblables à ceux qui sont garantis par son brevet........

Art. 33.

Quiconque, dans des enseignes, annonces, prospectus, affiches, marques ou estampilles, prendra la qualité de breveté sans posséder un brevet délivré conformément aux lois, ou après l'expiration d'un brevet antérieur, ou qui, étant breveté, mentionnera sa qualité de breveté ou son brevet sans y ajouter ces mots : sans garantie du Gouvernement, sera puni d'une amende de 50 à 1,000 fr. En cas de récidive, l'amende pourra être portée au double.

(1) La durée du Brevet court du jour du dépôt de la demande à la Préfecture, aux termes de l'article 8 de la loi du 5 juillet 1844.

La loi n'a point réservé à l'Administration le droit d'accorder des délais pour le payement des annuités ou pour la mise en exploitation des inventions ou découvertes.

Les questions de déchéance sont exclusivement de la compétence des tribunaux civils.

Le Ministre ne peut donc accueillir aucune demande tendant, soit à obtenir des délais pour le payement de la taxe ou la mise en exploitation des inventions ou découvertes, soit à être relevé d'une déchéance encourue.

Brevet d'Invention sans garantie du Gouvernement. 1

Le Ministre Secrétaire d'État au département de l'Agriculture, du Commerce et des Travaux publics,

Vu la loi du 5 juillet 1844;

Vu le procès-verbal dressé le 20 Octobre 1868, à 3 heures 40 minutes, au Secrétariat général de la Préfecture du département de la Seine et constatant le dépôt fait par le Sr Nestlé

d'une demande de brevet d'Invention de Quinze années, pour la fabrication d'une farine lactée alimentaire destinée aux enfants en bas âge, aux personnes faibles et aux valétudinaires.

Arrête ce qui suit :

Article premier.

Il est délivré au Sr Nestlé (Henri), représenté par le Sr Genty, à Paris, boulevard Beaumarchais, 95, sans examen préalable, à ses risques et périls, et sans garantie, soit de la réalité, de la nouveauté ou du mérite de l'invention, soit de la fidélité ou de l'exactitude de la description, un brevet d'Invention de Quinze années, qui ont commencé à courir le 20 Octobre 1868, pour la fabrication d'une farine lactée alimentaire destinée aux enfants en bas âge, aux personnes faibles et aux valétudinaires.

Article deuxième.

Le présent arrêté, qui constitue le brevet d'Invention, est délivré au Sr Nestlé pour lui servir de titre.

A cet arrêté demeurera joint un des doubles de la description déposés à l'appui de la demande, la conformité entre les pièces descriptives ayant été dûment établie.

Paris, le Cinq février mil huit cent soixante-neuf.

Pour le Ministre et par délégation :
Le Directeur du Commerce intérieur,

Figure 3.3 Infant food patent of Henri Nestlé. *Source:* FR82860. Archives INPI.

von Liebig's approach were the elimination of malt flour, the use of sweetened condensed milk, the baking of wheat flour and the preparation of a powder instead of a liquid. Some of the elements mentioned were already known, for instance the production of the condensed milk (Borden, 1856) and the use of baked flour and milk. The combination of the ingredients and the dried form were 'new' and probably also 'inventive'. The words *sans garantie du Gouvernement* are printed on the document shown in Figure 3.3. It means that the patent application was not examined by the authorities. The administration did not guarantee the process described by Nestlé, but provided only the delivery. Its value had to be proved in court. Since its early days *farine lactée* gained momentum and formed the basis of a very successful food company (Rowan, 2001).

Another infant food patent was filed 14 years later in 1882 by William Horlick. After approximately one year it was granted in the USA (Horlick, 1883), as can be read in the heading of Figure 3.4. von Liebig was mentioned in the state of the art of this patent; the French invention of Nestlé was probably unknown to USPTO. The process designed by Horlick is shown in Figure 3.2. Unlike Nestlé, no baked goods were used. There were many similarities between the approaches of von Liebig and Horlick, such as the ingredients applied and the conversion of wheat starch by means of malt flour into dextrin. The main difference was that Horlick produced an infant food in a dry form that was soluble in water when required for use. This was achieved by using a special vacuum pan. Part of the US patent is shown in Figure 3.4. The full text of the first claim is provided as it represents the 'heart' of the invention. The process is described in detail, which is relevant for industrial use. The product quality was enhanced by removing husks and insoluble matter (press and sieve). The dried form was a very useful improvement, which was crucial in the business approach of the Horlicks. The range of the product was enlarged tremendously. It was even used during polar expeditions, probably for its high calorie density. Mount Horlick, located on Antarctica, is named after the inventor. The product still exists in a somewhat adapted formula and is currently produced by GlaxoSmithKline.

Only two years later, in 1884, Meyenberg filed a patent application for preserving milk (Meyenberg, 1884a). The equipment is described in the document and a number of drawings illustrate the invention. Sterilization of milk in hermetically sealed cans was carried out by steam under pressure at a temperature exceeding 100°C. During the sterilization process the cans were continuously agitated by using a revolving reel. In Figure 3.5 two claims are mentioned; the invention is described in more detail in the second (dependent) claim. Meyenberg filed another application a few months later; its title was 'Process of preserving milk' (Meyenberg, 1884b). This patent claims the process can preserve milk in cans after it has been condensed in a vacuum. By making use of this invention, it was possible to preserve condensed milk without the aid of sugar – what would come to be called evaporated milk. In an earlier patent Borden (1856) claimed already that he could produce a concentrated milk by evaporation *in vacuo* without the admixture of sugar or other foreign matter. Despite this claim, records show that Gail Borden manufactured only sweetened

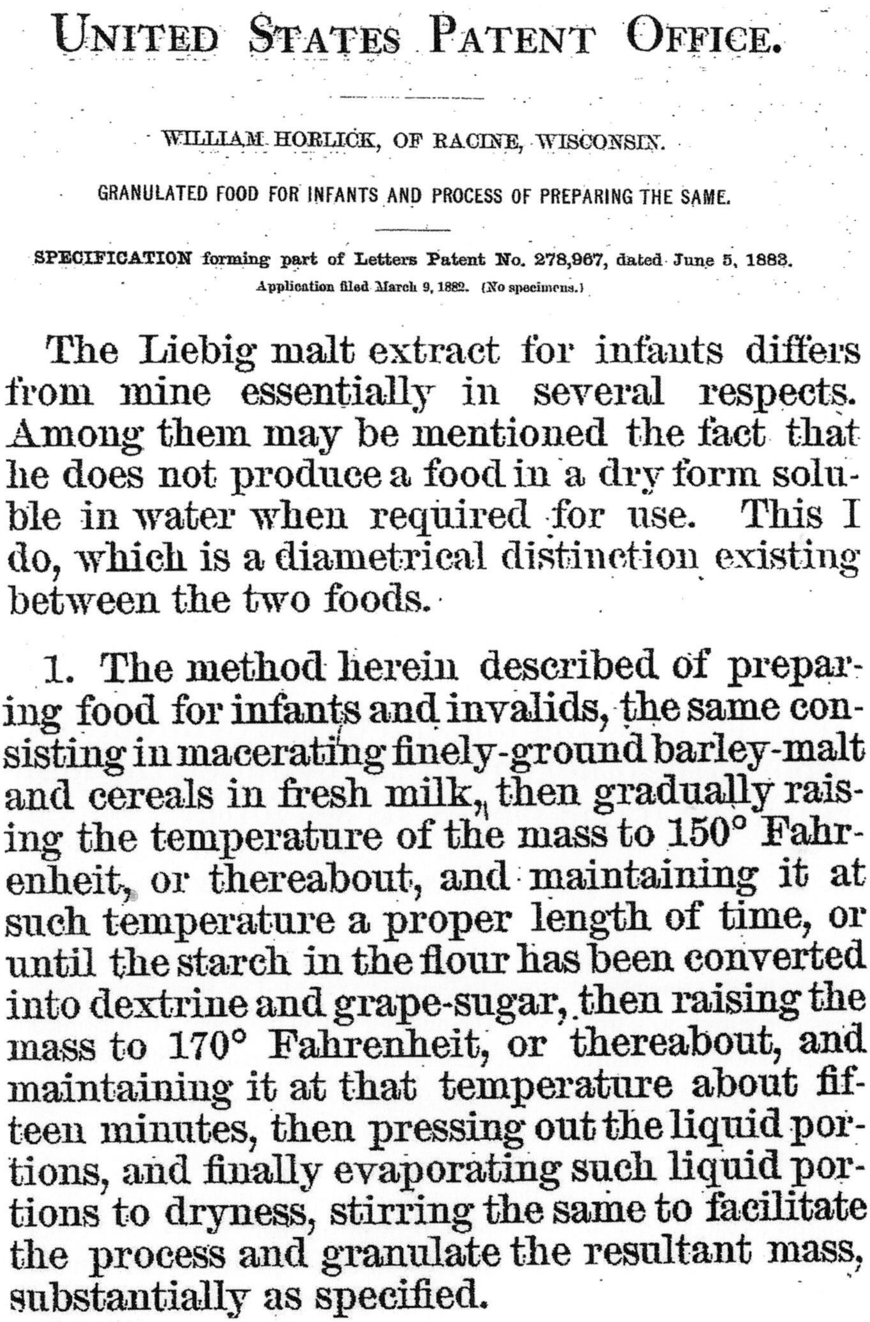

UNITED STATES PATENT OFFICE.

WILLIAM HORLICK, OF RACINE, WISCONSIN.

GRANULATED FOOD FOR INFANTS AND PROCESS OF PREPARING THE SAME.

SPECIFICATION forming part of Letters Patent No. 278,967, dated June 5, 1883.

Application filed March 9, 1882. (No specimens.)

The Liebig malt extract for infants differs from mine essentially in several respects. Among them may be mentioned the fact that he does not produce a food in a dry form soluble in water when required for use. This I do, which is a diametrical distinction existing between the two foods.

1. The method herein described of preparing food for infants and invalids, the same consisting in macerating finely-ground barley-malt and cereals in fresh milk, then gradually raising the temperature of the mass to 150° Fahrenheit, or thereabout, and maintaining it at such temperature a proper length of time, or until the starch in the flour has been converted into dextrine and grape-sugar, then raising the mass to 170° Fahrenheit, or thereabout, and maintaining it at that temperature about fifteen minutes, then pressing out the liquid portions, and finally evaporating such liquid portions to dryness, stirring the same to facilitate the process and granulate the resultant mass, substantially as specified.

Figure 3.4 Infant food patent of William Horlick. *Source:* Patent: 278,967, 1882. United States Patent and Trademark Office (USPTO) with permission from GSK.

UNITED STATES PATENT OFFICE.

JOHN MEŸENBERG, OF ST. LOUIS, MISSOURI.

APPARATUS FOR PRESERVING MILK.

SPECIFICATION forming part of Letters Patent No. 308,421, dated November 25, 1884.

Application filed February 16, 1884. (No model.)

I claim as my invention—

1. The combination of the revolving frame adapted to receive cans, the outer inclosing-cylinder, and the air and steam pipes, arranged and operating substantially as and for the purpose set forth.

2. The combination of the revolving frame H, cylinder I, surrounding the frame, steam-pipes U within the cylinder beneath the frame, air-pipes entering one end of the cylinder, and a door at the other end of the cylinder, through which the frame passes, as set forth.

Figure 3.5 Evaporated milk patent of John Meyenberg. *Source:* US 308,421 (1884). United States Patent and Trademark Office (USPTO).

condensed milk (Hunziker, 1946). The company founded by Borden marketed this type of condensed milk from 1856. The missing link in the approach of Borden was the sterilization step to ensure the long shelf life of evaporated milk.

In spite of the excellent bacteriological quality of evaporated milk and sweetened condensed milk, its application in infant food had two major disadvantages (Apple, 1987). Most physicians of the 19th century had rejected the use of sweetened condensed milk due to the fact that a significant quantity of sugar had to be added for its preservation. According to the American standard, this product contained 45% sugar and 28% milk solids. The other disadvantage was the digestibility of the casein milk protein. However, boiling improved the digestibility of raw milk and made the curd particles smaller and softer. The studies of Brennemann (1929) favoured the use of evaporated milk in infant food. The end-product could be easily prepared; only (boiled) water and some sugar needed to be added. Its application in the USA peaked in the first part of the 20th century (Fomon, 2001).

N° 16,016 **A.D. 1896**

Date of Application, 20th July, 1896

Complete Specification Left, 20th Apr., 1897—Accepted, 19th June, 1897

Backhaus's Improved Treatment of the Milk of Cows or other Domesticated Animals.

An Improved Treatment of the Milk of Cows, or other Domesticated Animals, to Render it Suitable as Food for Infants and Children.

claim is :—

1. The treatment of the milk of cows or other domesticated animals consisting in separating the said milk into cream and skim-milk, subjecting the skim-milk to the combined action of a ferment capable of converting caseïne into soluble albumen, and a ferment capable of precipitating caseïne, rendering the said ferments inoffensive after due action, and adding cream and milk-sugar after separation of the precipitated remainder of caseïne from the liquor, substantially as and for the purpose described.

Figure 3.6 Infant food patent of Alexander Backhaus. *Source:* GB 16,016 (1896) With permission from Nutricia Advanced Medical Nutrition.

The last invention to be discussed in this section was filed, inter alia, in Great Britain (Backhaus, 1896). Backhaus was both a professor and an entrepreneur. He filed in the USA, France, Germany and Belgium too. This patent can be characterized as a ‘process patent’. The words ‘treatment’ (Figure 3.6) or ‘method’ are often used in the claims of this kind of patent. As an alternative a product claim might have been an option. For example, the content of the range of the residual casein in the liquid infant food could have been used. These days a product claim is preferred as this claim can be enforced more easily compared to a process or manufacturing patent. The state of the art of the invention is human milk. It is stated in the British patent that ‘it has not hitherto been possible to make the percentage of casein and soluble albumen correspond with that present in human milk’. The procedure followed by Backhaus is explained in more detail in Figure 3.2. The digestion of the casein by enzymes is carried out under slightly alkaline conditions. The remaining residual casein is converted by rennet into insoluble casein, which can be easily removed. In the US patent the adjustment of the albumen/casein ratio is highlighted in more detail and the claims were slightly adapted. Subsequently, the liquid was standardized by adding lactose and cream (with some casein) to imitate the composition of human milk. The scientific work of Backhaus was very instrumental for the start of Nutricia as an infant food supplier. Backhaus was also a forerunner in the development of humanized infant food, which is the standard nowadays.

3.3 Present patent procedure

As was shown in the previous section, the inventor Backhaus filed a patent application in his native country but also in various foreign countries. This can be fairly time-consuming and costly due to differences in national legislation and language. Over a hundred years later the world is a lot more interconnected; suppliers of ingredients operate in many countries. The patent scenery has also changed, from a national to a more international approach.

The simplest approach is to file a patent in only one country, for instance in The Netherlands (NL), as indicated in Figure 3.7. A search report, which describes the

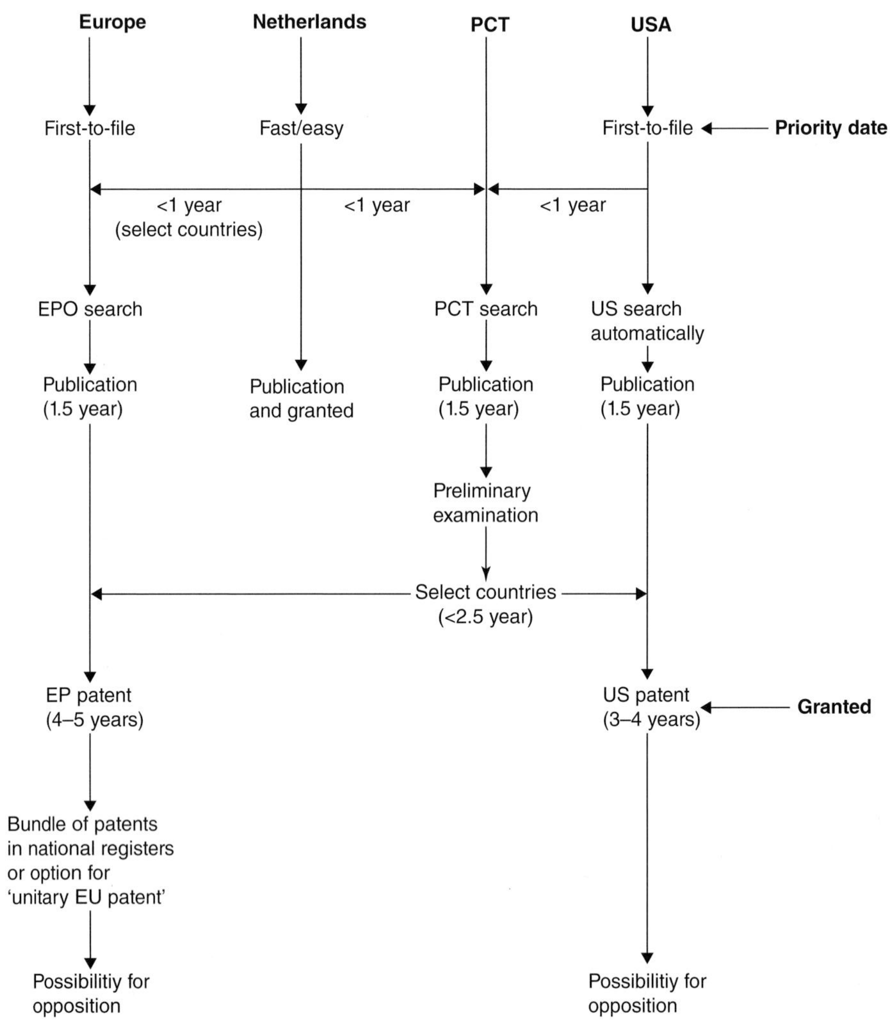

Figure 3.7 Survey of patent procedures. EPO, European Patent Office; PCT, Patent Co-operation Treaty.

state of the art, is delivered by the NL Patent Office. If the applicant takes no action and pays the fees, this NL patent is automatically granted. It can be characterized as 'fast/easy'. The advantage is that the idea is registered and a priority date is created; but its rights remain uncertain as there is no examination on the merits. The strength has to be proven in court. Similar situations occur in other countries, for instance *Gebrauchsmuster* (Germany) and 'Innovation Patent' (Australia). Usually these innovations do not meet the inventive threshold. The procedures are short, but internationally they are less valuable. For this reason they are hardly used for the ingredient market. In the following part of this section the focus will therefore be international patent applications, where a true examination takes place. As indicated in Figure 3.7 the NL document then needs to follow a more extensive procedure. Action must be taken within 1 year.

To make filing more convenient, international organizations are established in which a single patent application can be designated to various states at the same time. In Figure 3.7 different options are shown; two nations are mentioned as an example as well as procedures involving a regional patent office (European Patent Office – EPO) and the world organization (World Intellectual Property Organization – WIPO). In December 2012, the EPO had 38 member states, and in case of WIPO 185 countries signed a treaty called the Patent Co-operation Treaty (PCT). WIPO is a United Nations agency dedicated to the use of intellectual property as a means of encouraging innovation and creativity. The aim of the two organizations is to harmonize and thus simplify filing, searching and publishing international patent applications. The patents/applications of these organizations can be recognized by the abbreviations EP and PCT/WO respectively.

A patent application is published 1.5 years following its first filing date. At that time its contents will become known to the public. In many jurisdictions a search is carried out during the 1.5-year period. The search report will be published after 1.5 years, preferably together with the publication of the patent application (PCT/WO and EP systems). In the search report documents of particular relevance are shown. If the examiner questions whether the claimed invention is novel and/or inventive the documents cited are indicated by X and/or Y. Nowadays, the patent systems mentioned earlier allow the examiner to explain the objections in a search opinion that accompanies the search report. The applicant also has the opportunity to obtain a search report well before publication takes place. The original risk assessment can be updated regarding whether to patent or to keep it secret. At that stage withdrawal is a possibility and there will be no trace left in the public domain.

After publication the examination procedure is carried out, which should ultimately result in a refusal or grant. Dependent on its complexity it may take 4–5 years before the patent is granted after its filing date. If the PCT/WO path is chosen, there is the option for the application to undergo a preliminary examination to assess the patent chances. The period is limited to 2.5 years, as indicated in Figure 3.7. During this period the applicant should be able to evaluate the invention and select the countries that are most appropriate from a marketing perspective. The actual grant procedure is still in the hands of the PCT contracting states, which assume control 2.5 years after the first filing date. In case of the EP system the procedure ends in a package of the national patents. This situation is likely to change with the EP 'unitary patent'. In case of an unitary patent central EP courts can deal with infringement and validity matters. The situation will then be more comparable to that of the USA. This is an attempt to harmonize and reduce costs within the EP region. Finally, there is also a possibility that third parties question the allowance of a patent in a pre-grant or post-grant procedure. As indicated in Figure 3.7 the EP region allows the possibility of a post-grant legal challenge, also called an opposition. The post-grant opposition is limited to 9 months. This lengthy and costly procedure is followed only if major interests are at stake.

3.4 Infant foods in the 21st century

Compared to the 19th century the economies of scale in the 21st century dairy industry have increased tremendously. An advantage of the present size of operation is that minor constituents can also be produced in an economical way. This opportunity affects amongst others the use of dairy ingredients in infant food. Ingredients from milk were and still are a substantial part of infant formulas. The present situation is reflected in the information sheets. Infant or baby food is mentioned under the heading 'Application' in a large proportion of them.

Patents and patent applications are currently more structured than in the 19th century. The main elements of a patent are the abstract, description, examples and claims. The description includes the state of the art and contains an extensive explanation of the invention. It can be used as a rich source of information and sometimes offers a reference list. The examples show in some detail how the invention is carried out, in order to comply with the principles of enabling disclosure. The claims are the crucial part of a patent and determine the scope of the exclusivity. Once granted, third parties should not produce or offer materials within the scope of these claims, or they are at risk of being accused of infringement.

Nowadays, most patent applications are not the work of one individual as was the case in the 19th century, but are usually an inventor team effort. New tools in chemical analysis and in isolation procedures are responsible for the progress made. Companies often try to achieve an optimal coverage by applying for a series of patents. In this way they try to create an impenetrable wall around their intellectual property. It is called a *cluster* of patents. An example is the minor milk constituent osteopontin. An overview of a cluster of patents/applications including the abstracts is provided in Table 3.4 (status: October 25, 2012). As indicated by the priority dates, they cover a period of about 10 years (2000–11). The main aims of the applicant, Arla Foods, are to produce osteopontin as efficiently as possible and to humanize infant food. Surprisingly, dental and wound healing applications were also invented as a result. Obviously this a rather costly approach, which requires a

Table 3.4 A cluster of patents and patent applications regarding osteopontin filed by Arla Foods (status: October 25, 2012).

EP 1 244 702 (B1) and US 7,259,243 (B2)
Process for isolation of osteopontin from milk
Abstract: A process of isolation of milk osteopontin from a material containing milk osteopontin by optionally mixing the milk material with a calcium source and separate the osteopontin containing phase from the rest of the milk material by pH adjustment.
Priority date: January 7, 2000
Granted: April 26, 2006

EP 1 667 537 (A1) and WO 2005/025333 (A1) and US 2005/0287247 (A1) and US 2010/0034958 (A1)
Infant formula
Abstract: Infant formula containing osteopontin and use of osteopontin as supplement for infant formulas or baby food.
Priority date: September 18, 2003
Patent pending

EP 1 689 350 (B1) and WO 2005/053628 (A1)
Use of osteopontin in dental formulations
Abstract: Use of osteopontin for reducing plaque bacterial growth on tooth enamel and dental formulations containing osteopontin.
Priority date: December 1, 2003
Granted: January 30, 2008

EP 1 973 560 (A2) and WO 2007/068252 (A3)
Bovine osteopontin formulations for the improvement of the wound healing process
Abstract: Topical formulations containing bovine osteopontin for improved wound healing. The wound can be infected or inflamed or it can be a diabetic wound.
Priority date: December 16, 2005
Patent pending

WO 2012/117119 (A1)
Method for isolating osteopontin using feeds containing CMP or casein species
Abstract: The present invention pertains to a method for isolating osteopontin from a milk-derived feed containing caseino macropeptide and/or free beta casein, such as e.g. a feed based on milk serum or sweet whey. Particularly, the present method involves the use of a narrow window of pH and specific conductance of the milk-derived feed, which surprisingly has proven to provide a very efficient isolation of osteopontin from chemically complex feeds.
Priority date: March 3, 2011
EP procedure not started

WO 2012/117120 (A1)
Method for isolating osteopontin using concentrated feeds
Abstract: The present invention pertains to a method for isolating osteopontin from milk-derived feeds having a high concentration of protein. Particularly, the present method involves the use of a narrow window of pH and a specific conductance of the milk-derived feed, which surprisingly has proven to provide a very efficient isolation of osteopontin from chemically complex feeds.
Priority date: March 3, 2011
EP procedure not started

Source: Courtesy of Arla Foods.

long-term strategy and a substantial R&D budget. This strategy is also highlighted by two new patent applications, which were filed after the mentioned reference date (WO 2013/086459 - suppressing growth of cancer tumours and WO 2013/144247 - reducing oral related diseases). Some aspects of the patent cluster of Table 3.4 will be discussed below.

In Table 3.4 patents/applications are shown in various phases of the granting procedure. A granted patent is recognized by a letter 'B' or 'C' included in the patent number: for instance, US 7,259,243 B2. Worldwide, a patent application is indicated in the same way by an 'A'. The *'Examples'* mentioned in a patent are an important source of information. They describe the way an invention can be reproduced. Example 1 in Figure 3.8 illustrates nicely the extensive processing that is necessary for the production of osteopontin. The description does not have to be exhaustive, but it needs to be sufficient for those who are skilled in the art. Casein whey is applied as a starting material (0.002% osteopontin). The purity of the dried ingredient is still modest even after the intensive processing described in Figure 3.8 (2.3%). Usually several process variants are set out in the other Examples. The highest purity is approximately 45% as demonstrated in Example 8. More recent inventions of the same applicant show further improvement. The purity of osteopontin increased to a level close to 100%. At the same time a high yield can be achieved, as is indicated in patent application WO 2012/117120 (see Table 3.4). In Example 1 of this recent application, Whey Protein Isolate (WPI) was applied as a starting material. The purity level is close to that mentioned in the information sheet for the whey ingredient 'Osteopontin'.

In Table 3.4 two WO applications (WO 2012/117119 and WO 2012/117120) are mentioned, which have the same priority date March 3, 2011. Both describe a method of isolating osteopontin from milk derivatives. However, the dairy raw materials are quite different, which can be concluded from the protein content and protein composition. Due to the fact that there are differences in raw materials, it lacks the so-called *unity of invention*. In that case a separate filing is required.

The invention of the application of osteopontin in infant food is a result of a sophisticated chemical analysis. Human milk, bovine milk and commercial infant food formulas were investigated. Based on this finding, the addition of bovine osteopontin to infant food is claimed as an invention filed in several countries, which is called a patent family. Four of the patent applications are shown in the second item of Table 3.4. The priority date is September 18, 2003, so they already have a rather long history. As is demonstrated in Figure 3.9 a second US application was filed with the same priority date, but with a different publication date and number (US 2010/0034958 & US 2005/0287247). This is possible in the USA and is called a *continuation-in-part* (Eisenberg, 2000). Except for the publication dates and the numbers, the two applications are presumably the same. Only some minor clarifications are made in the text and the claims are further restricted to the amount of osteopontin to be applied in infant formulas. The intention of the applicant might be to continue uncertainty regarding the outcome of the granting process and to complicate 'designing around'.

United States Patent (10) **Patent No.:** **US 7,259,243 B2**
Sørensen et al. (45) **Date of Patent:** **Aug. 21, 2007**

PROCESS FOR ISOLATION OF OSTEOPONTIN FROM MILK

EXAMPLE 1

1000 kg of casein whey with pH 4.55, 0.53% protein, approximately 20 ppm osteopontin (0.002%), and 4.50% dry matter are concentrated in a spiral ultrafiltration plant with membranes having a nominal pore size so that the proteins do not pass through the membrane. Typical nominal pore size is 10,000 D. The temperature during the filtration is 50° C. and the mean pressure is 3.5 bars. Concentration is carried on until 900 kg of permeate, which do not contain osteopontin, are removed. During the concentration mean flux is 45.6 litres/m^2/h. 100 kg of retentate with pH 4.55, 3.86% protein, approx. 200 ppm osteopontin (0.02%), and 10.3% dry matter are left.

The 100 kg of retentate are pasteurized at 68° C. for 15 sec and cooled to 50° C. Hereafter pH is adjusted to 7.0 with 6% NaOH.

After being allowed to stand for 2 hours, microfiltration is performed on 0.1.4 μm ceramic membranes at 50° C. with a mean pressure of 4.0 bars. Diafiltration is carried out with 50° C. hot demineralized water pH adjusted to 7.0 until the conductivity in the permeate is below 100 μS. During the entire filtration the mean flux is 310 litres/m^2/h.

20,0 kg of retentate from the microfiltration are collected and cooled to 8° C. in ice water, with pH 7.0, 0.58% protein, approx. 1000 ppm osteopontin (0.1%), and 4.1% dry matter. The retentate is spray dried in a NIRO tower drier and 0.7 kg of powder with 13.6% protein, approx. 2.3% osteopontin, and 96.4% dry matter is obtained. The dry matter consists of 80% ashes, 27% calcium and 13.8% phosphorus.

Figure 3.8 Example of a patent as source of information. *Source:* Sørensen *et al.*, 2007.

Figure 3.9 Continuation-in-part in the USA

US 20100034958A1

(19) **United States**
(12) **Patent Application Publication** (10) **Pub. No.: US 2010/0034958 A1**
SØRENSEN et al. (43) **Pub. Date: Feb. 11, 2010**

(54) INFANT FORMULA

(75) Inventors: **Esben Skipper SØRENSEN**, Sabro (DK); **Hans BURLING**, Lund (SE); **Karen BØTTCHER**, Kibaek (DK); **Hans BERTELSEN**, Videbaek (DK); **Kristian ALBERTSEN**, Videbaek (DK); **Gitte GRAVERHOLT**, Aring;rhus N (DK); **Anders Steen JØRGENSEN**, Arhus C (DK); **Lotte SCHACK**, Risskov (DK)

Correspondence Address:
FINNEGAN, HENDERSON, FARABOW, GARRETT & DUNNER LLP
901 NEW YORK AVENUE, NW
WASHINGTON, DC 20001-4413 (US)

(73) Assignee: **ARLA FOODS AMBA**

(21) Appl. No.: **12/509,331**

(22) Filed: **Jul. 24, 2009**

Related U.S. Application Data

(63) Continuation of application No. 10/942.870. filed on Sep. 17. 2004.

(60) Provisional application No. 60/504.952. filed on Sep. 23. 2003.

(30) **Foreign Application Priority Data**

Sep. 18. 2003 (DK) .... PA 2003 01353

Publication Classification

(51) **Int. Cl.**
A23C 9/152 (2006.01)
A23C 11/00 (2006.01)
A23J 3/10 (2006.01)

(52) **U.S. Cl.** **426/657**

(57) **ABSTRACT**

Infant formula containing osteopontin and use of osteopontin as supplement for infant formulas or baby food.

US 20050287247A1

(19) **United States**
(12) **Patent Application Publication** (10) **Pub. No.: US 2005/0287247 A1**
Sorensen et al. (43) **Pub. Date: Dec. 29, 2005**

(54) **INFANT FORMULA**

(76) Inventors: **Esben Skipper Sorensen**, Sabro (DK); **Hans Burling**, Lund (SE); **Karen Bottcher**, Kibæk (DK); **Hans Bertelsen**, Videbæk (DK); **Kristian Albertsen**, Videbæk (DK); **Gitte Graverholt**, Arhus N (DK); **Anders Steen Jorgensen**, Arhus C (DK); **Lotte Schack**, Risskov (DK)

Correspondence Address:
Finnegan, Henderson, Farabow, Garrett & Dunner, L.L.P.
1300 I Street, N.W.
Washington, DC 20005-3315 (US)

(21) Appl. No.: **10/942,870**

(22) Filed: **Sep. 17, 2004**

Related U.S. Application Data

(60) Provisional application No. 60/504,952, filed on Sep. 23, 2003.

(30) **Foreign Application Priority Data**

Sep. 18, 2003 (DK) .... PA 2003 01353

Publication Classification

(51) **Int. Cl.**^{7} **A23L 1/30**
(52) **U.S. Cl.** **426/72**

(57) **ABSTRACT**

Infant formula containing osteopontin and use of osteopontin as supplement for infant formulas or baby food.

Figure 3.9 Continuation-in-part in the USA. *Source:* US 2010/0034958 and US 2005/0287247 available via Espacenet.

(12) INTERNATIONAL APPLICATION PUBLISHED UNDER THE PATENT COOPERATION TREATY (PCT)

(19) World Intellectual Property Organization
International Bureau

(43) International Publication Date
16 June 2005 (16.06.2005)

PCT

(10) International Publication Number
WO 2005/053628 A1

(51) International Patent Classification[7]: A61K 7/16

(21) International Application Number: PCT/DK2004/000835

(22) International Filing Date: 1 December 2004 (01.12.2004)

(25) Filing Language: English

(26) Publication Language: English

(30) Priority Data:
PA 2003 01777 1 December 2003 (01.12.2003) DK
60/525,828 1 December 2003 (01.12.2003) US

(71) Applicant *(for all designated States except US)*: **ARLA FOODS AMBA** [DK/DK]; Skanderborgvej 277, DK-8260 Viby J. (DK).

(72) Inventors; and
(75) Inventors/Applicants *(for US only)*: **BURLING, Hans** [SE/SE]; Hövdingavägen 15, S-22475 Lund (SE). **SØRENSEN, Esben, Skipper** [DK/DK]; Ranunkelvej 309, DK-8471 Sabro (DK). **BERTELSEN, Hans** [DK/DK]; Chr. Aagaardsvej 8, DK-6920 Videbæk (DK). **JØRGENSEN, Anders, Steen** [DK/DK]; Dyrehavevej 35, DK-8000 Århus C (DK). **GRAVERHOLT, Gitte** [DK/DK]; Narviksgade 4, DK-8200 Århus N (DK).

(74) Agent: **ZACCO DENMARK A/S**; Hans Bekkevolds Allé 7, DK-2900 Hellerup (DK).

(81) Designated States *(unless otherwise indicated, for every kind of national protection available)*: AE, AG, AL, AM, AT, AU, AZ, BA, BB, BG, BR, BW, BY, BZ, CA, CH, CN, CO, CR, CU, CZ, DE, DK, DM, DZ, EC, EE, EG, ES, FI, GB, GD, GE, GH, GM, HR, HU, ID, IL, IN, IS, JP, KE, KG, KP, KR, KZ, LC, LK, LR, LS, LT, LU, LV, MA, MD, MG, MK, MN, MW, MX, MZ, NA, NI, NO, NZ, OM, PG, PH, PL, PT, RO, RU, SC, SD, SE, SG, SK, SL, SY, TJ, TM, TN, TR, TT, TZ, UA, UG, US, UZ, VC, VN, YU, ZA, ZM, ZW.

(84) Designated States *(unless otherwise indicated, for every kind of regional protection available)*: ARIPO (BW, GH, GM, KE, LS, MW, MZ, NA, SD, SL, SZ, TZ, UG, ZM, ZW), Eurasian (AM, AZ, BY, KG, KZ, MD, RU, TJ, TM), European (AT, BE, BG, CH, CY, CZ, DE, DK, EE, ES, FI, FR, GB, GR, HU, IE, IS, IT, LT, LU, MC, NL, PL, PT, RO, SE, SI, SK, TR), OAPI (BF, BJ, CF, CG, CI, CM, GA, GN, GQ, GW, ML, MR, NE, SN, TD, TG).

Published:
— *with international search report*

For two-letter codes and other abbreviations, refer to the "Guidance Notes on Codes and Abbreviations" appearing at the beginning of each regular issue of the PCT Gazette.

(54) Title: USE OF OSTEOPONTIN IN DENTAL FORMULATIONS

(57) Abstract: Use of osteopontin for reducing plaque bacterial growth on tooth enamel and dental formulations containing osteopontin.

Figure 3.10 Designated countries. *Source:* WO 2005/053628 available via Espacenet. Reproduced by the permission of WIPO.

Osteopontin is a multifunctional protein involved in diverse biological processes (Mazzali *et al.*, 2002). For this reason, its scope is wider than infant food. Two patents/applications are shown in Table 3.4 regarding its use in wound healing and dental formulations. We will focus on the latter with number WO 2005/053628 A1 and titled 'Use of osteopontin in dental formulations'. A copy of the first sheet of this application is shown in Figure 3.10. The normal PCT procedures are followed; it is filed first in Denmark (DK) on 1 December 2003 and subsequently the international phase (WO) is started a year later. The WO application is published after 1.5 years (international publication date 16 June 2005). Under the heading 'Designated States' the various *countries* and regional organizations are mentioned. The countries are denoted by a two-letter code. It is only an option and does not mean that all countries will be used for a granted patent. Strategy and costs play an important role in the selection process. On 30 January 2008 the EP procedure had been started and a

WO 2005/053628
Use of osteopontin in dental formulations

PROPOSED CLAIMS:
1. Use of osteopontin for reducing or preventing plaque bacterial growth on tooth enamel.
2. Use of osteopontin according to claim 1 for reducing or preventing dental caries.
3. Use of osteopontin according to claim 1 for reducing or preventing calculus dentals.
4 Use of osteopontin according to claim 1 for reducing or preventing periodontal disease.
5. Use of osteopontin according to claim 1 for reducing or preventing demineralization of teeth.
6. A dental formulation containing osteopontin.
7. A dental formulation according to claim 6 in the form of a toothpaste, toothpowder, tooth gel, dental mouthwash, mouth spray or chewing gum.
8. A dental formulation according to claim 7 comprising about 5 mg osteopontin to about 1500 mg osteopontin per kg formulation.
9. A dental formulation according to claim 8 comprising about 100 mg osteopontin to about 1000 mg osteopontin per kg formulation.
10. A dental formulation according to claim 9 comprising about 200 mg osteopontin to about 500 mg osteopontin per kg formulation.
11. A dental formulation according to claim 10 comprising about 350 mg osteopontin per kg formulation.
12. Use of osteopontin for the manufacturing of a pharmaceutical composition for treating or preventing dental caries, calculus dentalis or periodontal disease.
13. A method for preventing or inhibiting plaque bacterial growth on tooth enamel by administration in the mouth of an effective amount of osteopontin.
14. Method of treatment or preventing of dental caries, calculus dentalis or periodontal disease by oral administration of an effective amount of osteopontin to a patient in need of such treatment.

EP 1 689 350 (B1)
Use of osteopontin in dental formulations

GRANTED CLAIMS:
1.Use of milk osteopontin for reducing or preventing plaque bacterial growth on tooth enamel.

3.A dental formulation according to claim 2 in the form of a toothpaste, toothpowder, tooth gel, dental mouthwash, mouth spray or chewing gum.
2.A dental formulation containing 5 mg of milk osteopontin to 1500 mg of osteopontin per kg of formulation.
4.Adental formulation to claim 3 comprising 100 mg of milk osteopontin to 1000 mg milk osteopontin per kg formulation.
5.A dental formulation according to claim 4 comprising 200 mg of milk osteopontin to 500 mg milk osteopontin per kg formulation.
6.A dental formulation according to claim 5 comprising 350 mg of milk osteopontin per kg formulation.

Figure 3.11 Proposed and granted claims for a patent application for the 'Use of osteopontin in dental formulations'. *Source:* WO 2005/053628 (reproduced by the permission of WIPO) and EP 1 689 350 (B1), both available via Espacenet.

patent was granted with number EP 1 689 350 (B1). The granted patent entered into force in five European countries.

In Figure 3.11 the proposed *claims* of WO 2005/053628, titled 'Use of osteopontin in dental formulations', are compared to the granted claims in EP 1 689 350 (B1). The EPO carried out an examination of the proposed claims, in which the earlier mentioned criteria 'new', 'inventive', 'useful' and 'sufficiency' were verified. The list of granted claims is considerably shorter than the proposed claims. The preliminary result is communicated with the applicant or their representative. The communication as part of the examination proceedings is open to the public through the 'Register' and accessible through a link[1]. It is a discussion between

[1] www.espacenet.com ('European Patent Register' → 'patent number' → 'all documents').

experts. Here the most important points of the examination will be mentioned. The examiner of the EPO investigated the oral healthcare literature before the priority date of this application, which was filed on 1 December 2003. Predominantly, two references were relevant that objected to the proposed claims 1–5 and 12–14. Proteins and peptides containing clusters of phosphoseryl residues were already used to reduce the level of dental caries (Reynolds, 1999). Particularly, the casein phosphopeptides were effective and inhibited enamel demineralization (see information sheet 'Casein phosphopeptides'). The second reference is more drug related and describes the use of an antifungal agent and antibiotic (Nonomura *et al.*, 2000). This invention relates to a periodontal treatment. Based on the information mentioned, the applicant restricted the claims to the technical problem that was solved by the invention. It means reduction of bacterial adherence and bacterial growth on enamel surfaces by dental formulations. Formulations in the form of a toothpaste, tooth powder, tooth gel, mouthwash, mouth spray or chewing gum were approved. Various concentrations of osteopontin were granted. In patent language, the concentrations in dental formulations reflected the functionality range of ostepontin. They refer to the granted claims 4, 5 and 6 respectively. Claim 1, the main claim, is restricted to 'milk osteopontin'; osteopontin is also available from other sources, which cannot be claimed. The last proposed claim, claim 14, is a medical treatment on the human body, which is excluded from patentability by law. This claim was rejected. Restriction of claims is a common phenomenon in the patent world, one in which examiner and applicant (or their representative) are involved. Claims are the centerpiece of a patent and they are weighed on a gold plate.

References

Abhinav S (2011) Doctrine of inherent anticipation under patentability criteria. Available at: http://ssm.com/abstract=1971168

Allen F (1994) *Secret Formula: How brilliant marketing & relentless salesmanship made Coca-Cola the best known product in the world.* Diane Publishing Company, pp. 26–8, 36, 48–9, 54, 69, 162, 368–9, 399–400, 404–17.

Apple RD (1987) *Mothers & Medicine – A social history of infant feeding 1890–1950.* University of Wisconsin Press, pp. 7–8, 23, 37–45.

Arla Foods (2013) Osteopontin variants for use in suppression or prevention of tumor growth and compositions containing such osteopontin variants [patent application], WO2013086459.

Arla Foods (2013) Nanoparticle aggregates containing osteopontin and calcium- and/or strontium-containing particles [patent application], WO2013144247.

Backhaus A (1896) Improved treatment of the milk of cows, or other domesticated animals, to render it suitable as food for infants and children. Patent GB189616016.

Barness LA (1987) History of infant food practices. *American Journal of Clinical Nutrition* **46**: 168–70.

Borden G (1856) Improvement in concentration of milk. US patent 15,553.

Bosworth D & Jobome G (1999) The measurement and management of risk in R&D and innovation. *International Journal of Technology Management* **18**: 476–99.

Bracken FJ (1956) The history of artificial feeding of infants. *Maryland State Medical Journal* **5**: 40–54.

Brennemann J (1929) The curd and the buffer in infant feeding. *Journal of the American Medical Association* **92**: 364–6.
Castilho SD & De Azevedo Barros Filho A (2010) The history of infant nutrition – review. *Jornal de Pediatria* **86**: 179–88.
Couzens PJ (2009) Intellectual property; protecting products and processes. In: Beckett ST (ed.), *Industrial Chocolate Manufacture and Use*. Blackwell Publishing, pp. 595–622.
van Dulken S & Thompson M (2001) *Inventing the 19th century: 100 inventions*. New York University Press, pp. 42–3.
Eisenberg HM (2000) Patent Law You Use – Continuation Patent Application. Available at: http://www.yale.edu/ocr/pfg/guidelines/patent/continuation.html
Fomon SJ (2001) Infant feeding in the 20th century: formula and Beikost. *Journal of Nutrition* **131**: 409S–420S.
Frischling GN & Orso KA (2005) Intellectual property litigation. *Biopharm International* **8**(8 Suppl.): 26–31.
Giuri P, Mariani M, Brusoni S *et al.* (2007) Inventors and invention process in Europe: Results from the PatVal-EU survey. *Research Policy* **36**: 1107–27.
Horlick W (1883) Granulated food for infants and process of preparing the same. US patent 278,967.
Hunziker OF (1946) *Condensed Milk and Milk Powder*. Published by the author, pp. 33–42.
Kellogg JH (1896) Flaked cereals and process of preparing same. US patent 558,393.
Kimura T & Tanaka Y (2010) Research on evaluation technique of patented invention using both technical value and economic value. In: *PICMET Proceedings, July 18–22, Phuket, Thailand*. Institute of Electrical and Electronics Engineers (IEEE), pp. 384–90.
von Liebig J (1866) *Suppe für Säuglinge*. Braunschweig: Druck und Verlag von Friederich Vieweg und Sohn.
Mazzali M, Kipari T, Ophascharoensuk V, Wesson JA, Johnson R & Hughus J (2002) Review: Osteopontin – a molecule for all seasons. *Quarterly Journal of Medicine* **95**: 3–13.
Meigs AV (1896) *Feeding in Early Infancy*. Philadelphia: WB Saunders.
Meyenberg J (1884a) Apparatus for preserving milk. US patent 308,421.
Meyenberg J (1884b) Process of preserving milk. US patent 308,422.
Nestlé H (1868) La fabrication d'une farine lactée alimentaire destine aux enfants en bas âge, aux personnes faible et aux valetudinaire. FR 82860 [patent].
Nonomura Y, Ikuta T & Matsumoto H (2001) An oral treatment/care agent [patent]. EP1206930.
O'Donnell RW, O'Malley JJ, Huis RJ & Halt GB (2008) *Intellectual Property in the Food Technology Industry*. Springer, pp. 3–22, 87–90.
Radbill SX (1981) Infant feeding through the ages. *Clinical Pediatrics* **20**: 613–21.
Reynolds EC (1999) Anticariogenic casein phosphopeptides. *Protein and Peptide Letters* **6**: 295–302.
Ritter D (2004) Switzerland's patent law history. *Fordham Intellectual Property, Media and Entertainment Law Journal* **14**: 463–95.
Rowan C (2001) Nestlé: Leadership through the years. *Food Engineering & Ingredients* May, pp. 22–6.
Seifert A (éd.) (2008) De la cuisine à l'usine. Les débuts de l'industrie alimentaire en Suisse, catalogue de l'exposition temporaire de l'Alimentarium, Musée de l'alimentation, une Fondation Nestlé, 7 mars 2008 – 4 janvier 2009, Vevey, 2008.
Tapscott D & Williams AD (2008) *Wikinomics*. Penguin Books Ltd.
Tetra Pak (2003) *Dairy Processing Handbook*. Tetra Pak Processing Systems AB.
USPTO (2011) *America Invents Act: Effective Dates*. United States Patent and Trademark Office.
World Intellectual Property Organization (WIPO) (2012) International Patent Filings Set New Record in 2011. Available at: http://www.wipo.org

Websites

European Patent Office (www.espacent.com).
Japan Patent Office (www.jpo.go.jp).
United States Patent and Trademark Office (www.uspto.gov).
World Intellectual Property Organization (http://www.wipo.org).

4 Reuse

In this chapter the focus will be on the reuse of dairy substances within the dairy industry itself. These dairy substances can be considered to fit the earlier definition of an ingredient: 'A substance of dairy origin that is used in the manufacture of a food' (see Chapter 1). In this chapter reuse is described as the effective utilization of milk. It can be related to the water part of the milk ('cow water') as will be described in Section 5.3.1.3. Here the focus will be on milk solids, examples of which are numerous in the dairy industry. For instance, in spray drier systems, milk powder particles as small as 1 μm can be extracted from the drying air that leaves the process (Westergaard, 2010). In this way pollution of the environment is avoided and the separated powder is reused again. In this chapter the focus will be on cheese. Due to the large volumes involved, cheese processing is an interesting illustration. Specific attention will be paid to the semi-hard Gouda and Edam cheeses. In turn, these cheeses can also be used again as an ingredient in, for instance, snacks, hamburgers or sandwiches (see Figure 7.14).

Three topics – cheese starter cultures, process residuals and cheese losses – will be discussed in this chapter. Some of the starters – cultivated on milk in today's Dutch cheese plants – come from dairies that have ceased to exist. However, these starters are still used through the application of modern technology. Various examples of the reuse of cheese process residuals are shown in the second section of this chapter. They require investment in equipment and reuse of residuals is particularly beneficial in case of large cheese plants. The process residuals are mostly derived from whey, while the cheese losses are directly related to the finished product. Cheese losses will be considered in the third section. The reduction of losses through changes in cheese-making technology is discussed, and reuse of the inevitable remaining cheese is highlighted.

From Milk By-Products to Milk Ingredients: Upgrading the Cycle, First Edition. Ruud de Boer.
© 2014 John Wiley & Sons, Ltd. Published 2014 by John Wiley & Sons, Ltd.

4.1 Cheese starter cultures

The starter culture serves two important functions in cheese manufacture. It prevents the growth of spoilage organisms by the conversion of lactose into lactic acid. Secondly, the starter culture together with chymosin and an indigenous milk enzyme (plasmin) breaks down the milk protein (casein) into peptides and amino acids. They are important for the development of the cheese flavour. Numerous cheese starter cultures are currently available. The three main types of starter are listed in Table 4.1 (Coffey *et al.*, 2001). The 'mixed strain starters' are complex mesophilic systems. These undefined starters have evolved naturally in industrial practice and are propagated for use in this environment. Some of these lactic acid producing bacteria are widely used in Gouda and Edam cheese manufacturing. They were selected from the best starter cultures of the numerous cheese-making plants during the 1930s. The starters of this 'Dutch heritage' are listed by codenames referring to historical plants, districts or cheese-makers. The second category of starters comprises the 'multiple strain starters', which are successfully applied in Cheddar cheese manufacture. This mesophilic starter system contains some well-characterized strains of lactic acid producing bacteria. The system is based on extensive research, in which rapid acid production, compatibility and resistance to bacterial viruses (bacteriophages) play an important role. A third category of starters are the 'defined strain starters'. They are applied in types of cheese that need a higher cooking temperature. Examples of the application of these 'defined strain starters' are in mozzarella and Emmentaler cheese manufacture. These thermophilic starters have a background in yoghurt manufacture and contain *Streptococcus* and *Lactobacillus* bacteria. 'Adjunct starters' have been developed to provide additional benefits with regard to flavour and texture. They are successfully used in, for instance, low-fat cheeses. Of the various starter cultures mentioned, the 'mixed strain starters' are identical to the original starters and therefore the best example of 'Reuse' of dairy ingredients.

Nowadays, 'mixed strain starters' are available to cheese plants in a frozen form. They are provided by specialized ingredient companies. An important advantage of a

Table 4.1 Microbiological risks and reuse of raw materials.

Starters
Mixed strain starters – mesophilic/naturally evolved – Gouda cheese
Multiple strain starters – mesophilic/composite – Cheddar cheese
Defined strain starters – thermophilic/composite – mozzarella cheese
Adjunct starters – composite with additional benefits (flavour, texture) – low-fat cheese
Bacteriophage
Contamination via whey – most heat-resistant lactococcal bacteriophage – inactivation 95°C/7 min
Clostridium tyrobutyricum
Contamination via raw milk – heat-resistant spores – inactivation 100°C/280 s

Source: Data from Atamer *et al.*, 2009 and Schreiber & Hinrichs 2000.

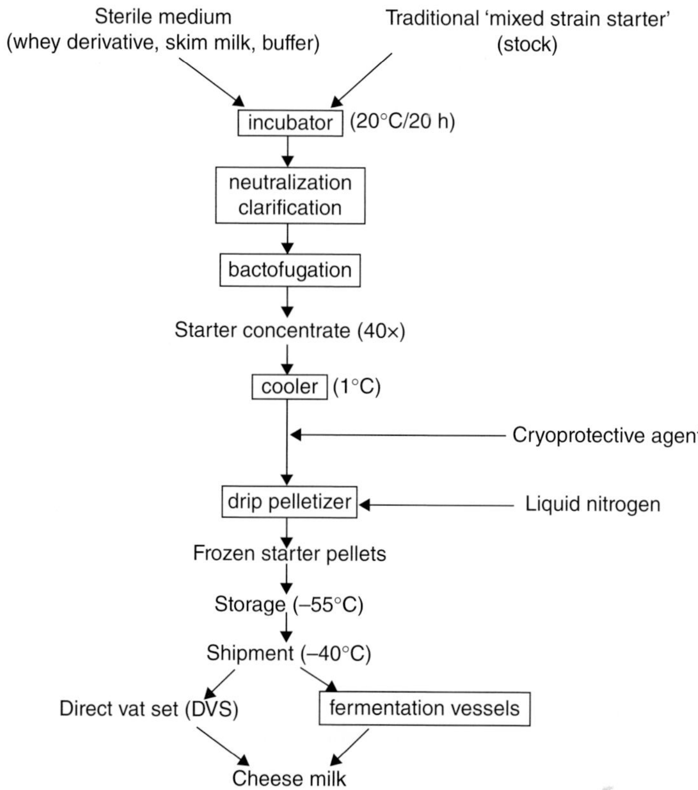

Figure 4.1 Scheme for the production of frozen 'mixed strain starters'. *Source:* Data from Stadhouders, 1974; Hup & Stadhouders, 1977; CSK Food Enrichment, 2001.

frozen starter culture is that the daily transfer of the culture in the cheese plant can be avoided. In this way the composition and performance of the starter culture remains stable. In Figure 4.1 the schematic production of frozen 'mixed strain starters' is shown (Stadhouders, 1974; Hup & Stadhouders, 1977; CSK Food Enrichment, 2001). The starter culture from stock (−196°C) is first incubated under appropriate conditions. Subsequently, the growth medium is clarified to dissolve insoluble compounds. In a following step the bacterial cells are harvested by bactofugation. The obtained concentrate of bacterial cells is cooled (1°C) and cryoprotective agents such as lactose are added. By using liquid nitrogen the concentrate is, for instance, converted to a convenient pellet form. The storage and distribution of these starter cultures takes place in deep-frozen conditions. In the cheese plant the frozen starter concentrate is directly inoculated into the cheese vat; the method is called 'direct vat set' (DVS). This procedure is usually applied in smaller facilities (Hull, 2004). In order to reduce costs, factory-derived starter cultures are widely used in large cheese plants. The 'starter' room is the heart of the cheese plant and extensive efforts are made to avoid any contamination. The frozen starter culture is inoculated in specially designed starter substrates or in skimmed milk heated at 95°C for 30 min. The bacterial culture grows in fermentation vessels under optimal conditions. The result is a bulk starter that is subsequently added to the cheese milk. In this case the number of transfers from stock to cheese milk are limited to three in order to minimize the risk of bacteriophage outbreaks.

In modern cheese plants many efforts are made to minimize or eliminate the risk of bacteriophage contamination. Use is made of special starter systems (Mills *et al.*, 2010), optimal operation procedures and separation of the whey processing from the cheese-making parts of the plant. Particularly, the reuse of whey constituents can be an additional threat for cheese manufacture. Due to a prolonged residence time of these constituents in the process equipment, contamination with bacteriophages is a serious hazard. Bacteriophages are viruses that can infect and replicate within the bacterial cells of the starter culture. They multiply much faster than the bacteria and impair the starter activity within a couple of generations. The result is a low-quality cheese and considerable financial losses.

Lactococcus lactis is one of the most important microorganisms present in the mesophilic starter cultures referred to earlier. Long-term control of bacteriophages in this microorganism is essential for the dairy fermentation processes. The difficulty in controlling outbreaks lies in the biodiversity and the dynamic state of the bacteriophage populations (Deveau *et al.*, 2006). Three *Lc. lactis* bacteriophage species are commonly found in dairy plants in Germany. They cause most of the fermentation failures (Atamer *et al.*, 2009). These three types of bacteriophages have a high thermal resistance and can survive standard pasteurization procedures. Inactivation of the most heat-resistant bacteriophage isolated from the aforementioned plants would require a holding time of 7 min at a temperature of 95°C. A reduction in the bacteriophage population by at least three log units could be achieved under these conditions. More intensive heat treatments of at least 40 min at 90°C or 95°C for 15 min are recommended for safe inactivation of all lactococcal bacteriophages in milk or similar media (Müller-Merbach & Hinrichs, 2006). Heat-resistant bacteriophages have emerged not only from mesophilic starter cultures but also from thermophilic lactic acid bacteria. Capra *et al.* (2013) investigated bacteriophages of both types of starter cultures and proposed a heating step of 90°C for 45 min. The most heat-resistant *Lc. lactis* bacteriophage was the determining factor. The recommended heat treatments are more intensive than earlier figures published by the International Dairy Federation (IDF) of 90°C for 15 min (Svensson & Christiansson, 1991). This reflects better monitoring of bacteriophages in the dairy industry and improved treatments for the safe reuse of dairy compounds.

Table 4.1 mentions another microbiological risk of reuse of milk solids, namely *Clostridium tyrobutyricum*. This anaerobic organism forms spores, which can be a threat to cheese-making. Accumulation of spores of *Cl. tyrobutyricum* together with some valuable milk solids, takes place during bactofugation and microfiltration of cheese milk (see Section 4.2). These spores can cause serious defects in semi-hard and hard cheeses, called 'late blowing'. Process residuals need to receive intensive heat treatment in order to kill the spores and be reusable as milk solids. A four log unit reduction can be achieved by a heat treatment at 100°C for 280 s (Schreiber & Hinrichs, 2000). Another option is the application of UHT treatment with temperatures ranging from 125 to 140°C for a few seconds (van der Berg *et al.*, 1989; Schreiber, 2001).

4.2 Process residuals

Yield and quality play a decisive role in the manufacture of cheese. In this section the focus will be on cheese yield. In Figure 4.2 regular milk solids containing materials for the manufacture of cheese are indicated. They are skimmed milk, milk protein concentrate (MPC) and cream. MPC can be added to the cheese milk to produce a constant quantity of cheese in the process line. A major part of the milk solids during the cheese-making process is released in the whey. Whey contains about 50% of the solids of the cheese milk; the other fraction is incorporated in the cheese. However, retention of the various milk constituents – protein, fat, ash, lactose – varies. (Waagner Nielsen, 2004). A typical value for the retention of protein is 75–78% of the total protein. The remaining part is discharged as whey protein in the cheese whey. The incorporation of fat in the cheese is higher, varying from 85 to 95%. The retention of ash compounds ranges from 35 to 40% for semi-hard and hard cheeses, the figure depending on the degree of acidification during cheese-making. The retention of citrate, which is linked to casein, is typically 10% in semi-hard cheese. Citrate as well as lactose are mainly released in the cheese whey; the retention of lactose ranges from 3% to 5% – the precise value

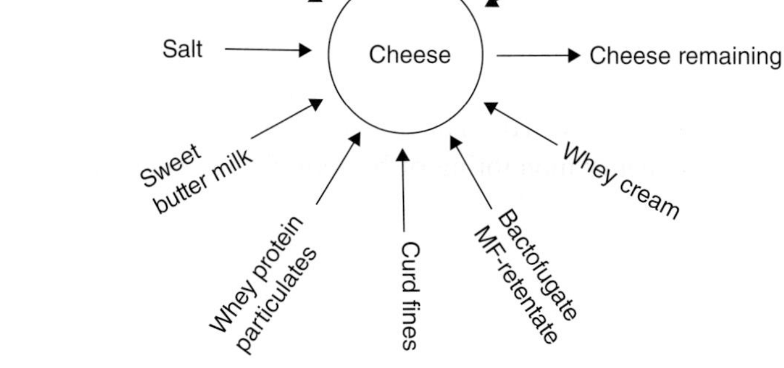

Figure 4.2 Options for reuse of raw materials in cheese manufacture. MF, microfiltration; MPC, milk protein concentrate.

depends on the type of cheese. The aforementioned figures result in the following composition of a young semi-hard cheese, for instance: 43% moisture, 2% salt and 55% milk solids. Protein and fat comprise approximately 93% of the milk solids. It can be deduced that the recovery of milk solids therefore focuses on the losses of those two constituents. In Figure 4.2 other sources of milk solids are mentioned too. They are bactofugate (or MF (microfiltration) retentate), curd fines, whey protein particulates and whey cream. These compounds are derived from the cheese manufacture itself. The indicated sweet buttermilk is an exception. This by-product of butter manufacture may also be useful as a raw material for cheese.

Bactofugate, or MF retentate, is derived from milk. The main aim is improvement of the microbiological quality of the cheese milk. Bactofugation removes a high percentage of bacteria, spores (including the anaerobic *Cl. tyrobutyricum*) and somatic cells by using a high gravitational force. The microorganisms and cells are more dense than the milk serum and are concentrated in the sludge. The sludge, or bactofugate, represents about 3% v/v of the milk. Besides the cells and spores some large, dense casein micelles are also removed. They may represent up to 6% of the total casein (Fox, 2000). By applying a UHT or a similar heat treatment, the loss of casein in the bactofugate can be prevented. After completion of this treatment, the bactofugate can be safely added back to the cheese milk. Adverse effects on the quality of Gouda cheese were not detected and no negative influence on cheese manufacture was mentioned by van der Berg *et al.* (1989). An alternative to bactofugation is microfiltration (MF) of the milk. MF is considered more effective than bactofugation (te Giffel & van der Horst, 2004). The MF process reduces the bacterial count and the aerobic and anaerobic spores by more than 99%. The pore size of the MF membrane and the diameters of the various milk constituents are key important factors. This process is applied to skimmed milk as there is an overlap in the diameters of fat globules and bacteria (see Table 6.1). Also the diameters of bacteria and casein micelles partly overlap. A MF membrane pore size of 1.4 μm retains the microorganisms effectively at the expense of only a limited retention of casein micelles. The MF retentate containing all of these microorganisms is mixed with the separated cream obtained at the beginning of the process. The mixture receives a UHT or similar heat treatment to inactivate particularly *Cl. tyrobutyricum*. After this treatment has been completed, the mixture is blended with the MF permeate (low-bacteria skimmed milk). In this way the cheese yield can safely be restored. The volume of the retentate can vary between 0.5 and 5% (Saboya & Maubois, 2000).

Despite the fact that MF removes *Cl. tyrobutyricum* spores more effectively than bactofugation, cheese manufacturers often prefer the latter as it is less complicated.

Part of the cheese curd, the so-called curd fines, are lost in the whey during processing. The losses thereof greatly depend on the firmness of the coagulum at the moment of cutting and stirring. If the coagulum is too weak, the losses will be high. If the curd is too firm, cutting and stirring need excessive force, which also enhances the losses. The amount of curd fines transferred to the whey varies from 0.2% to 1%,

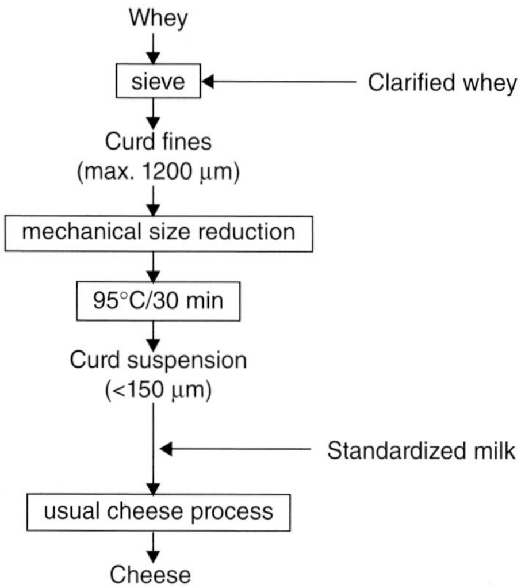

Figure 4.3 Reuse of curd fines in cheese. *Source:* Adapted from Burghout & Abeelen, 2004. NL1024367.

when expressed as a percentage of the amount of paracasein originally present in the curd (Walstra *et al.*, 2006). The size of the curd particles fluctuates from 5 to 1200 μm. They can be separated by, for instance, a sieve and are usually applied in feed or in process cheese. An alternative approach is the reuse of the curd fines in the cheese-making process itself (Burghout & van Abeelen, 2004). As shown in Figure 4.3 the curd fines are mechanically reduced in size, for instance by using a homogenizer, or a disc mill or colloid mill. In order to avoid settlement during coagulation, the particles should have a diameter of less than 150 μm. The comminuted particles can be suspended in whey or water, for instance in a concentration of 5% total solids (TS). After an intensive heat treatment (95°C/30 min) the bacteriophages and other contaminants are inactivated and the suspension can be added to the standardized cheese milk. Subsequently, the regular cheese-making process is carried out. The semi-hard cheeses enriched with curd fines cannot be distinguished from regular cheeses as far as flavour, texture and cross-section are concerned. The percentage of curd fines reused in the cheese varies from 50% to 80%. The additional output can be approximately 100 tonnes of cheese for a large plant with a capacity of 500,000 tonnes milk/year.

Another possibility to increase the cheese yield is by the incorporation of whey proteins. This can be carried out by raising the pasteurization temperature of the milk, as long as it does not negatively affect cheese quality. A variation on this theme is the inclusion of thermally modified whey protein concentrates in cheese. The treated whey proteins have here a dual purpose – to increase the yield and to be a suitable source for the manufacture of low-fat semi-hard cheese. As indicated in Figure 4.4 the whey proteins are concentrated approximately 20 times by using ultrafiltration (UF) starting from sweet cheese whey. Subsequently, the retentate is

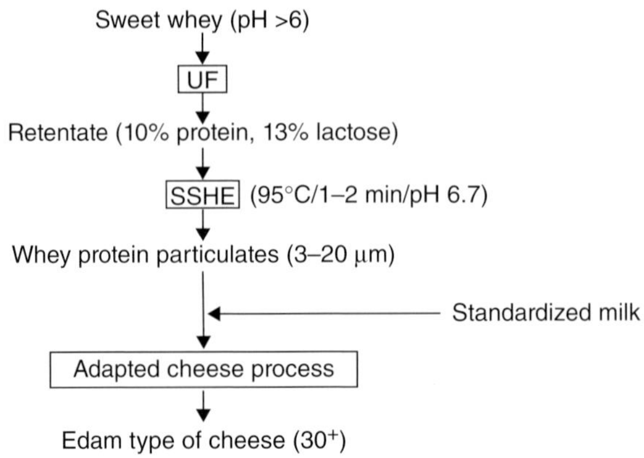

Figure 4.4 Incorporation of whey protein particulates in cheese. *Source:* Data from Schreiber *et al.*, 1998a, 1998b and Spiegel 1999.

treated in a specially designed scraped surface heat exchanger (SSHE). The whey proteins are denatured by a high heat treatment, while simultaneously applying a shear force. The size of whey protein aggregates can be chosen. Under carefully selected process conditions whey protein particulates (WPP) can be produced, which are similar in size to milk fat globules. Due to their size and structure WPP act as fat globules. To avoid complications during the heat treatment, the protein concentration should be maintained below its gelling point. If the lactose content in the retentate is kept low to medium, the whey proteins denature faster. Under these circumstances the protective effect of lactose is very limited. Small particles with a diameter of 3–20 μm and the desired structure can be obtained under high shear conditions at temperatures ≥ 95°C (Spiegel, 1999). The holding time is dependent on the pH of the whey as the denaturation reaction is delayed at lower pH values (Spiegel & Huss, 2002). The denaturation of the whey proteins should exceed 90% to maximize the incorporation of the MPP in the cheese. Under carefully chosen process parameters an inclusion level of 70% can be achieved (Hinrichs, 2001). One gram of protein binds about 4 g of water under optimal process conditions. The additional water has an impact on the texture of the cheese. For this reason the cheese-making process needs to be adapted (Schreiber *et al.*, 1998a, 1998b). The moisture content can be reduced by increasing the scalding temperature in order to improve the syneresis of the curd. The additional lactose originating from the WPP might also require more intensive washing. The theoretical amount of WPP is limited to 0.5 g per 100 g of milk. In that case the increase in yield could have a maximum of 15%. Beside an increase in yield, the creaminess of a 30^+ or low fat semi-hard cheese is enhanced and can be compared to a standard 40^+ or medium fat Edam cheese.

As indicated earlier, on average 10% of the fat is lost in cheese whey. In case of the semi-hard 48^+ or full fat Gouda cheese the loss is about 7%. The loss is caused by cutting and stirring of the coagulum. The coagulum is formed by a protein (paracasein) network, in which the fat globules are embedded. Compared to the paracasein part, fat globules close to the cut surface can transfer more easily to the whey. In a regular process this whey passes through a separator and the fat is recovered as whey cream. The risk of disturbing the acidification of the cheese-making process is avoided by applying a high heat treatment to the whey cream. Subsequently, this cream can be used again and added to the next batch of cheese milk. In this way the fat losses can be limited to about 1.5% of the fat originally present in the cheese milk.

The last of the dairy process residuals that will be discussed in this section is sweet buttermilk (see also information sheet 'MFGM buttermilk'). In the butter-making process the cream receives a high heat treatment of 85°C/15 s (Walstra *et al.*, 2006) to make the end-product more resistant to oxidative deterioration. In the following phase the cream is churned. During the churning process the milk fat globules are destabilized through mechanical agitation in the presence of air. The milk fat globule membrane (MFGM) migrates partially to the buttermilk. Buttermilk differs from skimmed milk due to the enhanced level of the highly surface-active phospholipids. These phospholipids have an amphiphilic character, which contributes to the emulsifying properties and water-binding properties of the buttermilk. For these reasons it is beneficial to use fresh buttermilk in the manufacture of cheese. However, this effect has to be weighed against a negative influence of the applied heat treatment (85°C/15 s) on the properties of the cheese.

Low-fat cheeses are usually characterized by a rubbery body and atypical flavour notes. The general goal in hard and semi-hard cheeses is to replace fat with moisture without adversely affecting the yield and quality of the product (Mistry, 2001). This can be accomplished by lowering the cooking temperature. Another method of increasing moisture retention is by incorporating sweet buttermilk in the cheese milk. Madsen *et al.* (1966) applied up to 30% of buttermilk. After 3 months of curing brick cheeses (a kind of American Cheddar), the addition of 20% and 30% buttermilk resulted in an undesirable quality as far as its flavour (bitterness) and body was concerned. Mayes *et al.* (1994) used various combinations of buttermilk and cream and concluded that homogenizing buttermilk and fresh cream slightly improves the sensory properties at an age of 23 weeks. An alternative approach is UF of buttermilk in which fat/phospholipids and protein are highly concentrated (Mistry *et al.*, 1996). The retentate was used in amounts of up to 5% in the cheese milk. At an age of 4 weeks the low-fat Cheddar cheeses were softer and the texture and body were of better quality than the control group. The quality differences are smaller at an age of 24 weeks. Additional trials with low-fat Cheddar cheese were carried out using various mixtures of (UF) milk and (UF) buttermilk (Turcot *et al.*, 2001, 2002). It was found that only 44% of the phospholipids in the starting milk were retained in the cheese curd. Sensory evaluation confirmed that the prepared low-fat cheeses had a soft texture. However, flavour defects – bitter and rancid – developed during extended storage. It was therefore recommended to consume the cheeses within 2 months. The age restriction is not needed if the dosage of the buttermilk is kept below 10% (HP Burghout, 2013, personal communication).

4.3 Cheese losses

Losses in the food industry are very product specific; for instance, meat, grain and dairy plants each have their characteristic losses. Russ and Meyer-Pittroff (2004) sought to quantify the losses or waste by introducing a *specific waste index*. It is defined as the mass of accumulated waste divided by the mass of saleable product. The specific waste index is limited to the processing part of the entire chain from farm to consumer. The index is on average 0.04 for hard and semi-hard cheeses. The figure may show some fluctuation, for instance depending on the cheese portion size. Sliced or diced cheese will result in greater losses than whole cheeses. The shipment conditions are not included in the index but they may play an important role in product losses as well. Shipment in tropical areas is more risky for the quality of cheese than in temperate regions. The attention in this section will again be focused on semi-hard cheese.

At the end of the 19th century the Dutchman Jan Hendrikzoon Eyssen struggled with the problem that his semi-hard, probably full fat, cheeses were not suitable for export to tropical areas. The shelf life was insufficient and as a consequence the product losses were high. Eyssen (1898) invented a modified cheese-making technique and introduced a tender, fully ripened, preserved cheese. Nowadays, this is referred to as 'processed cheese', which can be kept at ambient temperature. Proof of the success of this preserved Dutch cheese can be deduced from the 'royal' designation dated 1905 (Figure 4.5). The patent, in which no use is made of emulsifiers and preservatives, was granted to him in Great Britain in 1899. Unfortunately the inventor described the process only in general terms. Microbiological defects played an important role in his process and the need to expel air was emphasized. Cleaned, ripened cheese was subjected to a kneading step and transformed into a uniform pastelike mass. The cheese paste was packed in tins, which were hermetically sealed. A modern version of a kneading machine is an extruder.

Eyssen was clearly ahead of the famous James Lewis Kraft (1916), who was granted a patent for a similar approach nearly two decades later. Kraft described the process, which again uses no emulsifiers, in more detail. Instead of semi-hard cheese Kraft used Cheddar cheese. It was cut into pieces and subsequently heated in a kettle. Under proper agitation and continuous stirring the mass was heated to 80°C for 10–15 min. Subsequently, the hot 'sterilized cheese' was put in hermetically sealed containers and cooled. These first steps in avoiding cheese product losses under tropical conditions were followed by numerous improvements (Zehren & Nusbaum, 2000; Kapoor & Metzger, 2008; Lucey *et al.*, 2011). The introduction of emulsifying salts to process cheese manufacture appeared highly beneficial for further reductions of product losses. Defects such as oiling off, mealy body and faulty consistency could be dealt with more efficiently (Price & Bush, 1974).

The aspect of expelling air is of interest too. It is widely applied to extend the shelf life of packed natural cheese in order to reduce product losses. Nowadays, expelling air is developed further into 'modified atmosphere packaging' (MAP) of cheese

GB 11,207

'An improved process for keeping and packing cheese and the like'

Inventor: Jan Hendrikzoon Eyssen **Filed**: 17th May, 1898
Claim: Process for keeping and packing cheese without affecting the taste and the aroma in which the cheese in a ripe state is kneaded to a paste, the air expelled, the impurities removed and the paste packed into hermetically closed tins, substantially as described.

Figure 4.5 The granted patent claim (1899) and the 'royal' designation (1905) of Jan Hendrikzoon Eyssen. *Source:* Courtesy of Eyssen BV. (*For colour details please see colour plate section.*)

(Hotchkiss *et al.*, 2006; Khoshgozaran *et al.*, 2012; Singh *et al.*, 2012). MAP creates a preservative atmosphere. This is usually accomplished by creating a vacuum and then injecting a gas mixture into the foil-packed cheeses. The optimum atmosphere depends on the characteristics of the cheese type. Carbon dioxide is an important component of the gas mixture applied to hard and semi-hard cheeses. A major effect of CO_2 is the inhibition of mould growth on the surface of the cheese. Usually CO_2/N_2 mixtures (e.g. 25/75) are applied in which the N_2 acts as a filler. It prevents the collapse of the package and the formation of a vacuum as the CO_2 is absorbed by the cheese. The residual oxygen in the head space of the oxygen-impermeable package should be kept below 0.5% (Mortensen *et al.*, 2004). Besides MAP, the preservation of the natural ripened cheeses is extended by keeping the temperature below 7°C.

Products that are not sold through the regular sales channels have a lower value and are called cheese remainders. They are, for instance, residues from the cutting or grating of cheese. Others include cheeses with mechanical deformations or localized moulds, or cheeses with slight deviations in composition or sensory qualities. Some can only be used as animal feed. However, a considerable part is still suitable for human consumption. One widely used option is as a raw material for the process cheese industry. In Figure 4.6 another option is shown, in which the remaining cheese is reconstructed into a product that is in many aspects similar to the original cheese (Penders & Burghout, 2003). In this process the starting material is cleaned, the plastic coating removed and if needed the cheese is milled into pieces. The milling increases the surface area of the cheese and facilitates the following fusion process. An adhesion initiator is added and blended thoroughly with the cheese pieces. The adhesion initiators can be additives applied in regular cheese production such as rennet or $CaCl_2$. Another option is the use of skimmed milk powder, as is indicated in Figure 4.6. The casein part in the powder probably facilitates the adhesion of the particles. It reacts with the chymosin retained in the

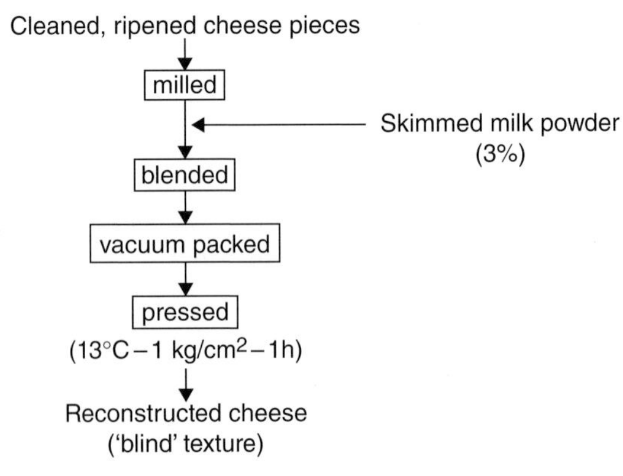

Figure 4.6 Reconstruction of natural cheese remainders. *Source:* Adapted from Penders & Burghout, 2003. NL1024528.

cheese particles and converts the casein into paracasein (Walstra *et al.*, 2006). After vacuum packing, pressing completes the fusion process and the cheese is put into its desired shape. The end-product can be used in several ways; differences from the original semi-hard cheese are that the cross-section is 'blind' (no 'eyes') and the shelf life is shorter.

References

Atamer Z, Dietrich J, Müller-Merbach M & Neve H (2009) Screening for and characterization of *Lactococcus lactis* bacteriophages with high thermal resistance. *International Dairy Journal* **19**: 228–35.

van der Berg G, Daamen CBG & Stadhouders J (1989) Bactofugation of cheese milk. *North European Food and Dairy Journal* **55**: 63–6.

Burghout HF & van Abeelen SJMH (2004) Cheese production which is otherwise conventional makes use of the curd particle-containing whey by comminuting to give still smaller curd particles which are added to the milk for incorporation in the final cheese. NL1024367 [patent].

Capra ML, Neve H, Sorati P *et al.* (2013) Extreme thermal resistance of phages isolated from dairy samples: Updating traditional phage detection methologies. *International Dairy Journal* **30**: 59–63.

Coffey A, Stokes D, Fitzgerald GF & Ross RP (2001) Traditional and molecular approaches to improving bacteriophage resistance of Cheddar and Mozzarella cheese starters. *Irish Journal of Agricultural and Food Research* **40**: 239–70.

CSK Food Enrichment (2001) Cryogenic blender enables CSK Food Enrichment to offer custom blends. *Powder and Bulk Engineering International* May, pp. 20–2.

Deveau H, Labrie SJ, Chopin M-C & Moineau S (2006) Biodiversity and classification of lactococcal phages. *Applied and Environmental Microbiology* **72**: 4338–46.

Eyssen JH (1898) An improved process for keeping and packing cheese and the like. GB 11,207 [patent].

Fox PF (2000) *Fundamentals of Cheese Science*. Aspen Publishers, pp. 50–1.

te Giffel MC & van der Horst HC (2004) Comparison between bactofugation and microfiltration regarding efficiency of somatic cell and bacterial removal. *Bulletin of the International Dairy Federation* no. 389.

Hinrichs J (2001) Incorporation of whey proteins in cheese. *International Dairy Journal* **11**: 495–503.

Hotchkiss JH, Werner BG & Lee EYC (2006) Addition of carbon dioxide to dairy products to improve quality: a comprehensive review. *Comprehensive Reviews in Food Science and Food Safety* **5**: 158–68.

Hull RR (2004) Factory-derived starter cultures for the control of bacteriophage in cheese manufacture. *Australian Journal of Dairy Technology* **59**: 230–5.

Hup G & Stadhouders JJ (1977) Starter culture production. US 4,053,642 [patent].

Kapoor R & Metzger LE (2008) Process cheese: scientific and technological aspects: a review. *Comprehensive Reviews in Food Science and Food Safety* **7**: 194–214.

Khoshgozaran S, Azizi MH & Bagheripoor-Fallah N (2012) Evaluating the effect of modified atmosphere packaging on cheese characteristics: a review. *Dairy Science and Technology* **92**: 1–24.

Kraft JL (1916) Process of sterilizing cheese and an improved product produced by such process. US 1,186,524 [patent].

Lucey JA, Maurer-Rothmann A & Kaliappan S (2011) Functionality of ingredients: emulsifying salts. In: Tamine AY (ed.), *Processed Cheese and Analogues*. Blackwell Publishing, pp. 110–132

Madsen FM, Reinbold GW & Clark WS (1966) Low-fat cheese. *Manufactured Milk Products Journal* **57**: 18–20, 22.

Mayes J, Urbach G & Sutherland BJ (1994) Does addition of buttermilk affect the organoleptic properties of low-fat Cheddar cheese? *Australian Journal of Dairy Technology* **49**: 39–41.

Mills S, Griffin C, Coffey A, Meijer WC, Hafkamp B & Ross RP (2010) CRISPR analysis of bacteriophage-insensitive mutants (BIMs) of industrial *Streptococcus thermophilis* – implications for starter design. *Journal of Applied Microbiology* **108**: 945–55.

Mistry VV (2001) Low fat cheese technology. *International Dairy Journal* **11**: 413–22.

Mistry VV, Metzger LE & Maubois JL (1996) Use of ultrafiltered sweet buttermilk in the manufacture of reduced fat cheddar cheese. *Journal of Dairy Science* **79**: 1137–45.

Mortensen G, Bertelsen G, Mortensen BK & Stapelfeldt H (2004) Light-induced changes in packaged cheeses – a review. *International Dairy Journal* **14**: 85–102.

Müller-Merbach M & Hinrichs J (2006) Thermische and hydrstatische Inaktivierung von Bacteriophagen. *Chemie Ingenieur Technik* **78**: 1723–30.

Penders JA & Burghout HF (2003) Cheese production by reducing ripened cheese to small pieces and molding under pressure includes addition of adhesion initiator, preferably dairy product, e.g. rennin, micellar casein, milk calcium, especially skim milk powder. NL 1024528 [patent].

Price WV & Bush MG (1974) The process cheese industry in the United States: a review. II. Research and development. *Journal of Milk and Food Technology* **37**: 179–98.

Russ W & Meyer-Pittroff R (2004) Utilizing waste products from food production and processing industries. *Critical Reviews in Food Science and Nutrition* **44**: 57–62.

Saboya LV & Maubois JL (2000) Current developments of microfiltration technology in the dairy industry. *Lait* **80**: 541–53.

Schreiber R (2001) Heat-induced modifications in casein dispersions affecting their rennetability. *International Dairy Journal* **11**: 553–8.

Schreiber R & Hinrichs J (2000) Rennet coagulation of heated milk concentrates. *Lait* **80**: 33–42.

Schreiber R, Neuhauser S, Schindler S & Kessler HG (1998a) Einbau von Molkenprotein Aggregaten in Schnittkäse – Teil 1: Prozeszoptimierung. *Deutsche Milchwirtschaft* **49**: 958–62.

Schreiber R, Post F & Kessler HG (1998b) Einbau von Molkenprotein Aggregaten in Schnittkäse – Teil 2: Einflusz von Grösze, Struktur und Zusammensetzung der Molkenprotein Aggregate. *Deutsche Milchwirtschaft* **49**: 1071–4.

Singh P, Wani AA, Karim AA & Langowski HC (2012) The use of carbon dioxide in the processing and packaging of milk and dairy products: a review. *International Journal of Dairy Technology* **65**: 161–77.

Spiegel T (1999) Whey protein aggregation under shear conditions – effects of lactose and heating temperature on aggregate size and structure. *International Journal of Food Science and Technology* **34**: 523–31.

Spiegel T & Huss M (2002) Whey protein aggregation under shear conditions – effects of pH value and the removal of calcium. *International Journal of Food Science and Technology* **37**: 559–68.

Stadhouders JJ (1974) Dairy starter cultures. *Milchwissenschaft* **29**: 329–37.

Svensson U & Christiansson A (1991) Methods for phage monitoring. *Bulletin of the International Dairy Federation*, no. 263, pp. 29–39.

Turcot S, Turgeon SL & St-Gelais D (2001) Effet de la concentration en phospholipide de babeurre dans le lait de fromagerie sur la production et la composition de fromages allégés de type Cheddar. *Lait* **81**: 429–42.

Turcot S, St-Gelais D & Turgeon SL (2002) Affinage de fromages allégés de type Cheddar fabriqués à partir de lait enriches en phospholipides. *Lait* **82**: 209–23.

Waagner Nielsen E (2004) Principles of cheese production. In: Hui YH (ed.), *Handbook of Food and Beverage Fermentation Technology*. Marcel Dekker, pp. 219–39.

Walstra P, Wouters JTM & Geurts TJ (2006) *Dairy Science and Technology*. CRC Press, pp. 485–7, 600, 635.

Westergaard V (2010) *Milk Powder Technology – Evaporation and Spray Drying*. GEA Niro, pp. 101–15.

Zehren VL & Nusbaum DD (2000) History of process cheese making. In: *Process Cheese*. Cheese Reporter Publishing, pp. 1–37.

5
Outside Constraints

Ingredients are often traded in a global market, and ingredients from milk are no exception to this pattern. The vast majority of dairy ingredients are supplied in a dried form to extend shelf life and reduce transport costs. To be successful, ingredients have to meet various requirements. Intrinsic qualities such as nutritional value, flavour and texture play a decisive role, of course; they are dealt with in Chapter 2 ('Milk Specifics') and Chapter 7 ('End User'). Other aspects are important too. In this chapter they will be indicated by the term 'Outside constraints'. The background of 'outside constraints' can be very diverse and even include import restrictions. We will focus here on religious concerns, food safety and sustainability. All three have an effect on technology.

The oldest aspect is religion, and Jewish and Muslim beliefs in particular are of great relevance. Dietary laws play a continuing role in the special food habits of adherents to these religions. Kosher and halal requirements will be discussed and the precautions that have to be taken during manufacture.

Apart from kosher and halal, food safety is crucial for all consumers. Various aspects of food safety will be dealt with. Elimination of food-borne pathogens is a key point. An outbreak of a disease, for instance via ingredients from milk, can be catastrophic for consumers. Moreover, it is extremely risky for manufacturers because of the adverse effect on their reputation. In this context tracking and tracing is essential and is a requirement for food safety (Raspor, 2005).

In the last part of this chapter sustainability will be discussed. The developments in this field are still undergoing rapid progress and they receive a lot of publicity. The general fear is that goods that are produced will deplete the earth's resources in the long run. We will focus here on resource efficiency in the 'dairy chain' and on greenhouse gas (GHG) emissions from the sector.

From Milk By-Products to Milk Ingredients: Upgrading the Cycle, First Edition. Ruud de Boer.
© 2014 John Wiley & Sons, Ltd. Published 2014 by John Wiley & Sons, Ltd.

5.1 Religious concerns

Kosher and halal dietary laws are very important to Jewish and Muslim communities. They are based on doctrinal rules that have no legal sanction (Sigman, 1993; Regenstein & Regenstein, 1992). Within their own communities, they are considered as laws. The kosher dietary laws are based on the Hebrew Bible, while the halal dietary laws can be found in the Koran. Both sources are interpreted and extended by scholars to address new issues and technologies (Regenstein *et al.*, 2003). Implementation is mostly carried out by food professionals. They are employed by kosher and halal certification agencies and examine the ingredients used, supervise the process and periodically inspect the production facilities. When permission has been granted, a symbol may be printed on the food packaging of the company concerned. The symbol is a registered trademark of a kosher or halal certification agency. Food companies pay a fee to enable these agencies to perform their activities. The symbol represents a necessary 'ticket to the market', and the market is greater than the number of actual believers. According to Regenstein *et al.* (2003) only one-fifth to one-third of the kosher consumers in the USA are Jewish. This can be explained by the quality image of kosher products.

In the following sections the focus will be on the manufacture of ingredients. Various aspects will be dealt with that play a role in the approval process for kosher and halal. Unlike meat products, dairy products can simultaneously be kosher and halal (Riaz & Chaudry, 2003).

5.1.1 Kosher

A meaningful explanation of the kosher dietary laws can be found in ethical considerations (Regenstein & Regenstein, 1979). A major concept of kosher laws concerns the separation of milk and meat. It is an example of the need to avoid cruelty; a mother and her offspring should not be killed at the same time. For Jews it means that dairy and meat products must be eaten separately. As a result, sandwiches containing a combination of cheese and meat are not permitted by their community.

The consumption of milk itself has limitations too. Historically, milk was only kosher if a so-called 'kosher witness' (rabbi) was present at the time of milking in order to make sure the milk was from a kosher animal. Unlike cows, camels are considered non-kosher animals, for instance. In the contemporary industrialized world, cow's milk production is highly regulated by food laws. For this reason, kosher agencies generally accept that cow's milk may be deemed kosher without a 'kosher witness' (G Buchbinder, 2013, personal communication).

The separation of milk and meat, as mentioned earlier, also has consequences for the use of additives. An example is rennet, which is the traditional cheese-coagulating enzyme. Rennet is an extract of calf's stomach. Its application results not only in non-kosher cheese, but also in non-kosher cheese whey products. Even the condensate from a cheese whey evaporator is considered non-kosher. Recycling of such a condensate in a plant may cause cross-contamination of the equipment via, for instance, steam heating or cleaning solutions. A plant can achieve a kosher certificate if the rennet is obtained from kosher slaughtered animals. Part of the procedure is supervision of the rennet extracting process by a rabbi. Also non-animal rennet alternatives are permitted. Examples of these are microbial rennet or biotechnology-derived chymosin using kosher equipment and ingredients. Another aspect of cheese-making is the scalding temperature. This is particularly important for the manufacture of Swiss and mozzarella types of cheese, which are not produced under rabbinical supervision. As a consequence, the cheese is regarded as non-kosher. The related cheese whey can only be kosher as long as the curds and the whey have not been heated above 49°C before the whey is drained off. By limiting the 'Jewish cooking' temperature (Regenstein & Regenstein, 2002), the transfer of a disqualifying taste from the cheese curd into the kosher whey can be avoided. Incidentally, this temperature is still subject of a debate between kosher agencies. However, there is no central organization whose opinion is binding on all agencies. To a great extent, these discussions are influenced by prominent Jewish scholars.

'Mixed' plants, in which kosher and non-kosher ingredients are produced on the same manufacturing line, need to separate their products clearly. It requires extra handling to comply with kosher regulations, and the equipment has to be 'kosherized' before kosher ingredients can be manufactured. Generally, the procedure for 'koshering' equipment is to leave it for 24 hours and then run boiling water through it until it overflows. This is a challenge for spray driers, which are among the most difficult pieces of equipment to 'kosherize'. Accurate planning is a key factor to overcome this problem in a cost-effective way.

Several kosher symbols can be found on packages of manufacturers. There are differences the way agencies interpret Jewish dietary laws, and some agencies tend to interpret the kosher laws more strictly than others. International recognition of the symbol is important for ingredient manufacturers as most of them act globally. Figure 5.1 shows a symbol of one of the 'mainstream' kosher organizations, the OU, or Union of Orthodox Jewish Congregations. Their supervision and certification programme dates back to 1924 (Sigman, 1993). Unless an ingredient is acceptable to the mainstream Jewish community, it is almost impossible to benefit from having a kosher ingredient for sale (Regenstein *et al.*, 2003). Kosher-Dairy is a further

Figure 5.1 A kosher dairy symbol of the OU. *Source:* Union of Orthodox Jewish Congregations trademark, with permission.

refinement. It is indicated as a 'D' and printed near the main kosher symbol on the food package. The symbol shows that the product is kosher and contains some milk or milk derivatives.

5.1.2 Halal

Halal is an Arabic term meaning lawful or permitted (Figure 5.2). As with kosher foods, numerous halal certifying agencies are active. The foundation of the Halal Certifying Agencies is more recent. One of the internationally renowned agencies is Ifanca (Islamic Food and Nutrition Council of America), which was founded in 1982. It is recognized by several regulatory agencies including those in emerging markets such as Indonesia and Malaysia. In Malaysia this agency is controlled by the state. Foods used in the earliest stages of human life are also certified. The Arabic signature of 'halal' can, for instance, be found on packages of infant foods. Agencies can closely cooperate with each other to make certification less expensive. The department of Islamic development in Malaysia has recognized foreign halal certifiers in 45 countries (JAKIM, 2012). The Halal Feed and Food Inspection Authority (HFFIA) is a private audit partner of the Malaysian organization for ingredients produced in The Netherlands.

The development of recognized agents was mainly initiated by the transition from rural societies to a modern society with its reliance on a supply of processed foods via supermarket chains and local grocery stores. Halal is derived from the Divine Commandment: 'Allah has prohibited it'. Among the basic reasons for prohibition are impurity and harmfulness (Regenstein *et al.*, 2003). The historical background to the prohibition of pork meat, by Jewish as well as Muslim communities, may stem from the risk of infection from pathogenic worms. The prohibition of alcohol by Muslims may be due to possible harmful effects on the nervous system. Codex Alimentarius (1997) also provides general guidelines for use of the term 'halal'; for instance, it gives a long list of forbidden foods of animal origin. Compounds derived from porcine materials like gelatin are also forbidden by Muslims. However, milk is one of the recommended foods for the Muslim population. As opposed to kosher,

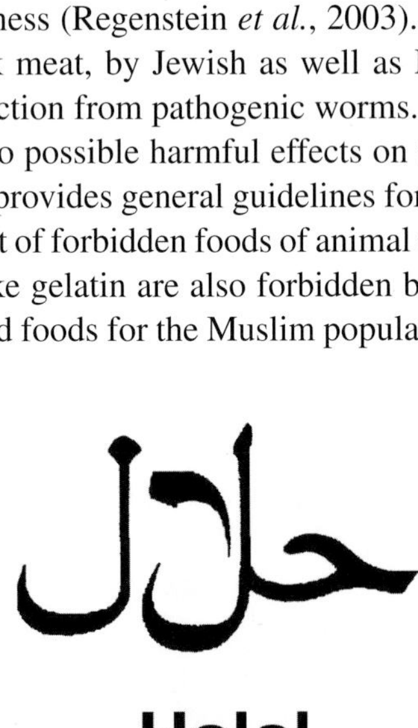

Figure 5.2 A characteristic signature found on halal products. *Source:* Courtesy of Halal Feed and Food Inspection Authority, the Hague.

there is no restriction on mixing meat and milk. For this reason an additional dairy symbol is not applied in halal certification.

The cheese and whey processing part of the dairy industry is a point of concern for halal certification too. The rennet enzyme has to be halal suitable. Bovine rennet produced from calves that have not been slaughtered according to Muslim requirements is not acceptable by most Muslims (Riaz & Chaudry, 2003). 'Allah' must be invoked by name at the time of slaughter (Regenstein *et al.*, 2003). It is an act of worship and obedience. However, rennet alternatives (microbial sources) are acceptable as long as the media in which these alternatives are cultured contain approved ingredients and cross-contamination of the fermentation equipment is avoided.

An overlap in requirements makes it possible for an ingredient to be both halal and kosher certified. Both bodies approve the use of microbial rennet and biotechnology-derived chymosin during cheese-making. A next step is, for instance, the enzymes used for the production of milk protein hydrolysates, which also need to be approved by the halal certifying agency. Double certification – halal and kosher – is here even more desirable as the markets are small and highly specialized. From a production point of view, it is hardly possible to manufacture each low-volume ingredient separately. 'Halal certification' of processed cheese ingredients, as mentioned in the information sheets, is even more complex. Not only the cheese but also the suppliers of packages and food additives – preservatives, emulsifiers, colours and thickening agents – need to be approved. The development of a global ingredient market still remains a challenge.

5.2 Safety

From a marketing point of view, food safety is a major concern in the ingredient business. Ingredients are generally distributed worldwide and they have to be safe for a long period of time. An extended shelf life at ambient temperature is therefore a key issue. A shelf life of 1–2 years is not uncommon and for this reason the ingredients are often supplied as a powder. Various safety aspects of powders will be dealt with in the following sections. About a quarter of the ingredients, referred to in the information sheets, are distributed in a solid, paste or oil form, including cheese ingredients (40–50% moisture). Their long shelf life is based on intensive heat treatment and formulation to inhibit growth and toxin production by spore-forming pathogenic bacteria (Glass & Doyle, 2005). A feared microorganism is *Clostridium botulinum*. Growth of pathogens can be controlled through the adjustment of pH, salt and use of antimicrobials. The anhydrous milk fat ingredients have an extremely low moisture content, often less than 0.5%. The quality and safety of these ingredients are ensured through standards formulated in the Codex Alimentarius (2010: Codex Stan 280-1973). In order to prevent the development of off flavours during storage, these standards focus in particular on the purity and stability of the lipids against oxidation and lipolysis.

5.2.1 General aspects

Not all aspects of food safety are mentioned in the information sheets; their focus is the microbiological part as is indicated under the heading 'Product analysis'. In Table 5.1 food safety is shown in a wider context. Together these factors are part of the 'dairy chain', which includes various aspects as the health of the cow, the production environment of the raw milk, the feed consumed and milk processing in the factory.

Zoonoses include, for instance, brucellosis, foot and mouth disease (FMD) and 'mad cow disease' (the scientific name for the latter is bovine spongiform encephalopathy (BSE)). They can have a large impact on the global distribution of ingredients. In case of pharmaceutical lactose, customers require a zero tolerance for BSE. Dual sourcing limited this risk and was the basis of cooperation between DMV FrieslandCampina and Fonterra in 2006, so BSE-free lactose could always be available for customers.

Dioxins are an example of environmental contamination via chemicals. It was caused by the operation of older waste incinerators in The Netherlands in the period 1989–90. The combusted chimney gases contaminated pastures in the vicinity and subsequently the silage and hay (Liem *et al.*, 1991). Residues of dioxins entered the milk through the cows. The maximum permitted limit has been set at 6 pg/g fat. Other chemical hazards such as residues of PCBs, herbicides, pesticides and radionuclides require continuous attention during quality control in the dairy industry.

The carry-over of various mycotoxins from cattle feed into milk has been reviewed by Fink-Gremmels (2008). Mycotoxins are toxic metabolites produced by moulds. The only mycotoxin for which a maximum permissible level in milk is established is aflatoxin M_1. In Europe the maximum acceptable level is set at 0.05 μg/kg milk. Due to the rumen flora of the cow the excretion levels in milk are low. Another important factor is the control of the raw materials for the production of feed concentrates. The feed plants play a decisive role in the elimination of this risk.

Residues of veterinary treatment are another concern. Particularly the use of antibiotics in the prevention or treatment of mastitis has been common practice in the dairy industry for several decades. The control of raw milk for the presence of antibiotics is a well-established routine. Any such contaminated milk has to be

Table 5.1 Safety aspects of milk production.

Zoonoses
Environmental contamination Feed contaminants Residues of veterinary treatment Adulteration Allergens Process control

kept on the farm. The background of this routine testing originated in possible antibiotic disturbance to fermentation processes in the factory (yoghurt, cheese). More recently, it has been debated if minor residues of antibiotics in food might affect microbial resistance to them in the human population. Oliver *et al.* (2011) reviewed the situation in the USA. They concluded that milk is one of the few sources that is free of antibiotic residues. Antimicrobial resistance among dairy pathogens, particularly those found in milk, is likely not a human health concern as long as the milk is pasteurized. As dairy ingredients are produced from pasteurized milk, the risk of antibiotic residues is extremely low in countries with a well-developed dairy industry.

A recent adulteration is the addition of the industrial chemical, melamine, to increase the nitrogen content of the milk. It results in an enhanced false protein content. Melamine itself is not a permitted food additive and contamination of milk carries a significant health risk in particular for infants. Moreover, the melamine added in China was of waste quality. Tens of thousands of infants were hospitalized and six died as a result of the contamination. The long-term effect of melamine on the health of consumers is unknown. One of the risks is kidney damage, particularly where the organ is still in development (Wei & Liu, 2012). The melamine scandal seriously damaged the image of the Chinese dairy industry. This substance will be monitored for by control systems in future.

The list in Table 5.1 ends with the items 'Allergens' and 'Process control'. Allergy will be discussed in Chapter 7 (see Section 7.3.1), while process control will be addressed in Section 5.2.3.

5.2.2 Microbiological aspects

About three-quarters of the ingredients shown in the information sheets are distributed in a dried form, of which spray-dried powders are by far the most common. Dehydration and subsequent packaging ensure an effective barrier to the uptake of moisture and oxygen. An optimum preservation of various powders is obtained at a water activity (a_w) close to 0.2 at 25°C (Efstathiou *et al.*, 2002). Water activity reflects water availability. Higher a_w values result in growth of moulds, yeasts and bacteria as well as other changes of both a biochemical (lipid oxidation, enzymatic reactions) or a physical nature (caking and stickiness). The aforementioned a_w value of 0.2 corresponds with an ideal moisture content of a powder. But the composition of the powder plays a role as well. The corresponding moisture content is, for instance, 4% for skimmed milk powder and 6% for caseinate powder (Schuck *et al.*, 2007). Different critical process parameters are involved in producing powders, such as the drying temperatures and the humidity of the air (Schuck *et al.*, 2008). Various equipment options are also available for powder manufacture (Westergard, 2010).

As opposed to sterilizing of liquid formulas, drying is not considered a controlled killing step. The consequence is that a powder is not a commercially sterile product.

Table 5.2 Frequency of microbiological tests based on powders indicated in the information sheets.

Test	Frequency	Reason
Standard plate count/aerobic plate count	40	General sanitation
Yeast and mould	38	General sanitation
Bacillus cereus	5	Presence of spore-forming bacteria
Thermophilic count	2	Survival after high heat treatment
E. coli	27	Hygiene indicator
Coliforms	20	Hygiene indicator
Enterobacteriaceae	15	Hygiene indicator
Salmonella	40	Pathogenic
Staphylococci (incl. CP, *aureus*)	31	Enterotoxin, pathogenic
Listeria	6	Pathogenic
Enterobacter sakazakii/Chronobacter	1	Pathogenic

CP, coagulase-positive.

This is clearly shown in the information sheets. Even if the powder is produced under optimal conditions, a low level of microorganisms will still be present.

A summary of the various microbiological tests is shown in Table 5.2. The referenced microorganisms are indicated in the information sheets. The standard plate count as well as the number of yeasts and moulds give an impression of the general hygiene in an ingredient plant. Particularly, the standard plate counts show large differences. The counts for acid casein have a warranted level of less than 30,000/g, while its derivative calcium caseinate has typically a value of 100/g. The latter is, for instance, used in clinical nutrition and infant foods. From a microbiological point of view these are vulnerable applications, especially if the finished product has not been subjected to an additional heat treatment. In the case of inhalation lactose the numbers are even more stringent ($\leq$10/g). This ingredient is directly applied in the inhaler and infection of the respiratory tract needs to be avoided at all costs. These low counts can only be obtained in special lactose production facilities in which the final step is carried out under controlled conditions.

In some cases specific organisms are mentioned, for instance *Bacillus cereus*. This microorganism is part of the product analysis of low-lactose and lactose-free skimmed milk powder. This information is useful in cases of liquid applications. Spores of *Bacillus cereus* can survive pasteurization and result in spoilage and disease, particularly if the finished products are kept at ambient temperature (Feijoo *et al.*, 1997).

The next categories mentioned in Table 5.2 are hygiene indicators: tests for coliform bacteria and the large bacterial family of *Enterobacteriaceae*. *Escherichia coli* is also a member of this family. Usually the indicator microorganisms are non-pathogenic themselves. Most European producers use *Enterobacteriaceae* as an indicator, while coliform bacteria are more commonly applied by US companies. These organisms are the result of contamination with faeces and soil via the exterior of the udder during milking. They enter the ingredients because of insufficient pasteurization or recontamination from inadequately cleaned or ill-designed equipment located downstream of the pasteurizer. Such contaminations should be kept to the absolute minimum, as faecal contamination is an important source of food-borne pathogens such as *Salmonella* spp. (Oliver *et al.*, 2005). The information sheet 'Demineralized whey powder' gives an example of a strict standard, namely a negative test result of each lot for *E. coli*, *Enterobacteriaceae* and *Salmonella* in, respectively, 10 g, three samples of 1 g and five samples of 25 g. This powder is an important ingredient for infant foods.

The last category in Table 5.2 comprises tests for target or pathogenic microorganisms. These are an important safety concern. As the public's health is at stake, standards are set by Codex Alimentarius and national governments. The regulations of the USA and EU are very often referred to as examples. In addition to these defined standards, specific microorganisms may be added as a result of negotiations between customers and suppliers. The most referenced organisms are *Salmonella*, *Staphylococcus aureus*, *Listeria monocytogenes* and the recently emerged *Enterobacter sakazakii* (*Cronobacter* sp.). Numerous strains of each microorganism can be distinguished. A specific strain of *E. coli* O157:H7 can, for instance, be added to this list. In contrast to most *E. coli* variants, which are hygiene indicators, this strain is able to produce toxins. The same is the case for *Staphylococcus aureus*. This organism plays a role in udder inflammation (mastitis). In the information sheets it is usually indicated as *Staphylococcus* CP (coagulase positive). An overview of the number of cases of various food-borne diseases reported in the EU-25 countries is given by Eurostat (2008). Papademas and Bintsis (2010) provide an overview of the food-borne disease outbreaks involving dairy products. None of them were related to dairy ingredients. Particularly, raw milk products (cheeses) constitute a risk. The key operation in the production of safe milk and milk ingredients is pasteurization, which inactivates the vegetative microorganisms. In a recent study the heat inactivation of six milk-borne pathogens in milk was studied. The pasteurization process was carried out under industrial conditions (Pearce *et al.*, 2012). All were inactivated below 68°C/15 s at a $\log_{10}$ reduction of 6.8. It was concluded that high-temperature short-time (HTST) pasteurization in New Zealand is carried out at up to 2°C above the minimum legal requirement of 72°C/15 s. Extra safety is obtained by automatically diverting the milk flow if the temperature falls below 72.5°C. The minimum legal requirement is slightly different for the USA, namely 161°F, or 71.7°C (Pasteurized Milk Ordinance).

Powders for infants and young children need special care. The main reason is that no additional heat treatment is carried out before they are used. Pathogens of greatest concern are *Salmonella enterica* and *Cronobacter sakazakii*. *Salmonella* infections can be linked to outbreaks of disease (Cahill *et al.*, 2008). Contamination usually results in diarrhoea and can, in some cases, be life-threatening. The incidence of *Cronobacter* infections appear to be low; however, the consequences can be severe. Outbreaks of *Cronobacter* can be especially linked to neonatal intensive care settings

(Codex Alimentarius Commission, 2008). Various countries have already begun to include *Cronobacter* spp. in the microbiological testing programmes for powdered infant formulas (Buchanan & Oni, 2012). Worldwide, the procedures are not yet completed, which may be why this organism is only mentioned once in the present information sheets.

5.2.3 Critical control points

Ingredients from milk need to be produced with 'due diligence', which means that the manufacturer has to take all reasonable precaution to avoid product contamination (Mettler, 1994). This can be translated into several procedures (Luning & Marcelis, 2009). One of the best known examples is Hazard Analysis Critical Control Points (HACCP). In this concept critical control points (CCP) are indicated. A CCP is defined as 'any factor that, if not controlled, provides a threat to the product acceptability and/or consumer safety'.

A number of CCPs are indicated in the flow schemes for the manufacture of sodium caseinate, MPI and lactoferrin (Figure 5.3). These are high quality ingredients with a standard plate count below 1000/g. The first CCP is a HTST treatment, which is carried out to inactivate the vegetative pathogens. This standard treatment is

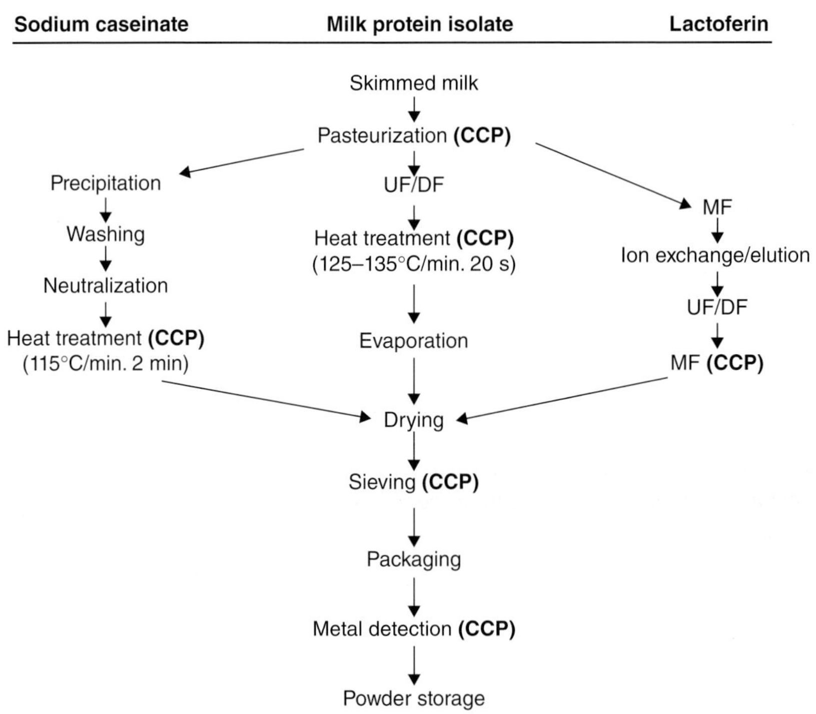

Figure 5.3 Examples of critical control points (CCP) of three ingredients from milk. DF, diafiltration; MF, microfiltration; UF, ultrafiltration.

followed by the indicated purification processes. Subsequently, more specific CCPs are located just before drying. Acceptability and safety aspects need to be in balance. Sodium caseinate allows intensive heat treatment. One of the reasons to select 115°C/> 2 min is the inactivation of surviving FMD viruses (Cunliffe & Blackwell, 1977; Cunliffe *et al.*, 1978; Haas, 2004). It is not a public health threat but the virus is highly contagious to cattle and thus represents a major risk for non-infected countries. Due to the presence of whey proteins, MPI is not a heat-stable ingredient. A less intensive heat treatment is therefore applied. Ultra high heat treatment (UHT) is shown in Figure 5.3, with the selected conditions 125–135°C/≥ 20 s. The most heat labile of the three ingredients is lactoferrin (Kussendrager, 1994). Additional heat treatments are avoided, being replaced by microfiltration. In this way the biological activity of lactoferrin is maintained. The following CCPs are situated after drying. Sieving and metal detection are the same for these three ingredients. The target is the removal of foreign matter, which of course needs to be absent in powders. These CCPs are introduced to avoid any harm to consumers and to minimize complaints.

Sometimes codes are developed for specific foods of vulnerable groups. An example is the 'Code of hygienic practice for powdered formulae for infants and young children' (Codex Alimentarius Commission, 2008). Different measures are recommended that focus on the prevention of post-drying contamination. Some of them are the adequate separation of wet and dry zones in a factory. Other preventive measures are a positive air pressure to avoid the entry of unfiltered air into the high-hygiene areas. Implementation of environmental programmes is a further step to minimize the risk of contamination. Additionally, it is of great importance that sufficient documentation is available to facilitate product traceability. In spite of the recommendations, the key issue is the commitment of management to secure the safety of their ingredient processes. Risk assessment (Schothorst *et al.*, 2009) and quality costs are topics that require exchange of opinions on a regular basis. Customer audits are also a valuable tool in checking that commitment, being beneficial for consumers as well as suppliers.

5.3 Sustainability

As opposed to food safety, which is part of well-established practice in the dairy industry, sustainability is still a work in progress. As a general principle sustainability is the capacity to maintain a certain process or state indefinitely. Milk has been used by humans for centuries, so it seems to meet that definition. However, over time the world has become much more densely populated, the milk flow from farm to consumer ('dairy chain') has become more complex, and the local dimension has diminished to a large extent. It is understandable that the consumer no longer sees the overall perspective. This is particularly the case for ingredients. They are used in all kinds of food products, as evidenced by the information sheets in this book. Furthermore, ingredients from milk find their way into a global market, which increases the complexity too. An attempt will be made to provide some background information regarding certain aspects of sustainability. We will focus on resource efficiency and GHG emissions; the latter has a more direct link with climate change.

5.3.1 Resource efficiency

The efficient use of scarce resources is an important issue today. As will be shown, major improvements have been achieved in this respect. We will concentrate in this section on the dairy cow, which is the basis of the dairy industry, the utilization and potential production of energy, and the use of water in the whole 'dairy chain'.

5.3.1.1 The dairy cow

The most valuable asset of cow's milk is protein. FrieslandCampina announced recently a new milk price system, which will take effect on 1 January 2014. The warranty price of the company will be based on protein : fat : lactose in the ratio 10 : 5 : 1. It reflects the value of milk protein in payment for the milk. Milk protein as well as other animal proteins are highly beneficial for an adequate human diet. Due to growth of the population, shortages of valuable animal proteins are expected in certain regions of the world in the near future. More of the existing sources need to be converted into food in order to be able to feed the world (FAO, 2011). In the context of shortage, the efficiency of dairy cows in producing edible protein is frequently questioned.

Shortage and sustainability are drivers for research regarding sustainable protein sources. They might offer an alternative for the dairy cow in the future. An example of a recent approach is indicated in Figure 5.4, in which the protein concentrate 'rubisco' is prepared from various plant materials (van de Velde *et al.*, 2011a). Rubisco (ribulose bisphosphate carboxylase/oxygenase) makes up to 50% of the protein fraction. It is abundantly present on Earth. In the technology as described materials such as grasses, leaves and beet tops could be directly converted into a colourless, tasteless protein concentrate. The powder obtained has an attractive amino acid profile, which makes it outstanding among the vegetable proteins. Rubisco also has potential as a functional ingredient due to its interesting foaming and gelling properties (Barbeau & Kinsella, 1988; van de Velde *et al.*, 2011b). The proposed technology consists of eight process steps (van de Velde *et al.*, 2011a). Maceration is one of the options to release the content of the vegetable cells with their rigid walls. The discharged green juice goes through a mild heat treatment. Part of the available protein is coagulated and separated. The purified rubisco protein is dechlorophyllated by activated carbon and subsequently dried. Some aspects of this process need substantial research, for instance scaling up to production level and valorization of the remaining cake, coagulated protein and chlorophyll-loaded carbon. This 'novel protein' project is quite a challenge, particularly for the field of agricultural side-streams. The future will reveal how it might affect the dairy industry.

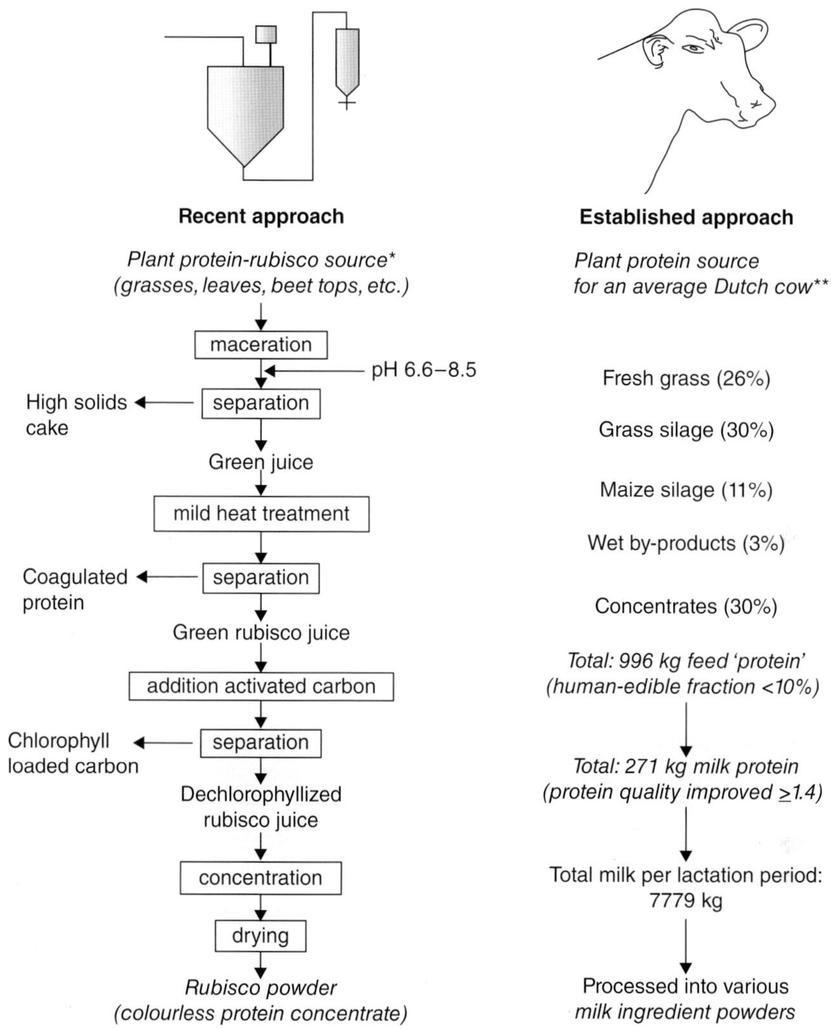

Figure 5.4 Plant materials as a source of protein production. *Sources:* *van de Velde *et al.* (2011a); **van Hooijdonk (2013).

Cows belong to the so-called 'established approach'. These ruminants are able to convert large quantities of human-inedible materials, such as grasses and herbaceous legumes, crop residues and by-products of the food- and fibre-processing industries, into highly nutrient-dense milk products (van Hooijdonk & Hettinga, 2013). In addition, they consume feed grains and oilseed products, for which are viewed as being in direct competition with humans. To some extent they compete for land with vegetable crops. In Figure 5.4 a diet of an average Dutch cow is given, in which concentrates have a share of 30%. These concentrate pellets consist of 75% food waste and 25% cereal grains, which means that less than 10% of the cow's feed is edible for humans. Furthermore, cows upgrade crude feed proteins to high-quality food proteins by at least a factor of 1.4 (ACM vanHooijdonk, personal communication, 2012). These conclusions support an earlier US study, which stated: 'Thus, diverting grains from animal production to direct human consumption would, in the long term, result in little increase in total food protein and would decrease average dietary quality and diversity' (CAST, 1999).

Over time enormous progress has been made in the efficiency of producing milk through selective breeding as well as by applying feed concentrates and fertilizers. Currently there are cows with a lifetime production of ≥100,000 kg of milk. An example of how this large quantity of milk converts into some valuable ingredients is given in Figure 2.2. A long horizon is needed for the profitable breeding of cows with a high lifetime production (VanRaden, 2004). The percentage of these exceptional cows is still very low in the Dutch cattle population. Their share rose from 0.4 to 1.8% during the last decade (Veepro Holland, 1999–2012). Compared to the population average, of 3.5 lactations and a lifetime production of about 35,000 kg of milk, the difference is large. In the case of a 100,000 kg cow, the feed needed for the first 2 years of maturation can be spread over more milk. From an environmental point of view this dilution of the so-called maintenance effect is favourable. It results in reduction of the amount of feed as well as reduced manure and greenhouse emissions per milk unit.

Originally the milk of cows was meant to feed only the calf. This situation has gradually changed and nowadays a high share of the milk production is available for food applications. Calves use only a modest part of the available milk solids (JAM Maas – FrieslandCampina Nutrifeed, 2013, personal communication). Some of the calf population is reared for future milk production, while the rest are destined for calf meat. The latter can be considered a by-product of milk production. If a cow produces 10,000 kg of milk per lactation period (about 1300 kg of milk solids), the following figures can be calculated:

White meat production (milk replacer)	~14% of original milk solids
Rearing calves (raw whole milk)	~3% of original milk solids
Rearing and rosy meat production (milk replacer)	~2% of original milk solids

An example of the composition of a milk replacer is shown in Table 7.5. The formula consists of about 80% of various milk solids. Colostrum is fed during the first few days, after which a milk replacer can be given during the subsequent 8 weeks (rearing, rosy meat). The production of a high-quality white meat requires the feeding of milk solids over a longer period. This practice reflects the relative high amount of milk solids used. Particularly, the low iron content of milk solids is responsible for the pale colour of this type of meat (Ngapo & Gariépy, 2006). The production of veal has been in decline for several decades now (Sans & de Fontguyon, 2009). One of the reasons for the change is the decreased availability of calves in Europe. A gradual switch from white to rosy meat resulted from animal

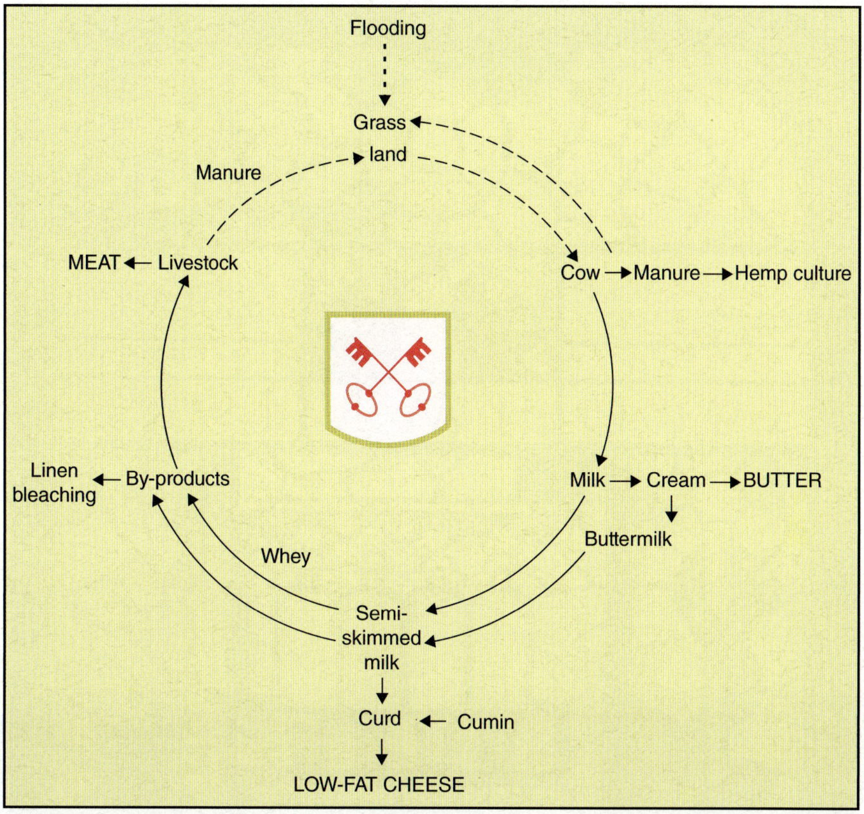

Plate 1.5 An example of a small cycle, depicting a farm area in the Middle Ages centred on the city of Leiden in The Netherlands.

From Milk By-Products to Milk Ingredients: Upgrading the Cycle, First Edition. Ruud de Boer.
© 2014 John Wiley & Sons, Ltd. Published 2014 by John Wiley & Sons, Ltd.

Plate 1.9 Artistic impression of the production and application of lactose dated 1950. Reproduced courtesy of Hollandsche Melksuikerfabriek-HMS.

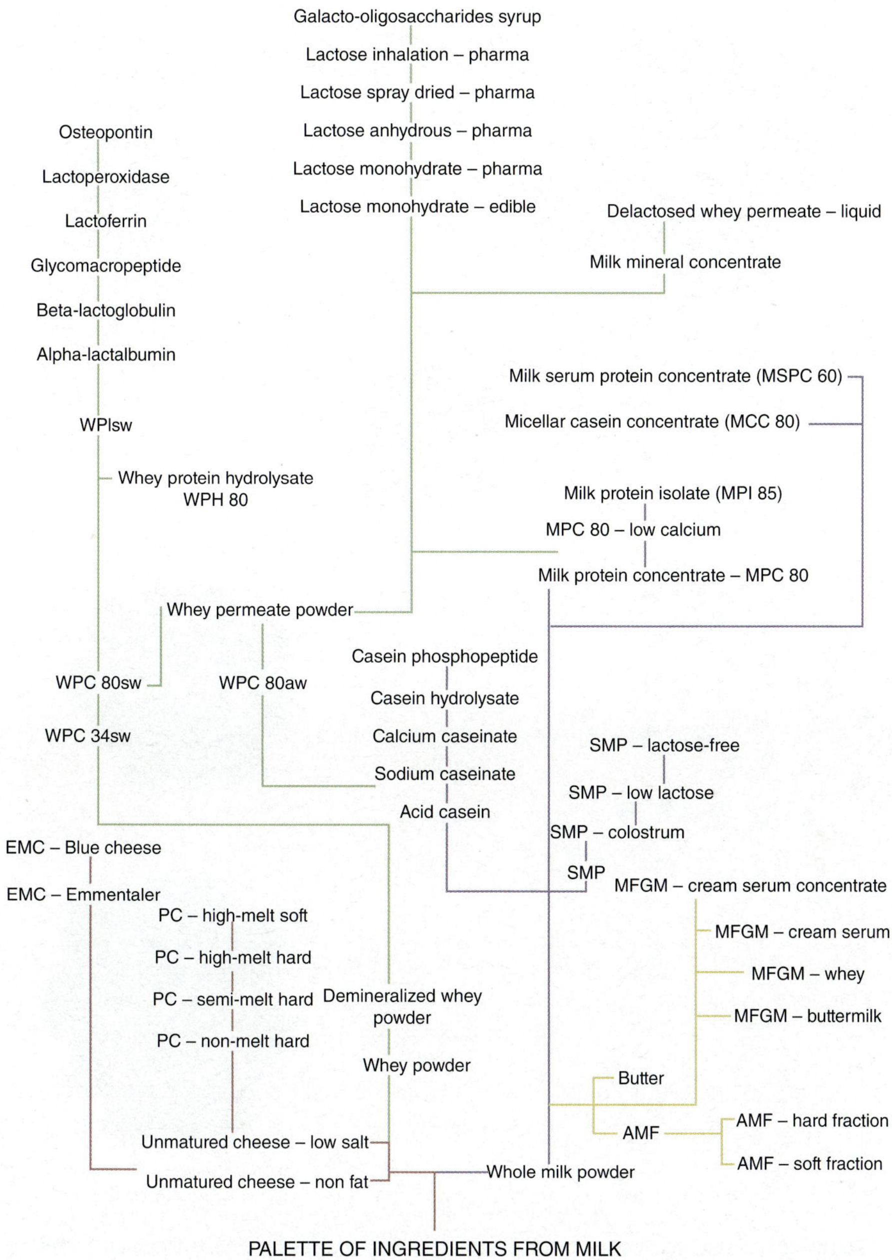

Plate 2.1 A palette of ingredients obtained from milk. AMF, anhydrous milk fat; EMC, enzyme modified cheese; MFGM, milk fat globule membrane; MPC, milk protein concentrate; PC, processed cheese; SMP, skimmed milk powder; WPC, whey protein concentrate; WPI, whey protein isolate.

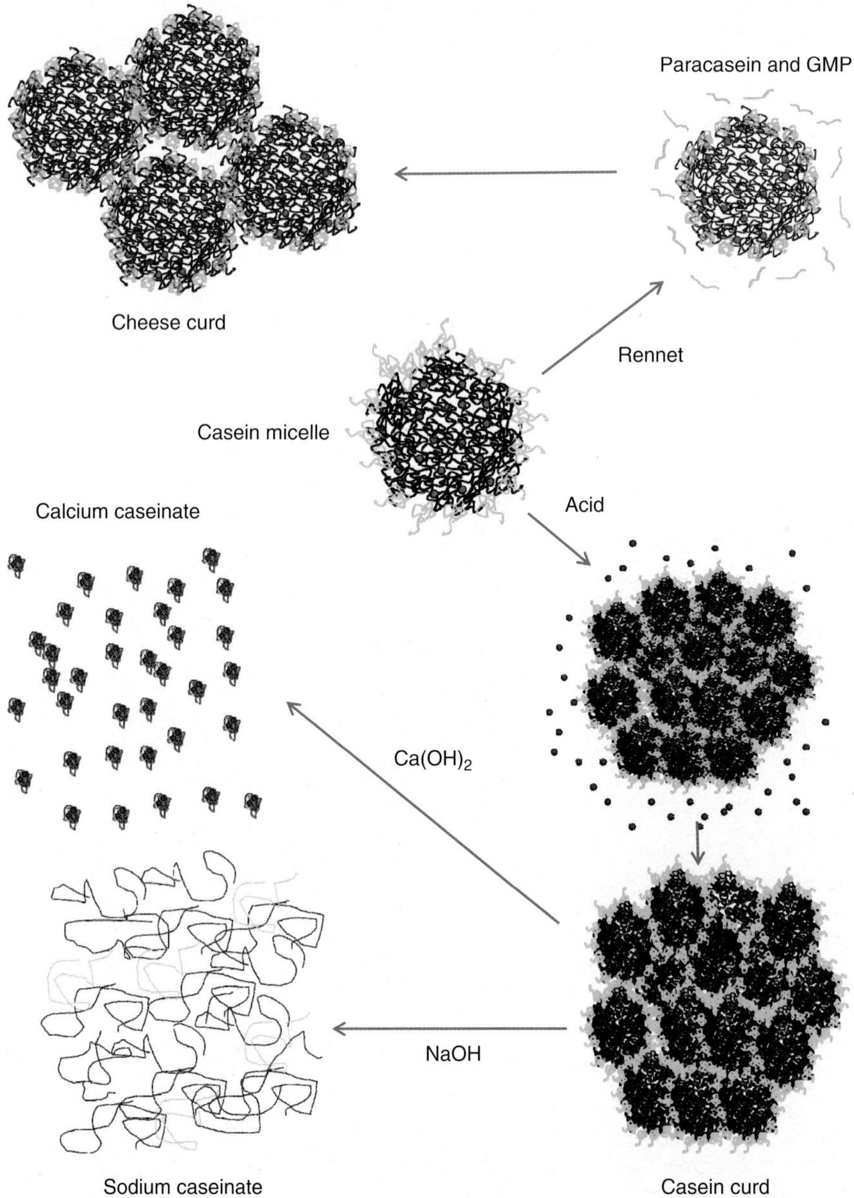

Plate 2.3 Casein micelles and modification. *Source:* Courtesy of FrieslandCampina DMV.

Plate 2.9 Tomahawk shape crystal of α-lactose monohydrate. The crystal faces are indicated by Miller indices (adapted from CCF brochure).

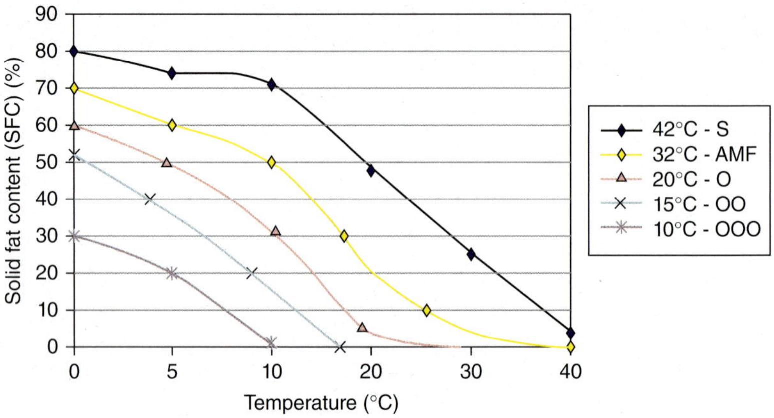

Plate 2.12 Melting curves of various milk fat fractions. *Source:* Boer, 2008. Reproduced with permission from the author, F. Boer. AMF, anhydrous milk fat; S, stearin; ○, ○○, ○○○, different olein fractions.

GB 11,207

'An improved process for keeping and packing cheese and the like'

Inventor: Jan Hendrikzoon Eyssen **Filed**: 17th May, 1898
Claim: Process for keeping and packing cheese without affecting the taste and the aroma in which the cheese in a ripe state is kneaded to a paste, the air expelled, the impurities removed and the paste packed into hermetically closed tins, substantially as described.

Plate 4.5 The granted patent claim (1899) and the 'royal' designation (1905) of Jan Hendrikzoon Eyssen. *Source:* Courtesy of Eyssen BV.

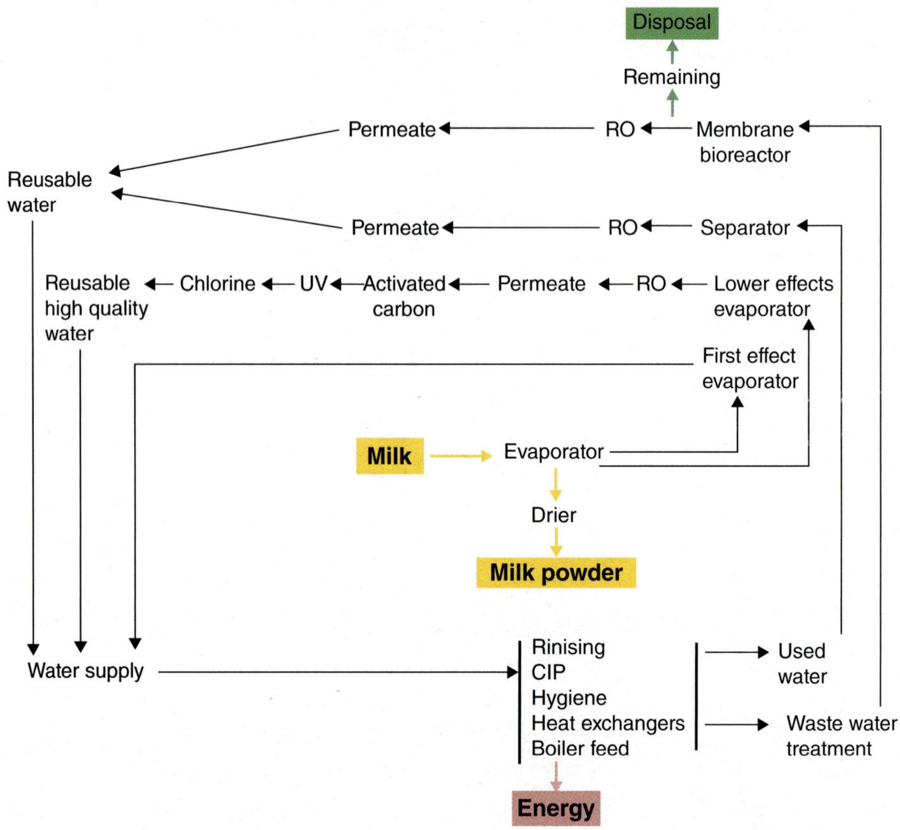

Plate 5.6 Scheme for recycling of water in a milk powder plant. RO, reverse osmosis.

welfare concerns, which influenced the type of feed given to the animals and favoured group housing.

Animal welfare is a complex and sensitive subject and difficult to define. McGlone (2001) emphasized a multi-disciplinary approach in defining animal welfare, including level of productivity, behaviour, physiology, health and anatomy of the animal. The perception of the public drives the animal welfare issue too and slogans such as 'the cow in the pasture' are popular in The Netherlands nowadays. Animal welfare is an important subject in high-producing cows such as the worldwide dominant Holstein-Friesian breed. The resources an animal has are limited and, as a result, if output is increased through one biological process, such as producing more milk, other functions such as fertility, maintenance, immune defence, etc. will be affected (Oltenacu & Broome, 2010). In cows with a lifetime production of 100,000 kg milk the health problems such as lameness, mastitis and infertility are well balanced against a high milk yield. It is a good example of careful management and in my opinion it provides guidance for the future.

5.3.1.2 Energy

Another key aspect of resource efficiency is the energy consumption in the 'dairy chain'. The total energy use in the life cycle of milk is on average 7.5 MJ/kg milk (IDF, 2009). The farm represents approximately 40% of the total energy use and the dairy plant is estimated to be about 60%.

Feed production and electricity consumption are the main factors at the farm. The cow's diet has a large impact, because the manufacture of feed concentrates are more energy consuming than the production of fodder for grazing and silage. The differences between New Zealand and the United Kingdom in Table 5.3 reflect the effect of agriculture practices (Saunders *et al.*, 2006). The electricity on the farm is used for milking, manure handling, fencing, etc.

The energy use for the dairy plants encompasses processing as well as packaging. Each accounts for about 20%, and the remaining relates to transport and cleaning. The total energy consumption of dairy plants depends on the end-product, as indicated in Table 5.3 (FAO, 2010). The figure for cheese of 7.7 MJ/kg is high as on average 1 kg of cheese can be produced from 10 kg of milk. The manufacture of skimmed milk powder is also very energy intensive and uses approximately 10 MJ/kg. The background is that 10 kg of water have to be removed for each kilogram of powder. The energy needed per kilogram of water removed is very dependent on the process. Reverse osmosis (RO) is the most efficient unit operation with 0.02 MJ/kg removed water; however, the low degree of concentration limits its application. Spray drying requires by far the most energy, approximately 4 MJ/kg removed water. The commonly applied evaporators require about 15 times less energy than spray driers (Ramirez *et al.*, 2006). The manufacture of skimmed milk powder is therefore carried out in such a way that evaporation of water during spray drying is maximized.

Table 5.3 Energy use and energy contribution in MJ per kg product in various stages.

Stage	Energy use	
Energy use at farm level[a]		
New Zealand	3.0 MJ/kg milk	
United Kingdom	6.5 MJ/kg milk	
Energy use in processing at plant level[b]	**MJ/kg of product**	**MJ/kg of milk**
Milk	0.56	0.53
Cheese	7.7	0.77
Whey	0.019	0.02
Skim milk powder	10	0.93
Energy use per unit operation[c]		
Reverse osmosis (RO)	0.02 MJ/kg removed water	
Evaporation (MVR - TVR)	0.1–0.4 MJ/kg removed water	
Spray drier	3.4–4.9 MJ/kg removed water	
Energy use per mode of transport[d]		
Air	8.1 MJ/kg.1000 km	
Road	0.5 MJ/kg.1000 km	
Rail	0.2 MJ/kg.1000 km	
Sea	0.1 MJ/ kg.1000 km	
Net energy contribution of by-products		
Lactose conversion into ethanol	0.3 MJ/kg milk[e]	
Manure fermentation into methane	1.8 MJ/kg milk[f]	

MVR, mechanical vapour recompression; TVR, thermal vapour recompression.
[a]Saunders *et al.* (2006).
[b]FAO (2010).
[c]Ramirez *et al.* (2006).
[d]Rizet (2003).
[e]Mawson (1990, 1994).
[f]van Tilburg *et al.* (2008).

Milk powder is usually transported to areas where there is a lack of milk. The distance between production and consumption can be thousands of kilometres. In Table 5.3 the energy use per mode of transport is compared. Transport by air requires by far the most energy. Due to the long shelf life of powder, it is not an obvious choice. Container shipment by sea is usually adopted. This mode of transport is very attractive compared, for instance, to transport by truck. Shipment by sea is about five times less energy intensive than transport by road (Rizet, 2003).

So far the production of dairy products has all been about energy inputs, but some energy can be gained from milk by the conversion of by-products into ethanol or biogas. For obvious reasons costs play a decisive role. The conversion of lactose into ethanol, which can be used as a compound for vehicle fuel, is less attractive. An energy contribution of about 0.3 MJ/kg milk can be calculated, if a pre-concentration step is carried out (Mawson, 1990). More value added applications are now available for lactose. An expanding field is the use of dairy manure for the production of biogas/electricity (Krebbekx *et al.*, 2011). In theory, the amount of biomass energy

Table 5.4 Nutrient flow of three elements retained for maintenance, secreted in cow's milk and excreted in manure as faeces and urine.

Element	Maintenance	Milk	Manure
Nitrogen (%)	20	20	60
Phosphorus (%)	13	27	60
Potassium (%)	12	13	75
Estimated milk production	32 kg/day		
Estimated manure production	72 kg/day		

Source: Data from Hart *et al.*, 1997.

present in manure is sufficient to replace all the fossil energy consumed in the 'dairy chain' in The Netherlands (van Hooijdonk & Hettinga, 2013).

A substantial by-product of milk is the excreted faeces and urine. For reasons of convenience these two will be referred to collectively as 'manure'. A highly productive cow with a yield of 32 kg milk/day excretes about 72 kg of manure, as indicated in Table 5.4. Only 15–25% of the nutrients in the feed end up in the milk (Hart *et al.*, 1997). A key factor in managing manure is recycling the nutrients for improving the production of the farmland. With large herds there is an abundance of manure at one location, so reuse of all the nutrients on the site is often not possible. This phenomenon is common in the dairy industry. Half a century ago cheese plants released more whey than the surrounding farms could absorb as animal feed. It forced the industry to process the whey and add value to it.

An example of the composition of cow manure in The Netherlands is demonstrated in Table 5.5 (Zeeman, 1991). About 75–80% of the solids comprises fibrous materials, which are complex carbohydrates such as cellulose, hemicellulose and lignin. These materials are referred to as volatile solids (VS). The VS can be further degraded in an anaerobic digester. In the digester a fermentation process takes place,

Table 5.5 Average composition of cow manure in The Netherlands in percentages.

Volatile solids	7.5 (cellulose, hemicelluloses, lignin)
Nitrogen	0.44 (organic-N, NO_3-N, NH_4-N)
Minerals	
P_2O_5	0.18
K_2O	0.55
CaO	0.21
MgO	0.1
Na_2O	0.1
Cl	0.3
SO_3	0.18
Total solids	9.5

Source: Data from Zeeman, 1991.

which also occurs in the rumen of the cow. The retention time in the digester is, however, several times longer than in the bovine rumen. Specifically the retention time is extended from 0.5–3 days in the cow to 10–20 days in the digester (Zeeman, 1991). The long residence time in the digester favours biogas production. Other beneficial effects of the anaerobic digester are a decrease in objectionable odours and GHG emissions as well as the availability of a high-quality fertilizer (digestate). Anaerobic digestion is a synergistic multi-stage process performed by a consortium of microorganisms of which hydrolysis is the first and methanogenesis is the final step (Li *et al.*, 2011). Biogas production from cow manure is dependent on the diet and quantity of milk produced. The yield varies from 208 to 268 norm litres (Nl) biogas/kg VS (Amon *et al.*, 2007). The main component of biogas is CH_4; it represents about 60–70% of the total. The other constituents are CO_2, NH_3 and H_2S. Due to the pre-digestion of the complex carbohydrates in the rumen of the cow, dairy manure has a modest potential to produce methane. The methane production of manure is about half that of a maize silage harvested at optimal conditions (Amon *et al.*, 2007). Mixing of manure and maize silage can be used to increase the biogas output. The system is called co-digestion. The downside of using maize silage may be that it deserves a better fate than co-digestion.

In Figure 5.5 various options are shown for the processing of manure in which use is made of an anaerobic digester. Farm and industrial settings are demonstrated separately, the latter below the dotted line. In case of off-farm use of dairy manure,

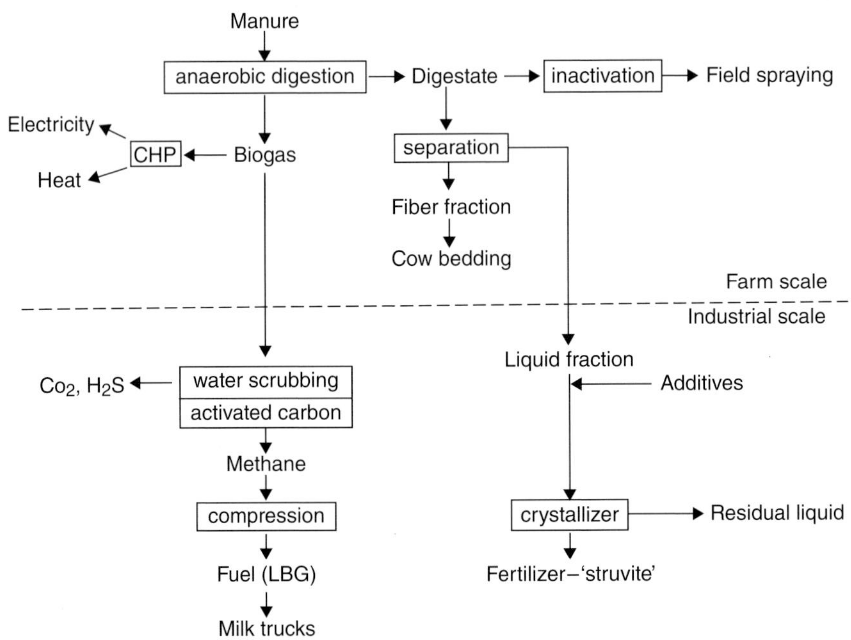

Figure 5.5 Some options for the utilization of dairy manure at farm and industrial scale. CHP, combined heat and power; LBG, liquefied biogas.

Table 5.6 Average production of electricity and bedding material of a dairy cow at a large farm scale.

Component	Output
Manure	57 kg/cow/day
Volatile solids (VS)	5.8 kg/cow/day
Methane	125.5–166.3 L/kg VS
Electricity	4.4 kWh/cow/day
Bedding material	0.019 m^3/cow/day

Source: Wang *et al.*, 2011.

biogas upgrading may be necessary. Figure 5.5 depicts the production of electricity at some large farms, as studied in Vermont (USA), where the average milking herd size is slightly above 1200 cows (Wang *et al.*, 2011). The biogas from the anaerobic digester is converted into electricity using a CHP (combined heat and power) generator. Thermal energy is produced as a by-product. The recovered combustion heat can be used for heating water, which in turn can be used for warming/cleaning of milking parlours and buildings. The digestate, which still contains an undigested fibrous fraction, runs through a screw-press separator. The solid fraction (~30% TS) is applied as cow bedding while the remaining fluid is used as a liquid fertilizer. This liquid digestate can be characterized as a N-K fertilizer and is comparable in composition to animal urine (Möller & Müller, 2012). Contamination of the environment and milk by pathogenic bacteria/viruses needs to be reduced to an acceptable level. Heat-inactivation of the digestate is, for instance, carried out at 70°C/1 h (Uemura *et al.*, 2008). Table 5.6 contains some data about the Vermont farms. The electricity produced was on average 4.4 kWh/cow/day. Even for farms of this size feasibility is still questionable. Given the present state of the art, grants and subsidies are necessary to justify the initial investments for an integrated anaerobic digester and electricity generator system. In order to accomplish the project customers pay an additional premium of $0.04/kWh. The electricity output of new farm systems is expected to be improved by up to 93% (Wang *et al.*, 2011). Better conversion of the VS material into biogas/electricity may be the basis for this improvement. The consequence is probably a reduction in volume of the bedding material.

Industrial-scale processing of manure can be achieved by purifying the biogas (Persson *et al.*, 2006). The removal of CO_2 increases the heating value of the biogas. A common technology for removing CO_2 and upgrading biogas to a purity of approximately 97% CH_4 is water scrubbing. The minor constituent H_2S is corrosive and should therefore be removed as well. Humidity and H_2S can be eliminated by using activated carbon. Subsequently, the purified biogas can be compressed and converted into liquid biogas (LBG). Similar to liquid petroleum gas (LPG), LBG can be used as a vehicle fuel. A 30 tonne milk truck can drive, for example, 630 km on the LBG fuel, which is processed from the manure of a cow with an annual milk production of 10,000 kg (PJ Roeleveld, Royal HaskoningDHV, 2013, personal comnuication). Besides biogas, the liquid fraction shown in Figure 5.5 requires attention. The elements N, P, K and Mg of the non-volatile substances have potential as nutrients (Table 5.5). An example of this is the manufacture of 'struvite' ($MgNH_4PO_4 \cdot 6H_2O$) or K-struvite ($MgKPO_4.6H_2O$). Both have a crystalline structure and are well suited to be produced in a crystallizer. A complicating factor is that the majority of the phosphate is present in a suspended form. This is caused by the elevated calcium levels in the digest. Struvite formation can be improved by adding EDTA (Zhang *et al.*, 2009, 2010). Approximately 90% of the total phosphorus is then released into solution. This facilitates the crystallization of struvite under the appropriate conditions. Struvite or its variants may be applied in the field of specialized fertilizers.

5.3.1.3 Water

Water is another important resource of the 'dairy chain'. Recently, the water footprint concept was introduced (Mekonnen & Hoekstra, 2010; Hoekstra, 2012). It is an indicator of water use in relation to the production of consumer goods. This is a complicated subject as various types of water can be distinguished: surface/ground water, rain water and polluted water. In this section the focus will be on the overall use of water in the 'dairy chain'. A tonne of milk requires about 1000 m^3 of water or 8 m^3 of water per kg of milk solids (Table 5.7). These figures are highly impressive. They are especially suitable for purposes of comparison. For example, the water

Table 5.7 The total water footprint of bovine products.

Milk producing countries (m^3/t milk)	
Weighted global average	1021
India	1078
Australia	800
USA	796
The Netherlands	528
Bovine products (global, m^3/kg solids)	
Milk	8
Beef	62
Bovine water needs (%)	
Feed (roughages, concentrates)	98
Drinking water	1.1
Service water	0.8
Feed mixing	<0.1
Allocation to milk ingredients (global, m^3/kg)	
Whole milk powder	4.7
Butter	5.6

Source: Adapted from Hoekstra, 2012 and Mekonnen & Hoekstra, 2010

requirement for beef solids is nearly eight times higher than that of milk solids. It reflects the unfavourable feed conversion efficiency for beef cattle.

The cited figure of 1021 m^3/tonne of milk is a global average; there are large differences between various countries, as indicated in Table 5.7. India and The Netherlands have respectively the highest and lowest water footprints. The latter needs about half the volume of water per unit of milk. Differences in climate and agricultural practice explain the difference. Water is mainly used by the dairy farm in the production of animal feed (roughages, concentrates), which represents 98% of bovine needs. High volumes of water are required for growing crops. In this context drinking water for cows is of minor importance. Compared to the overall figure for primary milk production, the use of water in dairy plants is very modest. The production of cheese and butter requires respectively 6 and 7–10 L of water per kg solids (IDF, 2009).

The overall water use of the ingredient whole milk powder is based on the global average. The global figure of 4.7 m^3/kg is the result of the method of allocation to whole milk powder and butter (Mekonnen & Hoekstra, 2010). This allocation is done on the basis of the market prices of these products. Water use can also be expressed in terms of protein, fat and calories (Hoekstra, 2012). Such an approach does not value all constituents of the highly nutritional milk system and for this reason the water use per kg solids is a preferred approach. An indication of the water use of an ingredient on the information sheet can be of interest for customers. A value of 2.5 m^3/kg is typical for whole milk powders produced in The Netherlands.

Water use is of interest not only in a global or regional perspective. Due to the large scale of dairy operations the importance of water use at a specific location, whether ground or surface water, cannot be underestimated. The reuse of water in new milk powder plants will be given as an example (Philardeau, 2012). In Figure 5.6 a flow scheme is presented, where the intake of external water is reduced to zero ('Zereau' plant). The key element in this process is the recycling of condensate, or 'cow water', which is released during the evaporation of milk into condensed milk. The water supply of the plant is made up of several sources. The most intensive processing is carried out to obtain reusable water of a high quality. This water is used to rinse equipment that has direct contact with the food product. The waste water stream also receives a thorough treatment to convert it into reusable or process water. This water can be applied for various purposes such as cleaning (cleaning in place, CIP) and hygiene. Accumulation of minor compounds (e.g. phenols, nitrogen compounds) in a closed-loop system is always a risk, which is very limited in the case of milk evaporator condensate (Verheyen *et al.*, 2009) Besides saving water a reduction of energy can also be achieved. Water is boiled in an evaporator in a sequence of vessels, each held at a lower pressure than the previous one. The condensate of the first effect has the highest pressure and temperature. Due to its high temperature it is suitable to be applied to heat exchangers in order to save energy.

A central piece of equipment in the purification of water is the RO 'polisher' (Vourch *et al.*, 2008). The type of membrane used in the RO polisher has a high salt

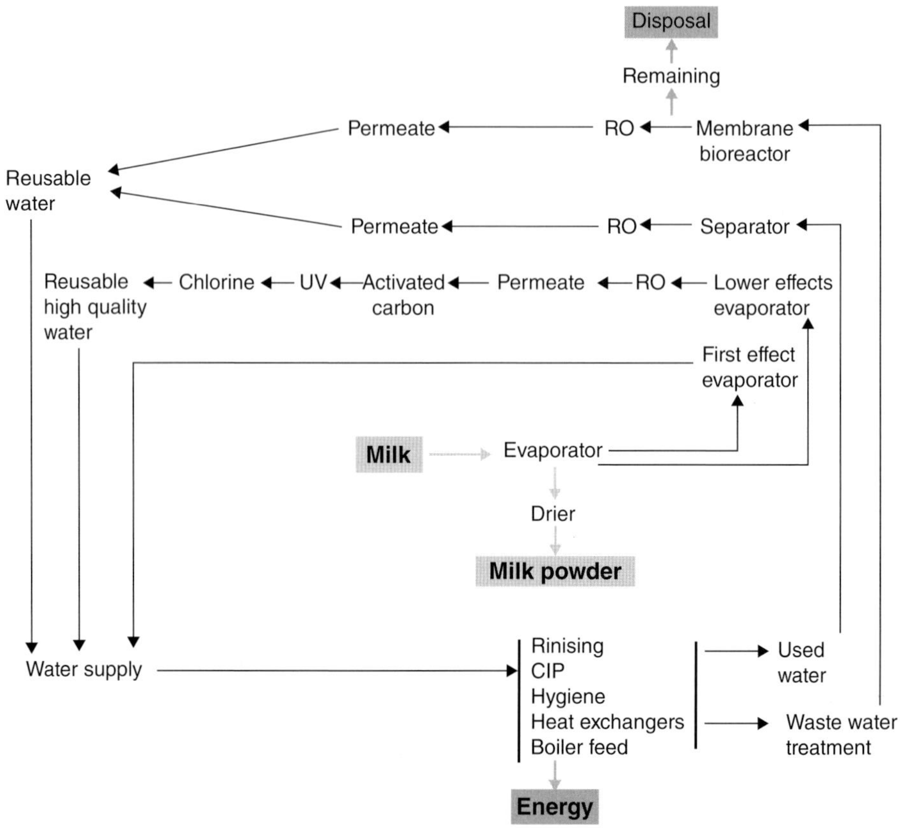

Figure 5.6 Scheme for recycling of water in a milk powder plant. RO, reverse osmosis. (*For colour details please see colour plate section.*)

rejection (99.7%). RO permeate is reused, while the RO concentrate is disposed of. As indicated in Figure 5.6, RO permeate can be further purified by using activated carbon. The microorganisms are inactivated by UV light and the addition of chlorine. An additional process is the membrane bioreactor, which is installed to reduce the chemical oxygen demand (COD: a measure of the amount of organic compounds in water) of the waste stream before it enters the RO polisher. The bioreactor is integrated in a membrane system. The organic matter is degraded under aerobic conditions. It is a relatively fast process requiring short retention times (Prazeres *et al.*, 2012). Extensive equipment is needed in order to reduce the external water supply to zero and to achieve a sustainable plant.

5.3.2 Greenhouse gas (GHG) emissions

Greenhouse gases trap the heat in the Earth's atmosphere and are largely responsible for climate change. The overall contribution of global milk production, processing

Table 5.8 Greenhouse gas (GHG) emissions from milk production, processing and transport (per kg milk of 4.0% fat and 3.3% protein; at farm gate).

Region (kg CO_2 eq.)	
Africa (sub-Saharan)	7.3
Asia (east, south)	2.0–4.5
Russian Federation	3.5
America (south and central)	3.3
Europe (west and east)	1.4
Oceania	1.4
America (north)	1.2
Global average	2.4
Farm: contribution of various GHG emissions (mixed farm system – temperate climate)	
Methane (CH_4)	52% (conversion factor 298)
Nitrous oxide (N_2O)	33% (conversion factor 25)
Carbon dioxide (CO_2)	15%
Plant: GHG emissions (kg CO_2 eq.) from processing, transport and packaging of various products (Europe)	
Fresh milk and cream	0.153
Fermented milk and cream	0.304
Cheese and whey	0.126
Skim milk powder and cream	0.157

Source: Data from FAO, 2010.

and transportation is estimated at 2.7% of total anthropogenic GHG emissions (FAO, 2010), and the largest GHG emissions are due to dairy farming. The latter is responsible for more than 80% of the GHG emissions of the total life cycle of milk (IDF, 2009). The causes of GHG emissions from the dairy farm are mainly enteric fermentation (CH_4), manure management (N_2O) and energy consumption (CO_2). The relative contributions of the three emission sources, shown in Table 5.8, are related to a mixed farm system in a temperate climate (FAO, 2010).

The figures provided by IDF show a similar distribution at the farm level. About 52% of GHG emissions are a result of enteric fermentation. Complex carbohydrates in feed (grass, silage, hay) are broken down through microbial fermentation in the rumen of the cow under anaerobic conditions. The final result is a highly nutritional milk; however, one of the by-products is methane, which is released to the atmosphere. About 33% of GHG emissions are estimated to be related to nitrous oxide (N_2O), which is mainly the result of manure management (storage and distribution) and use of nitrogen fertilizers. The remaining 15% is CO_2, which is caused by energy use at the farm, for instance by tractors and electricity. Methane and nitrous oxide are particularly potent GHGs, having an impact on global warming that is 25 and 298 times higher than that of the well-known CO_2 (IDF, 2009). Both conversion factors are used in the calculation of the CO_2 equivalents (CO_2 eq.). As opposed to farms, the GHG emissions at a factory level are modest. The production of cream and the ingredient skimmed milk powder results in the emission of 0.157 kg CO_2 eq./kg milk in Europe (FAO, 2010).

There are large differences between countries in the emission of GHGs, as shown in Table 5.8. The temperate climates of Europe, Oceania and North America clearly favour the production of milk and make for a relatively modest release of GHG, whereas the dry climate of Africa is less advantageous and results in about a five times higher GHG value per kg of milk (FAO, 2010). This is a consequence of the low digestibility of the grass in arid areas and the low milk production per cow. There is a significant relationship between milk production per cow and the total (post-farm gate) GHG emissions per kg milk. The emissions peak at 12 CO_2 eq./kg milk for cows producing less than 2000 kg milk per year and stabilize between 1.6 and 1.8 CO_2 eq./kg milk at a favourable annual production of more than 6000 kg (Vellinga *et al.*, 2013).

The development of an efficient dairy industry in temperate climates has a long history of gradual improvement. The carbon footprint per kg milk, including all GHG emissions from milk production on the farm, was 3.66 kg CO_2 eq./kg milk in 1944 and 1.35 kg CO_2 eq./kg milk in 2007 (Capper *et al.*, 2009). An enormous increase in milk production per cow was achieved – from 2074 kg milk/year in 1944 to 9193 kg milk/year in 2007. Breeding programmes formed the basis for the much improved yield per cow. Feeding practices also show a big difference, from pasture-based feeding regimes in 1944 to mixed feeding regimes in 2007. In addition to this study, two contrasting farm milk production systems were investigated in modern dairy countries (Flysjö *et al.*, 2011). An outdoor pasture grazing system in New Zealand (NZ) was compared to a mainly indoor housing system with a high input of feed concentrates in Sweden (SE). Notwithstanding differences in GHG emissions profiles, the total carbon footprint for both systems at the farm gate was relatively low, respectively 1.00 (0.60–1.52) for NZ and 1.16 (0.83–1.56) kg CO_2 eq./kg milk for SE. However, the large margins of error make it difficult to draw a firm conclusion. A better understanding of methanogenesis in the rumen of the cow is a promising approach for further reductions of GHG emissions (Dijkstra *et al.*, 2011).

The GHG value of food is an important aspect of sustainability. However, it cannot be viewed separately from its nutritional contribution. Both aspects are important for the future well-being of humans. The Nutrient Density to Climate Impact (NDCI) index was introduced to bridge both worlds in a single measure (Smedman *et al.*, 2010). The Nordic nutrition recommendations for 21 essential nutrients were used for the NDCI calculations. The NDCI value of milk was substantially higher than those for the alternative beverages used during a meal (soft drink, orange juice, beer, wine, bottled carbonated water, soy drink and oat drink). The very high nutrient density of milk has a significant effect on this outcome.

The allocation of GHG emissions to the various milk constituents is desirable in the case of ingredients. In Table 5.9 three approaches are set out, based on protein, energy value and total solids. All three represent a part of the nutritional value of milk. These calculations were based on a GHG emission of 1.4 kg CO_2 eq./kg milk, which is the average value for Europe as well as Oceania (FAO, 2010). The highest score is calculated if the entire GHG emission is attributed to the milk protein only, which gives 41.2 kg CO_2 eq./kg protein. The energy value may also be used to

Table 5.9 Allocation of GHG emissions to dairy ingredients.

GHG emission per kg milk: 1.4 kg CO_2 eq.
Milk composition: 13.3% total solids, 4.4% fat, 3.4% protein, 4.6% lactose, 0.2% organic acids, 0.7% minerals
Energy: 37.8 kJ/g fat, 16.8 kJ/g protein or carbohydrate

GHG emissions calculated per component(s):	
Protein	41.2 kg CO_2 eq./kg protein
Energy value	17.4 kg CO_2 eq./kg fat
	7.7 kg CO_2 eq./kg protein
	7.7 kg CO_2 eq./kg lactose
Solids	10.5 kg CO_2 eq./kg solids

allocate the GHG emission. Values can be calculated for the three main constituents of milk (fat, protein, lactose), as shown in Table 5.9. The total solids approach values equally all the constituents of the complex milk system, giving a value of 10.5 kg CO_2 eq./kg solids. It applies to major as well as minor constituents of the milk. This number can easily be used by customers to calculate the GHG emission of their finished products. An indication of the CO_2 eq. of an ingredient in the information sheet should cover the area of production.

References

Amon T, Amon B, Kryvoruchko V, Zollitsch W, Mayer K & Gruber L (2007) Biogas production from maize and dairy cattle manure – influence of biomass composition on the methane yield. *Agriculture, Ecosystems and Environment* **118**: 173–82.

Barbeau WE & Kinsella JE (1988) Ribulose biphosphate carboxylase/oxygenase (rubisco) from green leaves – potential as a food protein. *Food Reviews International* **4**: 93–127.

Buchanan RL & Oni R (2012) Use of microbial indicators for assessing hygiene controls for the manufacture of powdered infant formula. *Journal of Food Protection* **75**: 989–97.

Cahill SM, Kaye Wachsmuth I, de Lourdes Costarrica M & Karim Ben Embarek P (2008) Powdered infant formula as a source of Salmonella infection in infants. *Journal of Food Safety* **46**: 268–73.

Capper JL, Cady RA & Bauman DE (2009) The environmental impact of dairy production: 1944 compared with 2007. *Journal of Animal Science* **87**: 2160–7.

CAST (Council for Agriculture Science and Technology) (1999) *Animal Agriculture and Global Food Supply*. Task Force Report no. 135, p. 1.

Codex Alimentarius (1997) General guidelines for use of the term 'Halal'. CAC/GL 24-1997.

Codex Alimentarius (2010) Codex Stan 280-1973. Codex Standard for milk fat products, Revision 1999, Amendment 2006, 2010.

Codex Alimentarius Commission (2008) Code of hygienic practice for powdered formulae for infants and young children. CAC/RCP 66–2008.

Cunliffe HR & Blackwell JH (1977) Survival of Foot-and-Mouth disease virus in casein and sodium caseinate produced from the milk of infected cows. *Journal of Food Protection* **40**: 389–92.

Cunliffe HR, Blackwell JH & Walker JS (1978) Persistence of Foot-and-Mouth Disease virus in dried casein. *Journal of Food Protection* **41**: 706–7.

Dijkstra J, Oenema O & Bannink A (2011) Review: Dietary strategies to reducing N excretion from cattle: implications for methane emissions. *Current Opinion in Environmental Sustainability* **3**: 414–22.

Efstathiou T, Feuardent C, Méjean S & Schuck P (2002) The use of carbonyl analysis to follow the main reactions involved in the process of deterioration of dehydrated dairy products: prediction of most favourable degree of dehydration. *Lait* **82**: 423–39.

Eurostat (2008) *Food: From Farm to Fork Statistics*. European Commission, p. 166.

FAO (2010) *Greenhouse Gas Emissions from the Dairy Sector – A Life Cycle Assessment*. Food and Agriculture Organization of the United Nations.

FAO (2011) *World Livestock – Livestock in Food Security – Producing Enough Food*. Food and Agriculture Organization of the United Nations, pp. 78–87.

Feijoo SC, Cotton LN, Watson CE & Martin JH (1997) Effect of storage temperatures and ingredients on growth of *Bacillus cereus* in coffee creamers. *Journal of Dairy Science* **80**: 1546–53.

Fink-Gremmels J (2008) Mycotoxins in cattle feeds and carry-over to dairy milk; a review. *Food Additives and Contaminants* **25**: 172–80.

Flysjö A, Henriksson M, Cederberg C, Ledgard S & Englund JR (2011) The impact of various parameters on the carbon footprint of milk production in New Zealand and Sweden. *Agricultural Systems* **104**: 459–69.

Glass K & Doyle ME (2005) Safety of processed cheese – a review of scientific literature. Madison, WI: FRI Briefings, Food Research Institute, University of Wisconsin.

Haas B (2004) Inactivation of foot-and-mouth disease virus in milk. *IDF Bulletin* no. 392, pp. 77–83.

Hart J, Gangwer M, Graham M & Marx ES (1997) Dairy manure as a fertilizer source. Oregon State University Extension Service.

Hoekstra AY (2012) The hidden water resource use behind meat and dairy. *Animal Frontiers* **2**: pp. 3–8.

van Hooijdonk ACM (2013) Milk protein and food security – How cows convert human-inedible protein into high quality dairy protein. *IDF World Dairy Summit-Yokohama (JP)* IDF Bulletin 470.

van Hooijdonk ACM & Hettinga K (2013) Dairy in a sustainable diet: a question of balance. In: *Proceedings of the 'Sustainable Diets: Human Nutrition and Livestock' Conference*. Ulaanbaatar, Mongolia. FAO/MUST.

IDF (2009) Environmental/ecological impact of the dairy sector: Literature review on dairy products for an inventory of key issues & List of environmental initiatives and influence on the dairy sector. *Bulletin of the International Dairy Federation*, no. 436.

JAKIM (2012) *The recognized foreign Halal certification bodies & authorities*. Department of Islamic Development Malaysia (JAKIM)

Krebbekx J, Lambregts E, de Wolf W & van Seventer M (2011) Melk, de groene motor. *Agentschap NL*, April, pp. A–D.

Kussendrager KD (1994) Effects of heat treatment on structure and iron-binding capacity of bovine lactoferrin. In: *Indigenous Antimicrobial Agents of Milk*. IDF Bulletin, pp. 133–46.

Li Y, Park SY & Zhu J (2011) Solid-state anaerobic digestion for methane production from organic waste. *Renewable and Sustainable Energy Reviews* **15**: 821–6.

Liem AKD, Hoogerbrugge R, Kootstra PR, van der Velde EG & de Jong APJM (1991) Occurrence of dioxins in cow's milk in the vicinity of municipal waste incinerators and a metal reclamation plant in the Netherlands. *Chemosphere* **23**: 1675–84.

Luning PA & Marcelis WJ (2009) *Food Quality Management – Technological and Managerial Principles and Practices*. Wageningen Academic Publishers, pp. 283–306.

Mawson AJ (1990) Consideration of the fermentation of concentrated whey solutions for ethanol production. In: *Fermentation Technologies: Industrial Application.* Kluwer, pp. 264–70.

Mawson AJ (1994) Bioconversions for whey utilization and waste abatement. *Bioresource Technology* **47**: 195–203.

McGlone JJ (2001) Farm animal welfare in the context of other society issues: toward sustainable systems. *Livestock Production Science* **72**: 75–81.

Mekonnen M & Hoekstra AY (2010) The green, blue and grey water footprint of farm animals and animal products. *Value of Water Research Report Series* no. 48, UNESCO-IHE.

Mettler AE (1994) Present day requirements for effective pathogen control in spray dried milk powder production. *Journal of the Society of Dairy Technology* **47**: 95–107.

Möller K & Müller T (2012) Effects of anaerobic digestion on digestate nutrient availability and crop growth: a review. *Engineering in Life Sciences* **12**: 242–57.

Ngapo TM & Gariépy C (2006) Review: Factors affecting the meat quality of veal. *Journal of the Science of Food and Agriculture* **86**: 1412–31.

Oliver SP, Jayarao BM & Almeida RA (2005) Review: Food borne pathogens in milk and the dairy farm environment: food safety and public health implications. *Foodborne Pathogens and Disease* **2**: 115–29.

Oliver SP, Murinda SE & Jayaro BM (2011) Impact of antibiotic use in adult dairy cows on antimicrobial resistance of veterinary and human pathogens: a comprehensive review. *Foodborne Pathogens and Disease* **8**: 337–55.

Oltenacu PA & Broom DM (2010) The impact of genetic selection for increased milk yield on the welfare of dairy cows. *Animal Welfare* **19**(S): 39–49.

Papademas P & Bintsis T (2010) Food safety management systems (FSMS) in the dairy industry: a review. *International Journal of Dairy Technology* **63**: 489–503.

Pearce LE, Smythe BW, Crawford RA, Oakley E, Hathaway SC & Shepherd JM (2012) Pasteurization of milk: The heat inactivation kinetics of milk-borne dairy pathogens under commercial-type conditions of turbulent flow. *Journal of Dairy Science* **95**: 20–35.

Persson M, Jönsson O & Wellinger A (2006) Biogas upgrading to vehicle fuel standards and grid injection. *IEA Bioenergy*, task 37.

Philardeau T (2012) The water challenge in dairy. *IDF World Dairy Summit*, Nov. 5. International Dairy Federation.

Prazeres AR, Carvalho F & Rivas J (2012) Cheese whey management: A review. *Journal of Environmental Management* **110**: 48–68.

Ramirez CA, Patel M & Blok K (2006) From fluid milk to milk powder: Energy use and energy efficiency in the European dairy industry. *Energy* **31**: 1984–2004.

Raspor P (2005) Biomarkers: traceability in food safety issues – review. *Acta Biochemica Polonica* **52**: 659–64.

Regenstein JM & Regenstein CE (1979) An introduction to the Kosher dietary laws for food scientists and food processors. *Food Technology*, January, pp. 89–99.

Regenstein JM & Regenstein CE (1992) The kosher food market in the 1990's – a legal view. *Food Technology* **64**: 122–6.

Regenstein JM & Regenstein CE (2002) The story behind kosher dairy products such as kosher cheese and whey cream. *Cheese Reporter* **127**: 8, 16, 20.

Regenstein JM, Chaudry MM & Regenstein CE (2003) The Kosher and Halal food laws. *Comprehensive Reviews in Food Science and Food Safety* **2**: 111–27.

Riaz MN & Chaudry MM (2003) *Halal Food Production.* CRC Press, pp. 77–84.

Rizet C (2003) Energy consumed in freight transport: estimates from shipper surveys. IFSTTAR.

Sans P & De Fontguyon G (2009) Veal calf industry economics. *Revue de Médicine Vétérinaire* **160**: 420–4.

Saunders C, Barber A & Taylor G (2006) Food miles – comparative energy/emissions performance of New Zealand's agriculture industry. Report no. 285, Agribusiness and Economics Research Unit, Lincoln University, p. 117.

van Schothorst H, Zwietering MH, Ross T, Buchanan RL & Cole MB (2009) Relating microbiological criteria to food safety objectives and performance objectives. *Food Control* **20**: 967–79.

Schuck P, Mejean S, Dolivet A, Jeantet R & Bhandari B (2007) Keeping quality of dairy ingredients. *Lait* **87**: 481–8.

Schuck P, Dolivet A, Méjean S & Jeantet R (2008) Relative humidity of outlet air: the key parameter to optimize moisture content and water activity of dairy powders. *Dairy Science and Technology* **88**: 45–52.

Sigman SH (2004) Kosher without law: the role of non-legal sanctions in overcoming fraud within the kosher food industry. *Florida State University Law Review* **31**: 509–601.

Smedman A, Lindmark-Månsson H, Drewnowski A & Modin Edman AK (2010) Nutrient density of beverages in relation to climate impact. *Food & Nutrition Research* **54**: art. no. 5170.

Smith K (2008) *Dried Dairy Ingredients*. Wisconsin Center for Dairy Research.

van Tilburg X, Lensink SM, Londo HM *et al.* (2008) Technisch-economische parameters van duurzame energieopties in 2009–2010. Energy Research Centre of the Netherlands, ECN-E-08-090.

Uemura S, Ohashi A, Harada H *et al.* (2008) Production of biologically safe digested manure for land application by a full-scale biogas plant with heat-inactivation. *Waste Management & Research* **26**: 256–60.

VanRaden PM (2004) Invited review: selection on net merit to improve lifetime profit. *Journal of Dairy Science* **87**: 3125–31.

Veepro Holland (1999–2012) Average production figures of the black-and-white and red-and-white herdbooks. Available at: http://veepro.nl

van de Velde F, Alting AC & Pouvreau L (2011a) Process for isolating a dechlorophyllized rubisco preparation from a plant material. WO 2011/078671 [patent application].

van de Velde F, Alting AC & Pouvreau L (2011b) From waste product to food ingredient: the extraction of abundant plant protein RuBisCo. *New Food* **14**: 10–13.

Vellinga TV, Gerber P & Opio C (2013) Greenhouse gas emissions from global dairy production. In: de Jong P (ed.), *Sustainable Dairy Production*. Wiley-Blackwell, pp. 9–30.

Verheyen V, Cruickshank A, Wild K *et al.* (2009) Soluble, semi volatile phenol and nitrogen compounds in milk-processing wastewaters. *Journal of Dairy Science* **92**: 3484–93.

Vourch M, Balannec B, Chaufer B & Dorange G (2008) Treatment of dairy industry wastewater by reverse osmosis for water reuse. *Desalination* **219**: 190–202.

Wang Q, Thompson E, Parsons R, Rogers G & Dunn D (2011) Economic feasibility of converting cow manure to electricity: A case study of the CVPS Cow Power program in Vermont. *Journal of Dairy Science* **94**: 4937–49.

Wei Y & Liu D (2012) Review of melamine scandal: still a long way ahead. *Toxicology and Industrial Health* **28**: 579–82.

Westergard V (2010) Milk Powder Technology – Evaporation and Spray Drying. GEA NIRO, pp. 123–68.

Zeeman G (1991) Mesophilic and psychrophilic digestion of liquid manure. PhD thesis, Wageningen University.

Zhang T, Bowers KE, Harrison JH & Chen S (2009) Impact of calcium on struvite precipitation from anaerobically digested dairy wastewater. *ACS Symposium Series* **1022**: 123–34.

Zhang T, Bowers KE, Harrison JH & Chen S (2010) Releasing phosphorus from calcium for struvite fertilizer production from anaerobically digested dairy effluent. *Water Environment Research* **82**: 34–42.

6
Vital Membrane Processes

A collection of unit operations are applied in the manufacture of ingredients from milk. They are grouped into different categories according to their operation principle and/or purpose. In the information sheets 11 main unit operations are frequently mentioned:

- heat exchange, including pasteurization;
- centrifugal separation, commonly used for the separation of cream and fat;
- homogenization, applied for the reduction of particle size in a wet phase;
- precipitation, the key process step in the production of casein;
- crystallization, an essential step in the manufacture of lactose and milk fat fractions;
- enzymatic hydrolysis of various milk components;
- evaporation in order to concentrate all kinds of dairy liquids;
- drying, referred to in almost all ingredient process schemes;
- milling, particularly applied for size reduction of dried lactose particles;
- ion exchange/chromatography for removal of salts and isolation of proteins;
- membrane processes, including various pressure-driven processes and demineralization.

In the above list, the traditional dairy processes such as butter-making and cheese-making are not mentioned, since they are composed of different unit operations. Details of the traditional processes and the different unit operations can be found

From Milk By-Products to Milk Ingredients: Upgrading the Cycle, First Edition. Ruud de Boer.
© 2014 John Wiley & Sons, Ltd. Published 2014 by John Wiley & Sons, Ltd.

in various textbooks, some of which are referred to in the section 'Further reading' at the end of the chapter. Additional information regarding specific unit operations such as ion-exchange chromatography, enzymatic hydrolysis, precipitation and crystallization are provided in Chapter 7 ('End Users').

Membrane processes will be highlighted in this chapter as these have been of vital importance for innovation in the dairy industry during recent decades. These processes are mentioned in about half of the information sheets under the heading 'Description'. Maubois (1984) predicted the impact of novel technologies, in particular membrane processes, for the preparation of all kinds of ingredients, about 30 years ago. He introduced the term 'protein cracking'. 'Cracking', commonly used in the petroleum industry, describes the splitting of large hydrocarbon molecules into smaller molecules under the influence of heat, catalysts and solvents. A similar development took place in the dairy industry. Here the focus was on the most valuable milk constituent, its proteins. Unravelling of the numerous milk constituents was the main objective. In contrast to the chemical industry, the processing was carried out under mild temperature conditions, an approach into which membrane processes fit very well. Membrane processes are a highly fragmented unit operation and can be subdivided into six separation variants. They will be discussed in the following sections of this chapter. Emphasis will be placed on the manufacture of commodities from whey and skimmed milk. Various aspects of membrane processes will be discussed, such as their background, principles, dairy specifics, ingredients and by-products. Under the heading 'By-products', the integration of six unit operations, including membrane and ion-exchange processes, will serve to illustrate the complexity of today's technology.

6.1 Background

Artificial membranes were introduced for laboratory purposes in the 1930s (Lonsdale, 1982), but it took another three decades before this technique was ready for industrial use. The dairy industry was particularly interested in membrane techniques because of the processing of whey. Such techniques could be a solution for the industry in its search to add value to the whey stream and reduce waste water. The first electrodialysis (ED) plant for the demineralization of whey started operating in 1961 (Batchelder, 1987). The pressure-driven membrane processes were introduced in the dairy industry on a factory scale about 10 years later (Selitzer, 1972). The separation of whey proteins in their native form by ultrafiltration (UF) became a reality on a modest scale. The use of reverse osmosis (RO) for the concentration of whey started at the same time. Currently, nanofiltration (NF) is applied as an alternative for RO. The development of NF is described by Lindner and Kedem (2005). Microfiltration (MF) is used for the removal of bacteria. A variant of MF is used for protein fractionation purposes; it is called microfiltration fractionation (MFF). Advances in membrane processes used in dairy technology are reviewed by Pouliot (2008).

Table 6.1 Size of milk components.

Component	Size (nm)
Water	0.3
Cl, Ca ions	0.4
Lactose	0.8
Whey proteins	3–6
Casein micelles	20–400
Fat globules	100–10,000
Bacteria	200–8000
Somatic cells	6000–15,000
Cheese fines	200,000–2,000,000

As explained in Chapter 2, the composition of dairy liquids is complex. Table 6.1 gives the size of various constituents and residues present in milk and whey. This table is an extension of Table 2.3. The biggest particles – fines from cheese curd – are found in whey and can be seen by the naked eye. Bacteria and fat globules, which are smaller than somatic cells, show an overlap in size. The size of whey proteins in their native form is given (3–6 nm). They can be substantially larger after heat treatment, when aggregates are formed. This kind of treatment can thus influence the choice of membranes. Ions and lactose are similar in size.

The applications of various pressure-driven membrane processes are illustrated in Figure 6.1. Prior to membrane processing the fat is removed through centrifugation. In the case of MF this is necessary because bacteria and fat globules show a size overlap. Using MF extends the shelf life of pasteurized milk. Skim milk is also used for the concentration of casein micelles by MFF, because a fraction of the fat

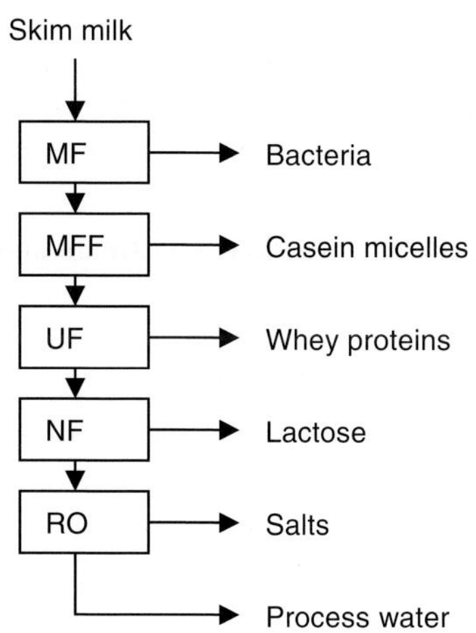

Figure 6.1 Overview of pressure-driven membrane processes for dairy technology. MF, microfiltration; MFF, microfiltration fractionation; NF, nanofiltration; RO, reverse osmosis; UF, ultrafiltration.

Table 6.2 Some characteristics of pressure-driven membrane processes.

	RO	NF	UF	MFF	MF	ED
Driving force	Pressure related					Electrical potential
Pore size (nm)	[a]	0.5–2	2–50	50–200	800–1400	[a]
Pressure (bar)	30–40	20–30	3–8	0.1–0.8	0.1–0.8	–
Temperature (°C)	10	10	10 or 50	50	50	Max. 40
Flow velocity (m/s)	2	2	2	4–6	4–6	0.1
	Spiral wound and tubular systems					Plate-and-frame

[a]Subject of debate.
RO, reverse osmosis; NF, nanofiltration; UF, ultrafiltration; MFF, microfiltration fractionation; MF, microfiltration; ED, electrodialysis.

globules is of the same size as the casein micelles. MFF separates whey proteins and casein micelles. The latter is utilized, for instance, as raw material for the manufacture of cheese. UF is used for the production of whey protein concentrates (WPC). In a next step, NF can be applied to concentrate and partially purify the lactose-containing permeate. In its final stage, RO is used for the separation of salts from water. This water is used as process water in many dairy plants.

Due to the nature of this book, emphasis in the next sections will primarily be on dairy issues rather than on engineering aspects.

6.2 Principles

Conventional filters (dead-ended) are used in raw milk lines, for instance. They separate suspended dirt particles of about 100 μm (Glimenius, 1980). Gravity is the main force for particle separation. The pressure-driven membrane processes referred to in Table 6.2 are an enormous advance on conventional filters. They operate in a cross-flow mode (feed flow is parallel to the membrane surface).

The operational principles of the indicated pressure- driven membrane processes are quite different. UF, MFF, and MF can each be seen as a sieving process. They can be described as flow through a bundle of capillaries according to the Hagen–Poiseuille equation[1]. This formula predicts a strong influence of pore radius on performance (permeate flux), which also depends on the membrane thickness and viscosity of the starting material. RO cannot simply be explained by the size differences between solvent and solute. The 'solution diffusion model' is useful for describing the behaviour of dairy liquids (Hiddink & de Boer, 1980). The principle

[1] Hagen–Poiseuille equation (Goulas & Grandison, 2008):

$$J = d^2 \Delta P \pi / 32\ \mu L$$

where J = permeate flux per unit area of membrane, d = capillary (pore) diameter, μ = dynamic viscosity, L = membrane thickness and ΔP = pressure difference.

of RO is based on reversing the solvent flow against the concentration gradient by means of the applied pressure. The pressures are highest for RO, because all components are concentrated. The osmotic pressure of milk and whey is 0.6–0.7 MPa, which is determined by the concentration of dissolved salts and lactose. The higher levels of protein and fat in milk have almost no influence on osmotic pressure. Casein solutions of 3.5% have an osmotic pressure of 0.003 MPa, whereas the osmotic pressure of a 10% whey protein solution is about 0.05 MPa (Jonsson, 1984; Goulas & Grandison, 2008). These values explain why UF and MFF operate at a far lower pressure than RO or NF. The concentration factor of whey through RO is therefore limited to five times at most, while the whey proteins through UF can be concentrated up to 40 times. In case of MF/MFF, ceramic (inorganic) membranes are commonly applied. The other processes (RO, NF, UF) make use of spiral wound modules with organic membranes (Da Costa *et al.*, 1993; Fane *et al.*, 1997). The modest investment costs, energy use and compact design of the spiral wound modules are their main advantages.

The pore size of a membrane is not an absolute value – there is always a pore size distribution. This complicates the prediction of the retention or rejection[2]. UF membranes are therefore also conveniently characterized by the molecular weight cut-off (MWCO). However, MWCO values are only a rough indication, since feed concentration, applied pressure, molecular characteristics, charge and absorption may have an impact too. NF membranes are characterized both by NaCl retention and MCWO; the latter is in the range of 150–300 Da. In MF/MFF, the pore size is listed by manufacturers; in case of MF, 1.4 μm is typical. In theory, an MF membrane with a pore size of 0.2 μm should be able to separate bacteria from skim milk, provided that the membrane has a fairly narrow pore size distribution and fouling is absent.

Reverse osmosis ideally retains all components other than water. The most common RO membranes used in the dairy industry have a retention of 0.997 for the reference salt NaCl (drink water application). UF retains macromolecules and particles of 2–50 nm. Ideally, the retention for whey protein is 1, while for lactose and salts it is 0. The choice of membrane is a compromise between performance expressed in flux (litres permeate/m^2 membrane surface/h) and protein yield. NF is categorized between RO and UF. NF and RO membranes have a comparable retention for multivalent ions, but NF shows a lower NaCl retention than RO membranes. Electrodialysis, a dialytic process in which an applied electric field is used to draw ions across a membrane, is able to remove salts to a substantially greater extent than NF. Electrodialysis membranes are homogeneous as opposed to the asymmetric structure of the pressure-driven membranes. MFF is able to retain casein micelles, but the separation of whey proteins is not complete and is affected by the

[2] Retention or rejection = c(feed) − c(permeate)/c(feed) × 100% where c = concentration.

process conditions (Samuelsson *et al.*, 1997). MF has a high retention (99.5%) for microorganisms when applied to raw milk (Madaeni *et al.*, 2011). Compared with pasteurization, the technique can remove a high percentage of spores. MF is often applied in combination with pasteurization to increase the shelf life of the milk.

Fouling of the membrane influences the performance of the plant. Minimization of fouling is essential to control the separation efficiency as well as the capacity. In this respect flow velocity is an important factor and is therefore indicated in Table 6.2. Control of the formation of the so-called 'secondary membrane' (fouling layer) is critical in industrial applications. Fouling can be reduced by equipment design and fluid flow practices. The flow velocity is relatively low in case of ED (plate-and-frame construction) and very high for tubular MF/MFF systems. In the latter, control of the 'secondary membrane' is most critical. According to Kromkamp (2005), this can even give rise to the following paradox: 'In case of complex suspensions as milk, the fractionation is not due to the size exclusion by the membrane, but instead due to segregation effects in the feed channel'. The background of the fouling phenomenon of MF/MFF of milk and the state of the art of these techniques in the dairy industry was reviewed by Brans *et al.* (2004).

In addition to minimization of fouling during processing, adequate cleaning is essential. Regular cleaning of the membrane equipment is an integral part of the process in order to restore the performance of the plant (IDF, 1995; D'Souza & Mawson, 2005). Disinfection is also an important step in the daily procedures of the production of ingredients with a high bacteriological quality. Cleaning and disinfection procedures are selected in such a way that adverse effects on membrane life are minimized.

In Table 6.2 the process temperature is also indicated. Besides fouling, microbial contamination is also important. Mostly, the choice is between an operating temperature of 10°C or 50°C. In both cases the growth of microorganisms is limited. For the ceramic filters, 50°C is always used; the elevated permeate flux compensates for the higher investment. In case of ED the maximum allowable temperature of 40°C is sometimes selected for reasons of plant output. As a consequence the running time is limited to, for example, 8 hours. Important considerations in this respect are the hygienic design of the equipment, the residence time, and the pH of the feed. Due to favourable conditions for bacterial growth at 40°C, a batch system is less advantageous than a continuous system. Microbial growth is, however, less critical for acid whey than for sweet whey.

6.3 Dairy specifics

Membrane processes in the dairy industry are mainly applied to skim milk and whey. The main differences between the two raw materials are the total solids (TS) content (9% vs 6% TS) and the protein content (3.3% vs 0.6%). The calcium content is 0.12% and 0.05% for skim milk and sweet whey, respectively. Approximately two-thirds of the calcium and half of the phosphorus are incorporated as colloidal calcium phosphate inside the casein micelles, as explained in Chapter 2. During acidification of skim milk – e.g. casein manufacture – colloidal calcium phosphate dissolves. Therefore, acid casein whey (pH 4.6) contains more than twice as much calcium as sweet Gouda whey (see Table 2.7).

Membrane performance is the focus of the next part. The permeate flux is closely related to the investment (in general, the higher the permeate flux, the lower the membrane surface needed and the lower the investment needed in equipment) and the retention is related to ingredient composition. To limit fouling of the membranes, the feed is normally pasteurized to inactivate microorganisms. Whey is carefully centrifuged to remove residual cheese fines and reduce the amount of fatty substances.

6.3.1 Reverse osmosis (RO) and nanofiltration (NF)

Tubular reverse osmosis membranes of cellulose acetate were used with a NaCl retention of 0.93 in Figures 6.2, 6.3 and 6.4. In Figure 6.2 the permeate flux is plotted against the osmotic pressure during concentration of pure water and various dairy liquids at 10°C. The effects of osmotic pressure (curve A), concentration polarization

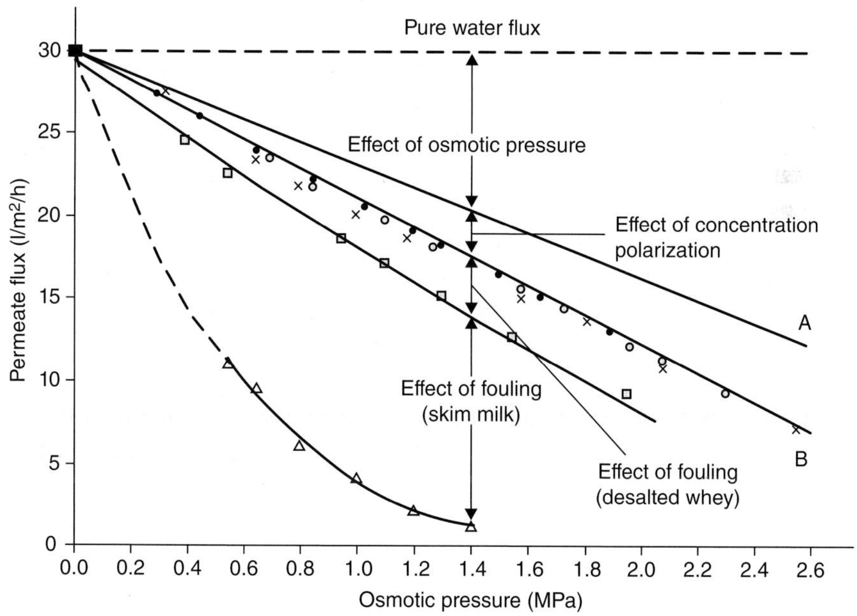

Figure 6.2 Permeate flux versus osmotic pressure of the bulk liquid for reverse osmosis of various liquids. The experimental fluxes are compared with those calculated assuming no concentration polarization and no fouling (curve A) and with those calculated assuming only concentration polarization (curve B). Process conditions: $T = 10°C$, $P = 4$ MPa, $v = 2$ m/s. • lactose, o UF-permeate, x sweet whey, □ desalted whey and △ skim milk. *Source:* Hiddink, Boer de & Nooij, 1980.

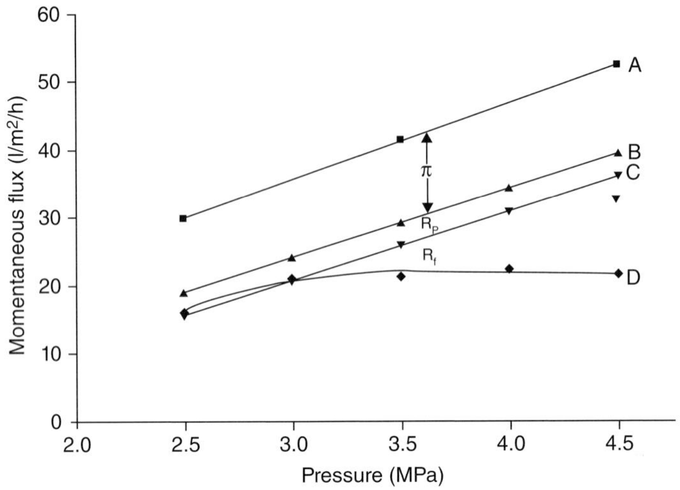

Figure 6.3 Permeate flux as a function of the applied pressure for reverse osmosis of skim milk in a batch system. Concentration factor 1.5; $T = 30°C$; $v = 2.6$ m/s; PCI membrane T2/15W. Curve A, pure water flux; curve B, permeate flux calculated without concentration polarization and fouling; curve C, permeate flux calculated without fouling; curve D, actual permeate flux. *Source:* Boer de & Hiddink, 1980.

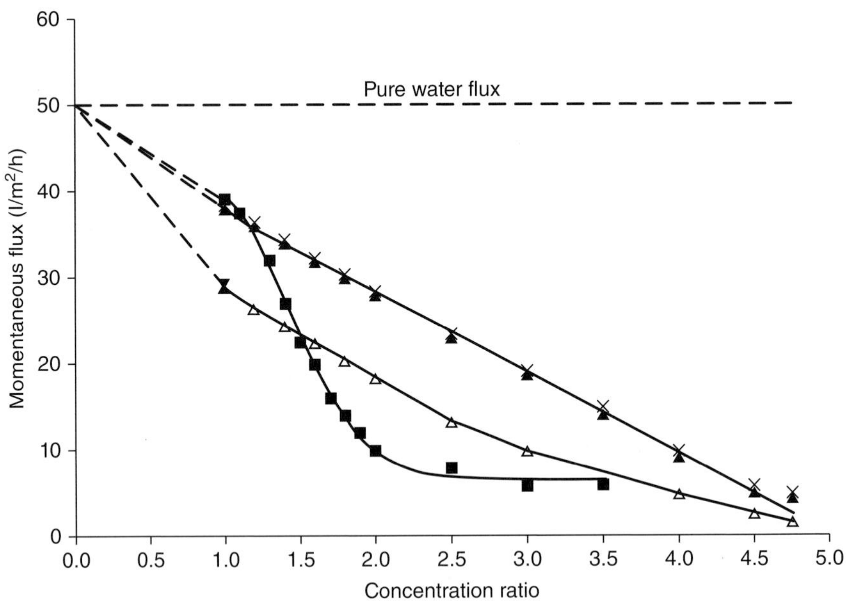

Figure 6.4 Permeate flux versus concentration ratio for reverse osmosis of Gouda whey with various pre-treatments. $T = 30°C$; $P = 4$ MPa; $v = 2$ m/s; PCI membraneT2/15W. ■ Gouda cheese whey pH = 6.6; Δ Gouda cheese whey acidified to pH = 4.6; ▲ Gouda cheese whey acidified to pH 6.0; X decalcified whey pH = 6.6. *Source:* Boer de & Hiddink, 1980.

(curve B) and fouling are indicated. The concentration polarization is the build-up of solutes at, or close to, the membrane surface, which results in an additional hydraulic resistance. Fouling is caused by deposition of colloidal material on the membrane surface, which increases the hydraulic resistance further.

The permeate flux of sweet whey, UF permeate and lactose follow the expected values as opposed to desalted whey. The 95% degree of desalting affects the stability of the whey protein, which may cause aggregation on the membrane surface. For skim milk, the differences between the expected and experimental fluxes are even bigger. The casein in the skim milk is responsible for considerable fouling. The high concentration and low diffusion coefficient of this constituent cause formation of a deposit on the membrane.

In Figure 6.3 four graphs are demonstrated for RO of skim milk at a process temperature of 30°C. The calculated resistances to permeate flux (R_m, membrane resistance; R_p, resistance attributed to concentration polarization; R_f, fouling resistance) are compared to the actual permeate flux. Under these conditions – higher temperature and flow velocity – the R_f value is lower than in Figure 6.2. The result is enhanced permeate flux. The optimum pressure has been shifted to a higher level, resulting in a further concentration of the skim milk.

In Figure 6.4 the permeate flux is plotted against the concentration factor for RO of Gouda whey. The whey had received various pre-treatments before RO started at a process temperature of 30°C. In contrast to a process temperature of 10°C a dramatic flux decline can be observed at a concentration ratio of Gouda whey (pH 6.6) exceeding 1.6. During RO of decalcified whey, this strong flux decline could not be found. Therefore, it can be assumed that during RO of whey, precipitation of calcium salts occurs on the membrane or in the membrane pores. If the pH of the Gouda whey is reduced to 6.0, the permeate flux does not decline and the whey can be concentrated five-fold. The flux of this acidified Gouda whey is about the same as that of decalcified whey. As the solubility of calcium phosphate increases at lower pH values, fouling is avoided. If the Gouda whey is further acidified to pH 4.6, the flux is lower than that at pH 6.0. As the calcium phosphate is completely soluble at pH 4.6, this lower flux is caused by the influence of pH on the solubility of whey proteins. At or near their isoelectric points, the whey proteins have a low net electrical charge, which favours aggregation and accumulation of whey proteins on the membrane surface.

As opposed to RO, NF is suitable for concentration as well as partial desalting. These membranes are characterized by a pore diameter close to 1 nm and have a slightly negatively charged surface. They exhibit a high permeability for monovalent ions (40–90%) and a low permeability for multivalent ions (5–20%). Figure 6.5 shows that using simulated milk ultrafiltrate (SMUF)/lactose (pH 4.6, 20°C), the transport of salt can be partially controlled by flux. At low flux when the contribution of diffusive transport is the most important, permeation of monovalent ions (sodium, potassium, chloride) is high. High flux levels, when transport by convection is most

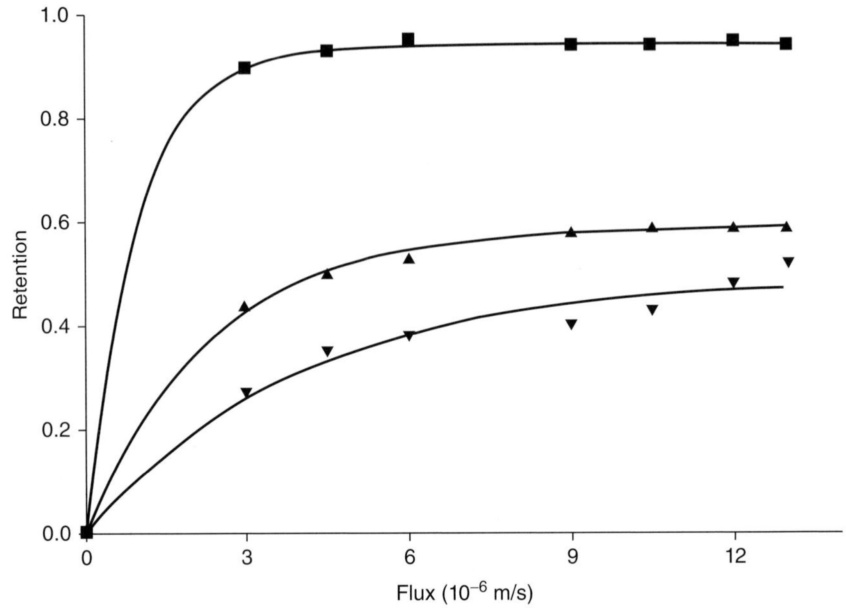

Figure 6.5 Retention versus flux of cations and anions during nanofiltration of simulated milk ultrafiltrate (SMUF + 4% lactose) at pH 4.6. ■ calcium + magnesium; ▲ anions; ▼ sodium + potassium. *Source:* Horst van der, Timmer, Robbertsen & Leenders, 1996. Reproduced with permission from International Dairy Federation.

pronounced, give the maximum retention. The retention of divalent ions (calcium, magnesium) is the highest, closely followed by a heterogeneous group of anions. To summarize, the best removal of salts takes place at a low flux level and thus at a low effective transmembrane pressure. The transport model on which Figure 6.5 is based is also applicable to industrial whey installations. In these situations the chances of fouling with protein or calcium phosphate are higher. For Gouda whey a reduction in pH from 6.6 to 5.8 is recommended in order to achieve both an increase in calcium phosphate solubility and a low protein fouling.

6.3.2 Ultrafiltration (UF)

Figure 6.6 shows UF experiments with Gouda whey at various pre-heating conditions (holding) and process temperatures. A constant but modest permeate flux is obtained if holding and UF are both carried out at 10°C. This matches with the results of the RO experiments (Figure 6.2). When the holding temperature is kept at 10°C and UF is carried out at 55°C, a fast flux decline occurs.

The fast flux decline can be attributed to precipitation of calcium phosphate at the UF membranes as the protective casein micelles are no longer present. Higher holding temperatures before UF result in a higher permeate flux. It seems that the precipitation of calcium phosphate during the holding step occurs in such a form

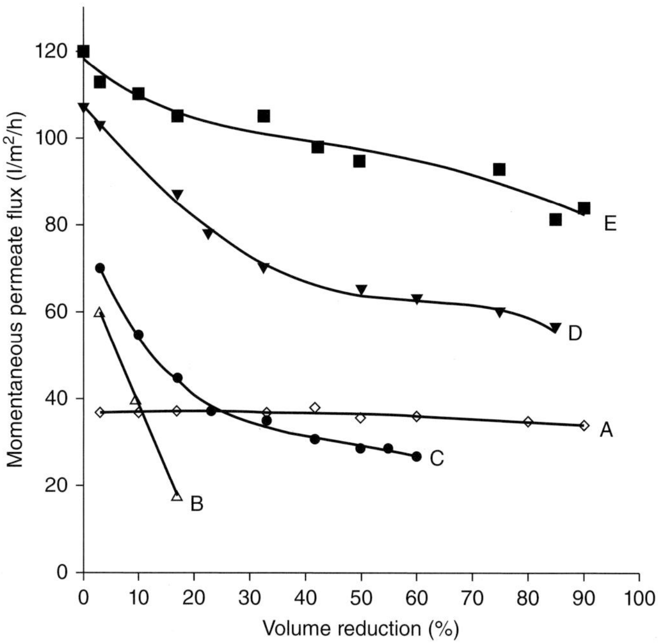

Figure 6.6 Effect of pre-heating on UF permeate flux of Gouda whey (5.5% TS). Batch system; tubular module; preheating for 30 min at the first temperature, ultrafiltration (UF) at second temperature. A,10/10°C; B, 10/55°C; C, 45/50°C; D, 55/50°C; E, 60/50°C. *Source:* Hiddink, Boer de & Nooij, 1980.

(probably as hydroxyapatite) that no blocking of the UF membranes takes place. The best result can be obtained if the holding temperature is about 10°C higher than the UF temperature. This pre-treatment should be limited to 60°C in order to avoid denaturation of whey proteins.

6.3.3 Microfiltration (MF) and microfiltration fractionation (MFF)

Figure 6.7 illustrates that MF/MFF can operate in a conventional mode or with co-current permeate flow. When MF/MFF operates in a conventional mode, high velocities are used to minimize concentration polarization and fouling. As indicated in Figure 6.7, this requires a high pressure drop ($P_{ri} - P_{ro}$). Permeate channels are usually operated at low (close to atmospheric) pressures. On the retentate side of the conventional system, the situation is quite different. The high pressure will cause fouling by compaction at the inlet side of the module, and will negatively influence the retention (rejection). The consequence of the low pressure at the outlet is that part of the membrane is used less effectively. Therefore, MF/MFF membranes quickly lose their separation characteristics and high permeate flux. To minimize concentration polarization and fouling, a high flow velocity should be combined with

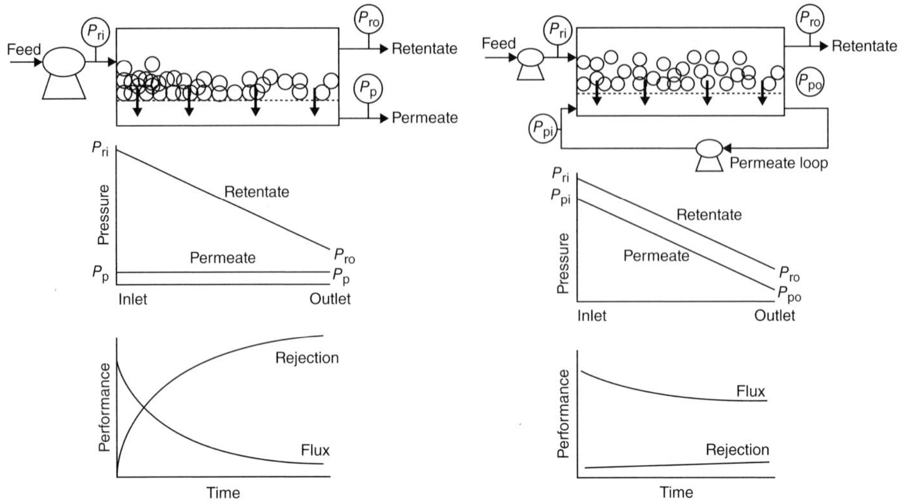

Figure 6.7 Conventional microfiltration (MF) (left) and co-current permeate flow (right). Cross-flow filtration showing profiles and expected behaviour of flux and retention. P_r = pressure retentate; P_p = pressure permeate; i = inlet; o = outlet. *Source:* Cheryan, 1998. Reproduced with permission from Taylor & Francis. Data from Sandblom, 1978.

a constant transmembrane pressure. This seems to be a contradiction. However, a simultaneous operation of a retentate pumping loop and permeate pumping loop solves this problem. This mode of operation is called 'uniform transmembrane pressure' (UTMP). Figure 6.7 is based on a patent granted on this method of filtration (Sandblom, 1978). Alternative ways of achieving UTMP are a progressive increase in the porosity of the membrane support layer or a decrease in thickness of the selective layer along the length of the module.

6.3.4 Electrodialysis (ED)

Electrodialysis reduces the salt content of whey – the removal of salts enhances its suitability in food applications. ED is an electrochemical process that makes use of direct current electricity as a driving force. Pairs of anion- and cation-permeable membranes are placed alternately in a stack. Under the influence of an applied voltage, cations (Na^+, K^+) will migrate towards the cathode, and anions (Cl^-) will move to the anode. A cation-permeable membrane will allow passage of cations but will repel anions, while the anion-permeable membrane performs the opposite. Between the membranes, spacers are placed at a distance of 1 mm or less. They prevent contact between the cation- and anion-permeable membranes and act as turbulence promoters, so that there is a mixing effect as the solution passes between the membranes.

The whey solids are split into two fractions through applying ED. One fraction is enriched in ionic species (concentrate or brine solution) and the other is depleted (desalted whey). Complete demineralization is impossible due to concentration polarization; the maximum achievable is approximately 90%. The ED plant can be run continuously up to about 60% demineralization. At a higher rate the whey has to be recycled through the stack and the capacity decreases substantially. The reduction in capacity is about fourfold, if one compares 90% with 50% demineralization. Desalting of whey poses problems with respect to fouling of membranes, such as deposition of calcium phosphate on the cathode side of the cation-permeable membranes and a layer of proteins at the surface of the anion-permeable membranes. For this reason, the brine solution is kept at a low pH and the current density is limited to about 17 mA/cm^2. At higher values splitting of water can occur, which leads to pH changes and operational problems. The electrodes at the end of the stack have separate rinse channels through which an acidified liquid is circulated to protect the electrodes from a chemical attack.

Table 6.3 shows the influence of the process conditions and the pre-treatment of Gouda whey on two important parameters, namely demineralization rate and 'power' efficiency. The change in temperature from 30 to 40°C corresponds to an increase in conductivity. It results in a higher demineralization rate of 60 g ash/m^2/h and has a positive effect on 'power' efficiency of 130 g ash/kWh. The increase of voltage per cell pair from 1.5 to 2.2 V results in a higher demineralization rate and has a negative impact on the 'power' efficiency. Obviously, the electric energy is used less economically at 2.2 V and partially dissipates into heat. All pre-treatments of Gouda whey (22% TS, pH 4.6 and decalcification) have a positive effect on both demineralization rate and 'power' efficiency. Pre-concentration increases the

Table 6.3 The influence of various process conditions and pre-treatments on the demineralization rate and 'power' efficiency during electrodialysis of Gouda whey (pH 6.6, 5.5% TS).

	Demineralization rate (g ash/m^2/h)	'Power' efficiency (g ash/kWh)
Process conditions:	70	710
Gouda whey/30°C, 1.5 V	+60	+130
Temperature: 30°C → 40°C	+40	−180
Voltage: 1.5 → 2.2 V		
Pre-treatments:		
Gouda whey/40°C, 2.2 V	190	630
Whey → concentrated whey (22% TS)	+90	+290
Whey → acidified whey (pH 4.6)	+110	+180
Whey → decalcified whey	+100	+150
Whey → simulated milk ultrafiltrate	+160	+160

Asahi Glass DW-XYO pilot plant; 40°C; reduction conductivity of 80%; concentrate (brine): 55 mS, pH 3.5 (de Boer & Robbertsen, 1983).
TS, total solids.
Source: Boer de & Robbertsen, 1983

conductivity of the whey, while acidification and decalcification convert the salts into the ionized form. The result of these pre-treatments is that the ions are more easily removed through ED. The best results were obtained in case of SMUF, the other whey constituents have apparently a negative effect on the demineralization rate. As far as the 'power' efficiency is concerned, the pre-concentration of the whey, a normal procedure in industrial processes, is the most positive.

6.4 Membranes and ingredients

Nowadays skim milk and whey, the by-products of butter and cheese, are converted into a variety of dried dairy ingredients. They are used in a wide range of food products because of their nutritional value, flavour and functional properties. In this section, the focus will be on how membrane processes contribute towards the addition of value to ingredients. In particular, the focus will be on purifying of milk proteins and whey proteins (UF and MFF) and desalting whey (ED and NF). The composition of the various ingredients obtained will be compared to the reference ingredients, skim milk powder and cheese whey powder.

6.4.1 Milk protein concentrate (MPC)

The concentration of milk proteins through membrane processes does not affect the original state of the milk proteins. If a UF process is chosen, both casein and soluble whey proteins are concentrated, as a consequence the ratio is about the same as in milk. The resulting product is called milk protein concentrate (MPC). When MFF is used on skim milk, a rather pure form called micellar casein concentrate (MCC) will result. In Table 6.4 the composition of MPC70, MPC80 and MCC85 is compared to skim milk powder.

By applying both UF and MFF the protein content is substantially increased, while the lactose content is greatly reduced and the fat content is slightly enhanced. The presence of calcium phosphate in the micelles results in a constant ash content in the various ingredients. These ingredients, of which MPC is the most important, are used for nutritional purposes (e.g. for athletes, aging population, infant formulas,

Table 6.4 The composition of dried milk protein concentrates and micellar casein compared to skim milk powder (%).

		Milk protein concentrates		Micellar casein
	Skim milk powder	MPC70	MPC80	MCC85
Protein	36	70	80	85
Lactose	51	16	6	1
Ash	8	8	8	8
Fat	1	2	2	2
Moisture	4	4	4	4

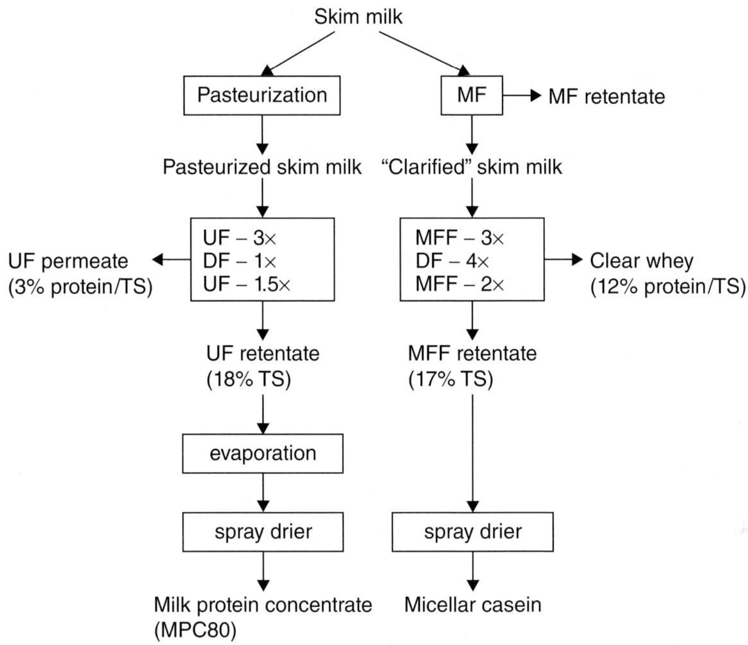

Figure 6.8 Examples of production processes for milk protein concentrate and micellar casein. DF, diafiltration; MF, microfiltration; MFF, microfiltration fractionation; UF, ultrafiltration. *Source:* Adapted from Novak, 1991 with Data from Schuck *et al.*, 1994.

etc.) and also in recombined cheese production. The more expensive MCC can be useful as an intermediate product for further fractionation of caseins.

In Figure 6.8 the process schemes for MPC and MCC are shown. A remarkable difference between the two processes is that in the case of MCC, heat treatments are avoided. Instead of pasteurization, MF (d = 1.4 μm) of the skim milk is applied to reduce the total bacterial count. Without heat treatment, the incorporation of some denatured whey proteins in MCC is prevented. A low-temperature pasteurization of the skim milk is used in the manufacture of MPC. The subsequent fractionation processes for MPC and MCC are, respectively, UF and MFF (d = 0.14 μm). A three-step process of concentration/diafiltration (DF)/concentration is applied.

Considerable purification can be achieved by either direct UF or MFF; however, the flux will drop to low values during processing and the pumping required will rise due to an increase in viscosity of the retentate. As continuous processes are normally applied in the dairy industry, a co-current, continuous DF step is carried out at about 10% protein content. This means that water is added to the retentate at the same rate as permeate is removed. The scheme of Figure 6.8 also shows that the DF step is more intensive in the case of MCC: the volumes of water are about four times bigger. Due to a partial retention of the whey proteins, the complete separation during MFF is still difficult (Samuelsson *et al.*, 1997). A disadvantage of DF is that the retentate is diluted. A final UF/MFF step is desirable to limit the drying costs and to avoid a too bulky powder. In case of MPC the retentate is concentrated in an evaporator to about 28% and is spray dried subsequently.

The liquid that passes the membrane in the two fractionation processes is quite different in appearance. In case of MFF a crystal clear microfiltrate is obtained, called 'ideal whey' or 'clear whey'. The product contains protein but includes no fat and rennet residues. It is an interesting starting material for WPCs with favourable functional properties such as foaming and gelling. The UF permeate from the MPC production, directly derived from milk, can be useful in the standardization of protein in skim milk powders.

6.4.2 Whey protein concentrate (WPC)

Due to increasing global cheese production, the amount of whey being processed is growing every year. As a consequence, the dairy industry is continually forced to develop new products and applications based on whey. Whey powder, which contains approximately half the solids of cheese milk, is still the main product as far as volume is concerned. An important application is the animal feed industry. The low protein content of whey powder limits its application in food. However, membrane processes (UF) make an increase of the protein content feasible and create value-added ingredients suitable for application in food. In Table 6.5 the composition of concentrated whey proteins is given, with whey powder as a reference. WPC 34 has about the same composition as skim milk powder (see Table 6.4). It is produced by five-fold concentration of the protein component through UF. Although the protein part consists of whey proteins only, WPC is useful as a skim milk substitute in many food applications. WPC 80 has a high protein (80%) and low lactose content, while whey protein isolate (WPI) is the purest form of the concentrated whey proteins. It combines a protein content of about 90% with a very low fat content. Both ingredients are used for nutritional (amino acid profile) as well as functional (solubility, foaming, gelling) reasons.

Figure 6.9 shows flow diagrams for the production of WPC 80 and WPI from cheese whey. In case of WPC 80 two membrane processes are combined. RO can be used to pre-concentrate the whey (about 11% TS) before UF takes place. In this way,

Table 6.5 The composition of various whey protein concentrates compared to whey powder (%).

	Whey powder	WPC 34	WPC 80	WPI 90
Protein	13	34	80	90
Lactose	75	53	6	4
Ash	8	7	3	3
Fat	1	3	7	<1
Moisture	3	3	4	3

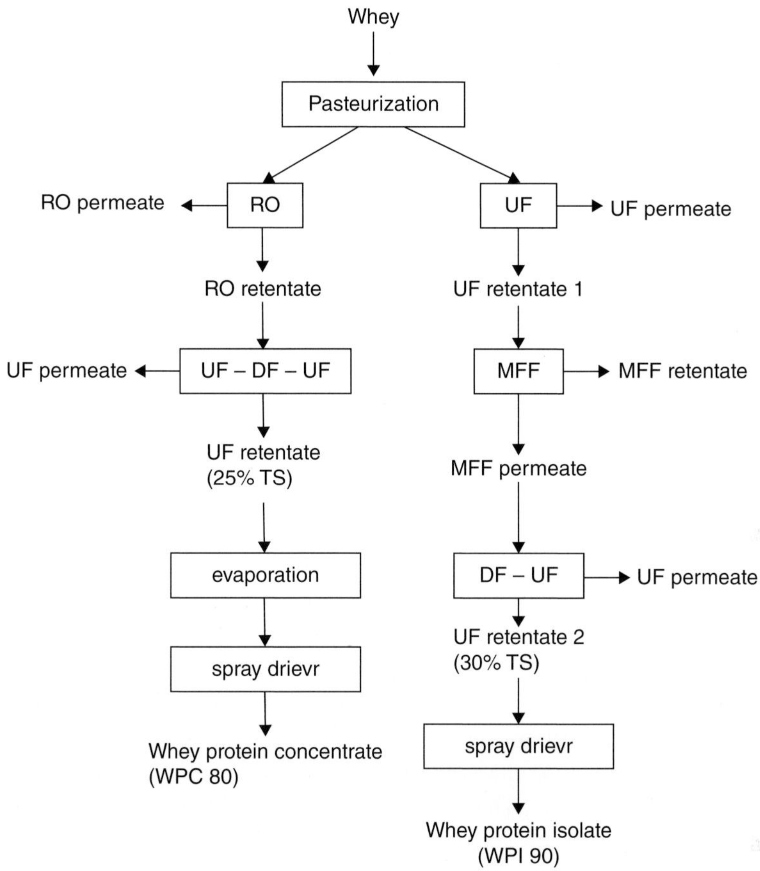

Figure 6.9 Examples of production processes for whey protein isolate (WPI) and whey protein concentrate (WPC) from cheese whey. DF, diafiltration; MFF, microfiltration fractionation; RO, reverse osmosis; UF, ultrafiltration.

part of the water is removed by RO and as such the UF process can be performed in a more efficient way. UF is carried out in a three-step process (UF-DF-UF) as referred to earlier. Compared to skim milk (MPC 80), the concentration factor for whey is substantially higher. The proteins in whey have to be concentrated up to 40 times to achieve the required purity. A continuous, industrial process for the manufacture of WPC 80 is usually marketed with 12 stages containing spiral wound membranes (Yee *et al.*, 2007). In such a process a proportion of the retentate is recycled in order to maintain the flow velocity and minimize membrane fouling. Also, a proportion of the combined permeate stream is recycled and mixed with fresh whey to minimize the effects of the variations in feed composition. An automatic controller is implemented to manipulate DF, retentate recycle and permeate recycle flow rates (Yee *et al.*, 2008). The DF water flow rate ensures that the WPC product is maintained within its desired specifications. To cope with long-term fouling, usually one of the stages is bypassed at the start and brought into operation at a later stage. After UF, a retentate of about 25% TS is obtained, which is evaporated to, for example, 35% TS and is spray dried subsequently.

The membrane processes UF and MFF (pore size 0.1 μm) are combined for the production of WPI. The process referred to is an alternative to the chromatography process. The whey proteins are partially concentrated by UF (3–10 times) before applying MFF. This serves two purposes. Firstly, it reduces the flow rate through the more expensive MFF system. Secondly, it improves the efficiency of fat removal by increasing its concentration and enhancing coalescence. The UF-retentate is subsequently sent to the MFF system, which removes a substantial amount of fat. Besides the fat, bacteria and some aggregated, denatured whey proteins are also separated. The MFF permeate, being the main product, is directed to a second UF/DF plant and further concentrated to the required protein content. Removal of this fat by MFF improves the flux of the second UF plant by about 50–100% and results in an ingredient with better functional properties. The retentate contains approximately 30% TS and is sent directly to the spray drier.

In Table 6.6 the composition of various whey protein concentrates is given for the UF-DF-UF system for producing WPC 80. The calculations were carried out using the indicated formula and the retentions were estimated for true whey proteins and fat at 1, and for lactose and non-protein nitrogen (NPN) at 0. The ash retention was estimated at 0.2 due to the fact that not all the mineral constituents are in the ionic state and bound to the protein.

During UF, the protein content increases gradually as the volume concentration ratio (VCR) becomes higher. The protein content of the retentate is 20% at a VCR of 40. The fat content increases also to about 8% of TS. This residual fat has a negative

Table 6.6 The composition of whey protein concentrates after applying UF-DF-UF in a continuous process.

Volume (L)	VCR/V_D	Protein (%)	Lactose (%)	NPN (%)	Ash (%)	Fat (%)	Total solids (%)	True protein/TS (%)
100	1	0.5	4.4	0.2	0.5	0.05	5.65	8.8
5	20	10	4.4	0.2	1.82	1	17.42	57.4
5	20/1.0	10	1.62	0.07	0.82	1	13.51	74
2.5	40	20	1.62	0.07	0.94	2	24.63	81.2

For the calculations, the following formulas were used (Cheryan, 1998):

$$\text{UF}: C_R = C_0(\text{VCR})^R; \text{DF}: C_R = C_0 e^{-V_D(1-R)}$$

where C_P is the solute concentration in the permeate; C_R is the solute concentration in the retentate; C_o is the solute concentration in the feed and R is $(1 - C_P)/C_R$. Retentions were estimated at protein and fat = 1, ash = 0.2, lactose and NPN = 0.

DF, diafiltration; NPN, non-protein nitrogen; TS, total solids; UF, ultrafiltration; VCR, volume concentration ratio; V_D, volume diafiltration.

Table 6.7 Removal of milk salts during ultrafiltration and diafiltration of Gouda cheese whey.

	TS (%)	Protein on TS (%)	Ash on TS (%)	Inorganic constituents (mg/100 g solids)					
				Na	Ca	Retention	Cl	P	Retention
Whey	5.8	13.4	8.2	724	676		1409	604	
Retentate VCR = 20; pH 6.6	17.3	68.3	3.7	324	442	+	281	160	−
Retentate V_D = 3.0; pH 3.2	11.7	85.0	0.6	9	24	−	84	48	+

Source: Hiddink, Boer de & Romijn, 1978. Reproduced with permission from Elsevier.
TS, total solids; VCR, volume concentration ratio; V_D, volume diafiltration.

effect on functional properties (e.g. foaming) and can result in the development of off flavours during storage. During DF (V_d = volume permeate/volume feed = 1.0), the lactose and NPN contents of TS are reduced by about 50%. The lactose content is 6–7% of TS in the final stage, while the NPN part of total protein is reduced to 0.3–0.4%. The decrease of the ash content during UF-DF-UF is more modest due to the retention of 0.2. The ash content is 3–4% of TS at the end of the process.

The removal of the individual minerals of WPC under various process circumstances is demonstrated in Table 6.7. This shows that chloride and phosphorus decrease strongly compared to sodium and calcium at a VCR of 20 and a pH of 6.6. The retention of chloride and phosphorus is even negative. The opposite happens at a subsequent DF step at pH 3.2. In this case, there is a steep decrease in sodium and calcium compared to chloride and phosphorus, and the retention of sodium and calcium is negative. This phenomenon can be explained by the Donnan effect. In this situation, the whey proteins play the role of ions that are unable to permeate the membrane. They have an electrical charge depending on the pH value. If the whey proteins have a negative charge (pH 6.6) the retention for the cations (Ca^{2+}, Na^+) is positive and there is negative retention for the anions (Cl^-, mainly PO_4^{2-}). Below the isoelectric point, the whey proteins are positively charged and the situation is reversed, with retention positive for anions and negative for cations. By applying a pH of 6.6 and subsequently 3.2 during UF/DF of cheese whey, an ash content of 0.6% can be achieved at a high VCR.

6.4.3 Demineralized whey

Although the protein content of cheese whey is modest, it is a vital factor in determining its value. A disadvantage is that this valuable protein source contains also a relatively high salt or ash content. The salts in the whey have a negative effect on flavour and are mostly a drawback for nutritional applications. For this reason, a good balance between protein and ash content is important. The protein/ash ratio will be calculated for various ingredients obtained by different processes based on ED and NF. Whey powder and skim milk powder will be used as reference ingredients.

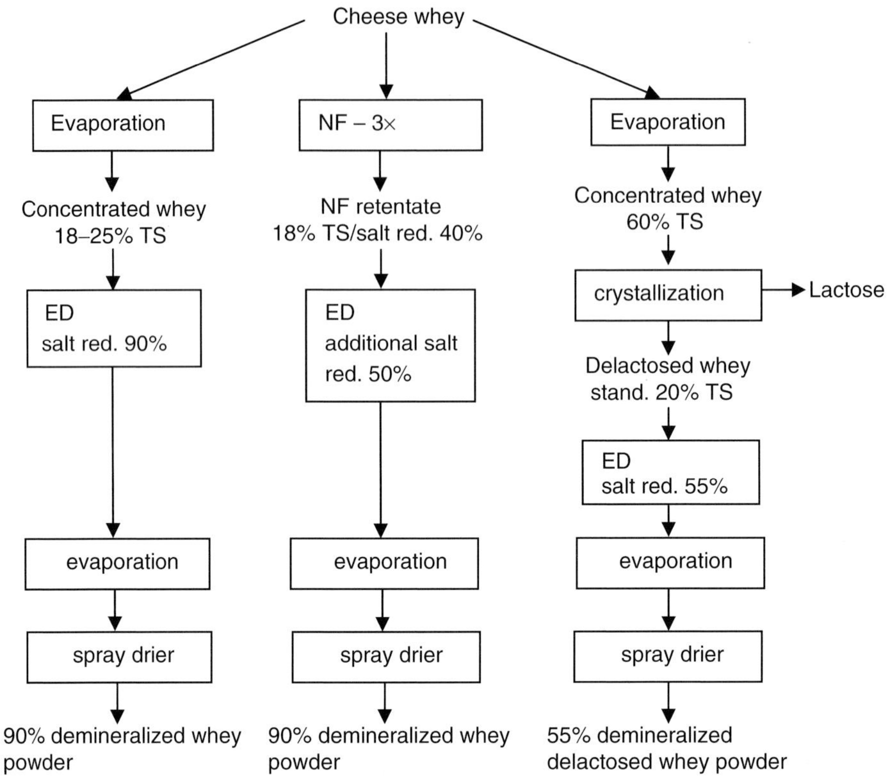

Figure 6.10 Examples of production processes for demineralized whey and demineralized delactosed whey. ED, electrodialysis; NF, nanofiltration; TS, total solids; red., reduction; stand., standardized.

Figure 6.10 shows three processes in which ED, a combination of NF/ED and lactose crystallization connected to ED are highlighted. Due to the concentration and subsequent removal of lactose crystals, the starting material for the latter has the highest ash content and consequently the highest conductivity. A demineralized, delactosed whey is obtained if the ash content is reduced by about 55%. This ingredient has about the same chemical composition as skim milk powder (Table 6.8). The protein/ash ratio is in both cases approximately 4.5. In case of the 90% demineralized whey, starting with a four times concentrated whey, an ingredient with a protein/ash ratio of 13 is produced. In this case the calcium and phosphate, which are valuable from a nutritional point of view, are reduced less than chloride, potassium and sodium. This can be explained by the fact, that calcium and phosphate are either bound to the protein or present in an insoluble state. The 90% demineralized product is suitable for use in infant formulas.

The latest development in processing is the combination of NF and ED. A three times NF concentration results in 40% ash removal and improves the efficiency of

Table 6.8 Upgrading of cheese whey by electrodialysis: powder composition (%).

	Whey powder	90% demineralized whey powder	55% demineralized, delactosed whey powder	Skim milk powder
Moisture	4.5	3.0	2.5	4.0
Fat	1.0	1.0	1.8	1.0
Lactose	74.0	82.0	54.0	51.0
Protein	12.5	13.0	30.2	36.0
Ash	8.0	1.0	7.0	8.0
Calcium	0.8	0.2	0.8	1.2
Magnesium	0.2	<0.1	0.2	0.1
Potassium	1.9	0.3	1.2	1.5
Sodium	0.8	<0.1	0.7	0.5
Chloride	1.6	<0.1	0.6	1.2
Phosphorus	0.6	0.2	0.8	0.9
Protein/ash ratio	1.6	13.0	4.3	4.5

ED for the production of a 90% demineralized whey powder. Adjustment of pH can be part of the ED process too. The phenomenon whereby calcium and phosphate remain in the retentate is even more pronounced in case of NF, because it is the monovalent ions that are chiefly removed.

Table 6.9 shows the composition of acid whey, a by-product of the precipitation of casein from skim milk by means of HCl, in addition to the composition of sweet cheese whey. The acid whey has a lower protein and higher ash content than sweet whey and is therefore less valuable. As is demonstrated in Table 6.9, acid whey can be upgraded by NF to achieve a similar composition as sweet whey. The protein/ash ratio improves from 0.9 to 1.6 after a four times concentration through NF at a pH of 4.4. Sweet whey can be converted into a more valuable product too. After a five-fold

Table 6.9 Upgrading of casein whey and cheese whey by nanofiltration – calculated on total solids (%).

	Acid whey (HCl)	Acid whey (VCR 4, pH 4.4)	Sweet whey (Gouda)	Sweet whey (VCR 5, pH 5.8)
Lactose	76	82	78	81
Protein	11	11	13	14
Ash	12	7	8	4
Calcium	2.0	1.8	0.8	0.7
Magnesium	0.2	0.2	0.2	0.2
Potassium	2.4	0.8	2.0	0.9
Sodium	0.6	0.2	0.8	0.3
Chloride	4.2	2.1	1.7	0.3
Phosphorus	1.2	1.0	0.6	0.5
Protein/ash ratio	0.9	1.6	1.6	2.8

VCR, volume concentration ratio.

concentration by NF at pH 5.8, the protein/ash ratio increases from 1.6 to 2.8. The mineral reduction with respect to the monovalent ions during NF of sweet whey is comparable with the ED product obtained through 60% demineralization.

6.5 By-products

Table 6.10 shows the composition of the by-products of five membrane processes. MF retentate, the by-product of the separation of microorganisms from milk, has a roughly similar composition to skim milk. This stream is normally recycled in the same plant. As opposed to the by-product from MF of skim milk, the ED by-product (a salt solution) is released as effluent. The total solids content is about 2%, of which ash represents about three-quarters and the remainder is NPN and lactose. The NF permeate is highly diluted. It contains approximately 0.3% TS of which ash makes up 85%. The costs of discharging NF permeate can be as high as 50% of the process costs. If feasible, it is preferable to carry out additional processing. More favourable is the situation in case of RO. By applying a polisher, RO permeate of a quality close to drinking water can be produced. This RO permeate can be reused as process water while the modest quantity of RO retentate (e.g. 20 times concentrated) can be discharged. The most problematic by-product is UF permeate. It contains a very high percentage of lactose and may cause serious pollution, mitigation of which requires further processing. In the following section various options for the processing of UF permeate will be discussed.

In Table 6.11 an estimate is calculated of the amount of some whey ingredients (WPC 34, WPC 80 and WPI 90) that can be produced from 10,000 kg of cheese whey. The amount of solids of UF permeate for each of the main ingredients is expressed as a percentage of the original whey solids. The quantity of whey powder that can be produced from 10,000 kg cheese whey is about 565 kg. In this situation no by-product is obtained. In the production of WPC 34 a large amount of UF permeate is released; the solids represent 72% of the original whey solids. The ingredients WPC 80 and WPI represent only one-tenth of the original whey solids. The consequence is that the by-product stream contains as much as 90% of the whey

Table 6.10 Composition of by-products from various processes.

		Based on total solids (%)			
	Total solids (%)	Lactose	Ash	Protein	Usage
MF retentate	9	50	8	37	Recycling
UF permeate	5	85	10	↑	Further processing
ED salt solution	2	20	75	NPN	Effluent
NF permeate	0.3	5	85	↓	Further processing
RO permeate	0.02	–	50		Process water

ED, electrodialysis; MF, microfiltration; NF, nanofiltration; RO, reverse osmosis; UF, ultrafiltration.

Table 6.11 Ultrafiltration of cheese whey: its main products and by-products (quantities based on 10,000 kg of cheese whey).

	Main product (quantity, kg)	By-product (% of whey solids)
Whey powder	565	0
WPC 34	161	72
WPC 80	62	89
WPC 90	48	92

WPC, whey protein concentrate.

solids. As UF is widely used in the dairy industry, a solution for this problem is vital for the various value added ingredients.

Currently, the most common solutions are the production of permeate powder or a combination of lactose and delactosed permeate manufacture (see information sheet 'Delactosed whey permeate'). The latter is mostly offered in a liquid form and contains residual lactose and minerals in particular. It is usually used as a form of 'molasses' in the local animal feed market. Recently, drying technologies became available for permeate powder and the even more hygroscopic delactosed permeate (Henningfield & Dimesen, 2002). The dried form provides more handling freedom. Despite this advantage, further upgrading of the UF permeate is desirable.

Compared to milk protein and lactose, applications for milk salts are less developed. Milk salts are a heterogeneous group containing chloride, phosphate, citrate, potassium, sodium, calcium, magnesium and trace elements. The current picture is that they have a negative value and end up in animal feed or as an effluent, as mentioned above. An exception to this rule is 'milk calcium' (see information sheet 'Milk mineral concentrate').

Figure 6.11 illustrates an example of the processing of this calcium supplement (J van Loon, 2009, personal communication). Initially, sweet whey permeate is concentrated by NF to about 22% TS. Via the NF permeate a substantial fraction of the monovalent ions are removed (4.5% of UF permeate solids). The NF concentrate is kept at 80°C/30 min and pH 7.2. Under these conditions various calcium phosphates and to some extent also citrate compounds are formed, which can be separated by centrifuge or UF. In Figure 6.11 the UF process is shown as it results in a greater yield due to supplementary inclusion of colloidal calcium compounds. By applying additional DF, the calcium content of the supplement increases to a maximum of 25% of total solids. Subsequently, this UF concentrate is spray dried. In the example shown in Figure 6.11, mass balance calculations of whey from semi-hard cheese indicate that only 2.5% of the solids from the original UF permeate are recovered in the supplement. Further processing of the decalcified UF permeate is a logical approach. Due to its low calcium content it can be concentrated by evaporating to 65–68% TS without deposits being formed excessively. An edible lactose yield of about 85% can be achieved. If the lactose yield is expressed in the original

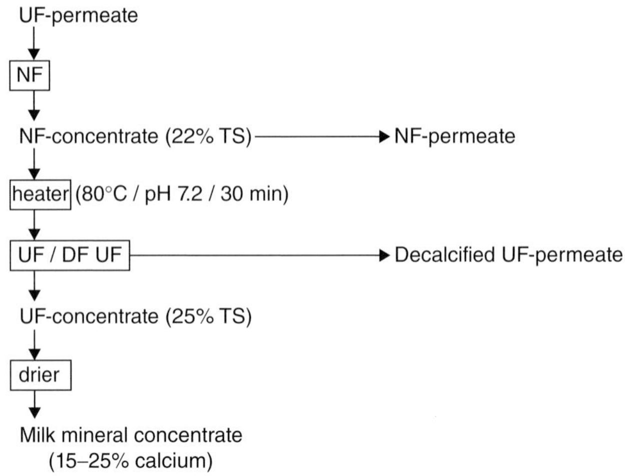

Figure 6.11 Process for the manufacture of a milk mineral concentrate from sweet UF permeate. DF, diafiltration; NF, nanofiltration; UF, ultrafiltration.

UF permeate solids, 70% of the solids are recovered. The lactose contribution is of crucial importance for the feasibility of the calcium supplement project. The residual 23% of the permeate solids, a mother liquor containing all kinds of milk serum substances, is used as a binder and cheap energy source in animal feed. The mother liquor contains, for example, a variety of interesting nitrogen substances in an enhanced concentration. This poses a promising starting point for further R&D work.

The yield and quality of lactose produced from UF permeate can be further improved. However, the processing becomes more complicated. de Boer and Robbertsen (1983) described the removal of calcium and riboflavin from two-fold concentrated UF permeate. The calcium was thoroughly removed by applying a weak acid ion-exchange resin, which replaced the calcium ions of the UF permeate by sodium. The riboflavin was reduced by means of a resin with weak and strong acid groups. A lactose of a high purity level was obtained after crystallization. These authors noted that the lactose crystals were larger than in a process without decalcification and removal of riboflavin. After having completed the process a mother liquor (minerals, NPN and residual lactose) remained as well as the spent regeneration fluids of both resin columns. Durham *et al.* (2004) and Durham (2009) went a step further and combined membrane processes (NF, RO) with ion exchange and ion exclusion, as demonstrated in the process scheme of Figure 6.12.

Ion exchange and ion exclusion make use of cation exchangers. In the ion exclusion process the resin is saturated with counter ions in the same composition as the UF permeate; in this way there is no net ion exchange between the resin and the feed. The permeate ions (K^+, Na^+, Cl^- and organic acids) are excluded by the resin. They pass through the column with the eluent (RO permeate) more quickly

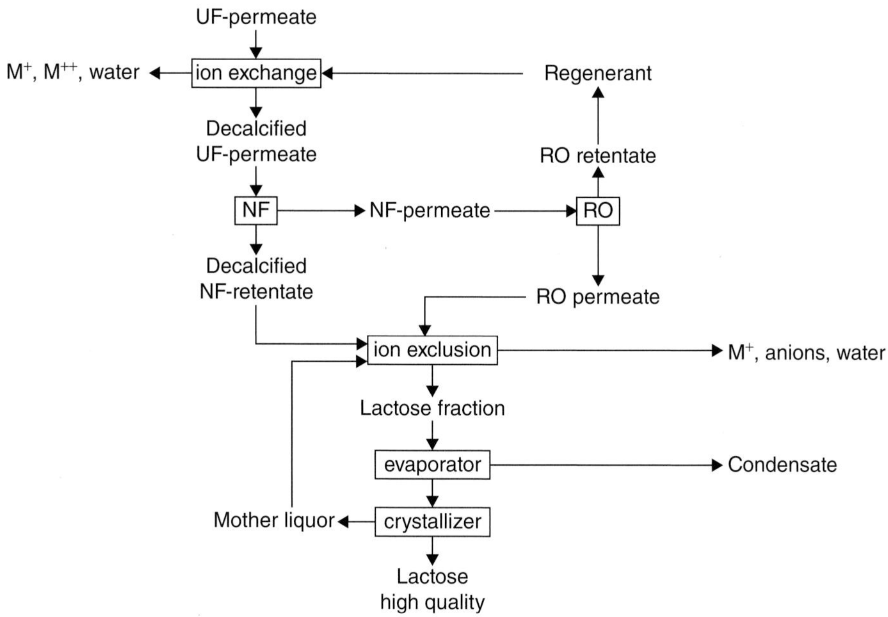

Figure 6.12 Lactose yield maximization from cheese UF permeate. M^+, M^{++}, mineral cations; NF, nanofiltration; RO, reverse osmosis; UF, ultrafiltration. *Source:* Adapted from Durham *et al.*, 2004.

than the non-ionic lactose fraction. Lactose is absorbed by the resin beads and therefore goes through the column more slowly. Hence, this procedure allows the lactose fraction to be separated from the mineral fraction. However, the ion exclusion process is adversely affected by the presence of the supersaturated calcium phosphate constituents present in the UF permeate. The 'blocking phenomenon' also causes fouling of equipment (evaporators, membranes). An ion-exchange step was introduced in the scheme of Figure 6.12, which reduced the calcium content by approximately 99%. The decalcified UF permeate improves the efficiency of ion exclusion and creates sharp peaks with a highly purified lactose fraction. The advantage of this additional step is that the mother liquor can be recycled through the ion-exclusion column. As a result almost all lactose can be used for the production of lactose or its derivatives (Durham *et al.*, 2004. Two streams remain after processing, namely the effluent of ion exchange – mineral cations (M^+, M^{++}), water – and the mineral-rich fraction of ion exclusion (M^+, anions, water). The ion-exchange effluent can be reused for the regeneration of the resin. Use is made of two phenomena. Firstly, UF permeate is relatively rich in potassium and sodium compared to calcium and magnesium (see milk serum in Table 2.1). Secondly, use is made of the difference in retention between the monovalent and divalent cations during NF of decalcified whey. The NF permeate is therefore particularly rich in monovalent ions, which can be concentrated by using an RO polisher. The RO concentrate is in turn utilized as regenerant. This approach makes the process more sustainable.

References

Bargeman G, Timmer M & van der Horst C (2005) Nanofiltration in the food industry. In: Schäfer AI, Fane AG & Waite TD (ed.), *Nanofiltration Principles and Applications*. Elsevier, pp. 305–28.

Batchelder BT (1987) Electrodialysis applications in whey processing. In: *Trends in Whey Utilization*. Bulletin IDF no. 212, pp. 84–90.

de Boer R & Hiddink J (1980) Membrane processes in the dairy industry. *Desalination* **35**: 169–92.

de Boer R & Robbertsen T (1983) Electrodialysis and ion exchange processes: The case of milk whey. In: Cantarelli C & Peri C (ed.), *Progress in Food Engineering*. Foster-Verlag, pp. 393–403.

Brans G, Schroën CGPH, van der Sman RGM & Boom RM (2004) Membrane fractionation of milk: state of the art and challenges. *Journal of Membrane Science* **243**: 263–72.

Cheryan M (1998) *Ultrafiltration and Microfiltration Handbook*. Technomic, pp. 268–71, 293–304.

Da Costa AR, Fane AG & Wiley DE (1993) Ultrafiltration of whey protein solutions in spacer-filled flat channels. *Journal of Membrane Science* **76**: 245–54.

D'Souza NM & Mawson AJ (2005) Membrane cleaning in the dairy industry: a review. *Critical Reviews in Food Science and Nutrition* **45**: 125–34.

Durham RJ (2009) Modern approaches to lactose production. In: Corredig M (ed.), *Dairy-Derived Ingredients – Food and Nutraceutical Uses*. Woodhead Publishing, pp. 101–44.

Durham RJ, Hourigan JA, Sleigh RW & Johnson L (1998) Process for the purification of nutrients from food process streams. US 6475390B1 [patent].

Durham RJ, Sleigh RW & Hourigan JA (2004) Pharmaceutical lactose: a new whey with no waste. *Australian Journal of Dairy Technology* **59**: 138–41.

Fane AG, Da Costa AR, Cao Z & Wiley DE (1997) Factors affecting performance of membrane processes – the spiral wound element design and operation. *Australian Journal Dairy Technology* **52**: 71–2.

Glimenius R (1980) Membrane processes for water, pulp and paper, food state of the art. *Desalination* **35**: 259–72.

Goulas A & Grandison AS (2008) Applications of membrane processes. In: Britz T & Robinson R (ed.), *Advanced Dairy Science and Technology*. John Wiley & Sons, Ltd., pp. 35–74.

Henningfield TD & Dimesen RA (2002) Process and plant for evaporative concentration and crystallization of a viscous lactose containing aqueous liquid. US 6790288B2 [patent].

Hiddink J & de Boer R (1980) Concentration of milk and whey by reverse osmosis. In: Linko P, Maelkki J & Olkku J (eds), *Food Processing Engineering* Vol. 1. Applied Science Publishers, pp. 561–71.

Hiddink J, de Boer R & Romijn DJ (1978) Removal of milk salts during ultrafiltration of whey and buttermilk. *Netherlands Milk and Dairy Journal* **32**: 80–93.

Hiddink J, de Boer R & Nooij PFC (1980) Reverse osmosis of dairy liquids. *Journal of Dairy Science* **63**: 204–14.

Hiddink J, de Boer R & Nooy PFC (1981) Effect of various pre-treatments on the ultrafiltration of sweet cheese whey at about 55°C. *Milchwissenschaft* **36**: 657–63.

van der Horst HC, Timmer JMK, Robbertsen T & Leenders J (1996) Nanofiltration for concentration and demineralization of cheese whey: a mass transport model. *International Dairy Federation Bulletin* no. 311, pp. 9–11.

IDF (1995) Fouling and cleaning in pressure driven membrane processes. *International Dairy Federation Special Issue* 9504.

Jonsson G (1984) Boundary layer phenomena during ultrafiltration of dextran and whey protein solutions. *Desalination* **51**: 61–77.

Kromkamp J (2005) Particle separation and fractionation by microfiltration. PhD thesis, Wageningen University.

Linder C & Kedem O (2005) Introduction/history of nanofiltration membranes 1960 to 1990. In: Schäfer AI, Fane AG & Waite TD (ed.), *Nanofiltration Principles and Applications*. Elsevier, pp. 1–31.

Lonsdale HK (1982) The growth of membrane technology. *Journal of Membrane Science* **10**: 81–181.

Madaeni SS, Yasemi M & Delpisheh A (2011) Milk sterilization using membranes. *Journal of Food Process Engineering* **34**: 1071–85.

Maubois JL (1984) Separation, extraction and fractionation of milk protein components. *Le Lait* **64**: 485–95.

Novak A (1991) Milk protein concentrate. In: New applications of membrane processes. *International Dairy Federation Special Issue* 9201, pp. 51–66.

Pouliot Y (2008) Review: Membrane processes in dairy technology – From a simple idea to worldwide panacea. *International Dairy Journal* **18**: 735–40.

Samuelsson G, Dejmek P, Trägårdh G & Paulsson M (1997) Minimizing whey protein retention in cross-flow microfiltration of skim milk. *International Dairy Journal* **7**: 237–42.

Sandblom RM (1978) Filtering process. US 4105547 [patent].

Schuck P, Piot M, Méjean S *et al.* (1994) Déshydration par atomization de phosphocaséinate natif obtenu par microfiltration sur membrane. *Le Lait* **74**: 375–88.

Selitzer R (1972) Crowley begins membrane processing of cottage cheese whey. *Dairy & Ice Cream Field* **155**: 34.

Yee KWK, Wiley DE & Bao J (2007) Whey protein concentrate production by continuous ultrafiltration: Operability under constant operation conditions. *Journal of Membrane Science* **290**: 125–37.

Yee KWK, Alexiadis A, Bao J & Wiley DE (2008) Effects of multi-stage membrane process designs on the achievable performance of automatic control. *Journal of Membrane Science* **320**: 280–91.

Further reading

Tetra Pak Processing Systems (2003) *Dairy Processing Handbook*. Tetra Pak Processing Systems, pp. 81–212, 356–58.

Walstra P, Wouters JTM & Geurts TJ (2006) *Dairy Science and Technology*. CRC Press, pp. 467–95, 583–639.

Westergaard V (2010) *Milk Powder Technology – Evaporation and Spray Drying*, 5th edn. GEA Niro.

7
End Users

A surprising number of new finished products based on ingredients from milk were launched globally during the last decade. According to Business Consulting 3A (2008), new food products containing whey protein concentrate (WPC) increased from 740 (period 2001–03) to 1179 (period 2004–06). In the same time frame the number of new products containing lactose doubled and reached 9434 in the period 2004–06.

The field of food applications is very broad and the numbers reflect the differences in food habits throughout the world. Ten categories of food will be distinguished and discussed in Section 7.3 about food products. Additional information regarding the use of concentrated and dried milk ingredients, whey ingredients, and milk fat and butter ingredients in all kinds of food products in the USA is also illustrated in the *Dairy Ingredient Applications Guides* (DMI, 1996). Numerous recipes are shown in these guides. The follow-up of the guides is more scientifically based and is described in many application monographs. Some of the monographs are product focused such as snacks and seasonings, bars and gels, or baked goods, while others are topic centred, for example weight management, immunity and gut health (USDEC, 1999 to present).

The scope of the applications of ingredients from milk is not limited to the food industry. By considering other industries the number of finished products increases further. Suppliers of ingredients approach this complex situation mostly through the use of a limited number of 'model recipes' to minimize the product development costs of the customer. The 'model recipe' includes both ingredients and the preparation. In these recipes the specific performance or cost benefits of the ingredients in a finished product can be demonstrated. The involvement of ingredient suppliers only partially explains the enormous diversification in end-products. In my opinion the main driving force behind the high numbers is 'business-to-business' communication. In a typical scenario the manufacturer of the finished

From Milk By-Products to Milk Ingredients: Upgrading the Cycle, First Edition. Ruud de Boer.
© 2014 John Wiley & Sons, Ltd. Published 2014 by John Wiley & Sons, Ltd.

product translates the results of the 'model recipe' to his or her product portfolio. The end-product then receives its 'unique selling point', so it can be distinguished from competitors' products. In this chapter the communication with the end users will be emphasized.

The development of ingredients require a long-term strategy due to the high investments in R&D and production facilities. Kramer (1991) describes the development of ingredients derived from skimmed milk, particularly after World War II, within DMV in The Netherlands. Until 1962 management hesitated regarding the addition of the milk sugar component; they considered the production of lactose an intermediate step. The final goal was the large-scale production of lactic acid. Based on market prospects and improvements in technology, the production of pharmaceutical lactose was selected in favour of lactic acid. Nowadays, this activity is successfully continued under the name DMV Fonterra Excipients. A similar example outside the dairy industry is Tate & Lyle, a private company with a long tradition in sugar refining. Bennett (2012) describes the change from a bulk low-margin product (refined sugar) to lower volume, higher margin food additives and ingredients. The company ceased using sugar technology, and today Tate & Lyle's focus is the production and marketing of ingredients. The low-calorie sweetener, sucralose, is one of the successful ingredients. The basis for this ingredient dates from 1966, when research concerning the versatility of the sucrose molecule to produce new ingredients was initiated. Both companies, DMV Fonterra Excipients and Tate & Lyle, are examples of successful ingredient businesses that adapted themselves over the course of time. Both have considerable know-how in the production as well as the application of their own ingredients from which end users can benefit.

Particularly in the early stages of the development of new ingredients the cooperation between the end user and the producer is vital. In the following part of this chapter we will focus on the requirements imposed by the end users and the field of application in which they are active.

7.1 Ingredient requirements

Requirements for ingredients will complement the more general demands with respect to religious concerns, safety and sustainability, which have been discussed in Chapter 5. Five requirements, namely nutrition/health, functionality, flavour, convenience and price, are shown in the outer edge of Figure 7.1. They are important considerations in the communication between the end user and the ingredient supplier, in which aspects regarding reproducibility are also involved. Since production aspects play an important role in this communication, the processes of various ingredients will be highlighted too. In the following sections, examples of ingredients and their processes are indicated in bold.

The applications of ingredients from milk are centred around human foods, animal feeds and pharmaceutical products, as shown in Figure 7.1. Some milestones in the development of ingredients for these industries are given as examples. Industrial

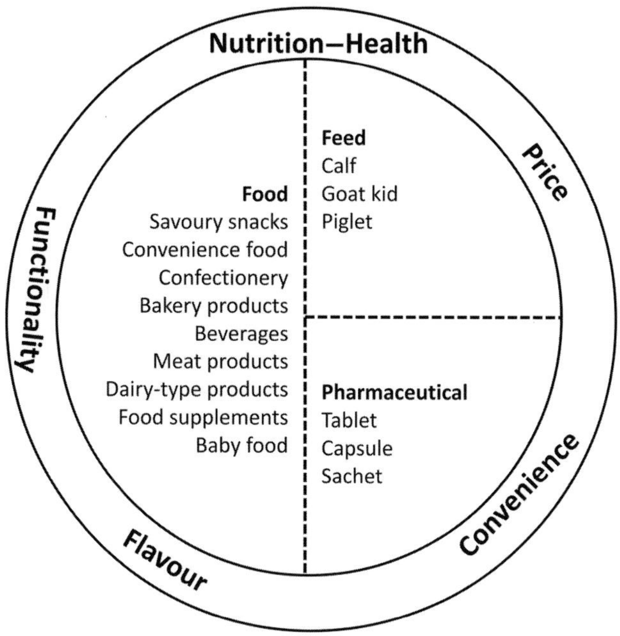

Figure 7.1 Ingredients requirements and applications.

tablet formation, making use of lactose as a filler, started about a hundred years ago. The development and application of caseinates coincided with the introduction of processed foods in the 1960s. Around that time the dairy industry started activities with milk replacers in the feed segment. Each of these industries has its own rules and standards; these are captured in respectively the Codex Alimentarius, the Code of Practice on Good Animal Feeding and various pharmacopoeias. The main objective of these rules and standards is to minimize risks to consumers' health. The human food segment is the most extensive of the three today. Collective terms are widely used, for instance in bakery products and confections, instead of individual products.

7.1.1 Nutrition/health

An important requirement for ingredients in this segment is that they improve the nutritional and/or health status of the application concerned. Milk ingredients are well known in this respect. In Figure 7.2 the two main constituents of dairy ingredients, protein and lactose, are distinguished as they play the most dominant roles in various applications. 'Milk calcium' is another interesting ingredient in human nutrition. Adequate 'milk calcium' intake is associated with multiple favourable health outcomes, like improved bone health, lowered blood pressure, weight management, and decreased risks for bacterial infections and colorectal cancer.

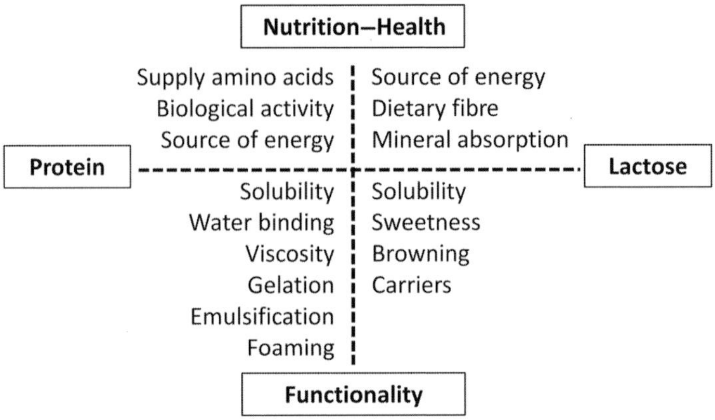

Figure 7.2 Aspects of nutrition-health and functionality of milk protein and lactose.

Milk protein is five to ten times the price of lactose and as a consequence it gives a higher added value. It is an excellent source of essential amino acids. The profile of whey proteins is slightly more favourable than that of casein based on the sum of the essential amino acids (see Table 2.5). Whey proteins also include a spectrum of proteins and peptides some of which show biological activity – meaning that they exceed the properties of a high-quality amino acid source. Lactoferrin (red protein) has well-documented antimicrobial and antiviral properties (Wakabayashi *et al.*, 2006; González-Chávez *et al.*, 2009). This protein may strengthen innate defence and plays a role in immune modulation. Another whey protein component that has recently emerged is caseinomacropeptide, or glycomacropeptide. Promising biological functions such as appetite control and impact on brain development are claimed (Thomä-Worringer *et al.*, 2006). The efficacy of these peptides needs to be further substantiated in human clinical studies. More examples of the bioactivities of milk proteins, such as osteopontin, lactoperoxidase and casein phosphopeptides, can be found in the information sheets.

In addition to being a source of energy, lactose has some specific benefits (Schaafsma, 2008). It can promote gut health (prebiotic or dietary fibre-like properties). Part of the lactose can escape digestion and will favour growth of beneficial bifidobacteria and lactobacilli. Like all non-digestible carbohydrates, which can be fermented by the intestinal flora, they will enhance the absorption of calcium and magnesium. A less favourable aspect of lactose is the high prevalence of lactose intolerance. However, undesirable effects in subjects deficient in β-galactosidase (lactase) can be prevented by careful use. In those cases, ingestion is limited to 11 g of lactose per day. As the level of dairy ingredients in food products is mostly low, negative effects can be avoided.

The role of lactose as a prebiotic attains a higher level in the lactose derivative, galacto-oligosaccharide (GOS). This causes growth of lactobacilli and bifidobacteria in the large intestine to be stimulated. It is a specific health effect, also called the 'bifidogenic effect' of GOS. The influence of GOS on bone mineralization is also of interest, as well as its impact on the immune system (Macfarlane *et al.*, 2008). Moreover, some promising effects have been mentioned in relation to inflammation, rheumatism and cancer.

In the evolution of the nutritional and health aspects of milk ingredients, the manufacturers of baby food have played a crucial role in which both whey proteins and lactose have been involved. Ion exchange and ion-exchange chromatography are important tools to fulfil the requirements of this sophisticated application. The latter has recently developed from a laboratory technique to become an industrial process. Three stages can be distinguished within this technology: demineralization, isolation of whey protein fractions (α-lactalbumin, lactoferrin) and hydrolysis of lactose into GOS connected to further purification of a mixture of saccharides. The principles of each stage will be discussed in the following part.

Ion exchange is applied to obtain whey with a **demineralization** degree of at least 90% as generally required for baby foods. It is a mature technology that has been utilized for decades (Houldsworth, 1980). A fixed-bed technique is applied involving use of ion-exchange resins. These resins are macromolecular composites, which can be constituted as acids or bases. Its physical form is a small sphere with a diameter of 0.3–1.2 mm. A cation-exchange resin removes positively charged ions and an anion-exchange resin removes negatively charged ions. If whey passes over a bed of a cation-exchange resin the cations in the whey (Ca^{2+}, Mg^{2+}, K^+, Na^+) are replaced by H^+ ions. Subsequently, the whey is charged on an anion exchanger, where the anions from the whey ($citrate^{3-}$, HPO_4^{2-}, NO_3^-, Cl^-) are adsorbed and displaced by OH^- ions. For the treatment of whey strong acid cation- and weak base anion-exchange resins are normally used. The latter can be regenerated more easily. A typical cycle for treating non-concentrated whey would be 2 hours of adsorption and 4 hours of regeneration. One litre of resin is able to demineralize about 10 L of whey. A continuous flow of demineralized whey can be achieved by using three pairs of ion-exchange columns in parallel. This so-called 'merry-go-round system' can be beneficial for the production of a highly reproducible ingredient.

The regeneration solutions need to be sufficiently concentrated (4–5%) to remove the adsorbed cations and anions and replace them with H^+ and OH^- ions. Weak base anion-exchange resins need 150% molar equivalent NaOH, and the strong acid cation resin requires 300% molar equivalent HCl. If in the latter case a countercurrent system is applied the usage of HCl can be reduced by 40%. A further reduction of regeneration solution, which represents 60–70% of the operational costs, can be achieved through the more complicated SMR process (Burling, 1979). In this system the whey anions are exchanged with HCO_3^- and subsequently the cations with NH_4^+. The desalted whey is conveyed to an evaporator, where NH_4HCO_3 is volatilized as NH_3, CO_2 and H_2O. More than 80% of the NH_3 and CO_2 is recovered by distillation to create new regeneration solutions and thereby reducing effluent problems.

Desalted whey, which is mostly adjusted to 90% demineralization, is frequently used as part of a liquid baby food formula. The liquid formulas are sterilized using UHT treatment. Adjustment of the pH level before UHT sterilization is important in order to achieve stable emulsification and thermal properties (McSweeney *et al.*, 2004). A pH of 6.8, as indicated in the information sheet for 'Demineralized whey', is close to the optimal pH value.

Besides desalting of whey by ion exchange, a combination of nanofiltration (NF) and electrodialysis (ED) can also be applied (see Figure 6.10). Reproducibility is again an important issue as baby food formulas based on demineralized whey powder can contain 50–60% whey solids. Outinen *et al.* (2010) investigated the effect of changes in cheese production processes on composition and functional properties of an 85% demineralized whey powder. Partially high-temperature treatment and/or ultrafiltration (UF) to standardize the protein content of cheese milk had only a modest effect. Despite differences in process history, the amino acid profile of the powders was similar. Five functional properties were investigated of which four showed no significant difference. The heat stability was only improved by the UF treatment of the cheese milk.

As set out above, some milk proteins have very specific biological functions. The processing is complicated and has to be carried out with special care in order to manufacture the ingredient in a reproducible way. For the isolation of **whey protein fractions** ion-exchange chromatography is mostly applied. The aim is isolation of a specific whey protein, while the remaining part of the whey leaves the chromatography column. The process can be divided into four steps: loading, rinsing, elution and regeneration. In this process, particular use is made of the amphoteric character of the proteins (see Section 2.3.3). The pH level of the starting material influences which type of protein is adsorbed during the loading process. β-Lactoglobulin and α-lactalbumin are positively charged at a pH of 4.0. The proteins are captured under these conditions on the strong cation-exchange resin. The whey protein isolate (WPI) fraction will be isolated if elution is carried out with a NaOH solution (Figure 7.3). By changing the elution conditions and the use of a sodium acetate solution of pH 4.9, α-lactalbumin will be removed. An α-lactalbumin-depleted WPI is left (Doultani *et al.* 2004). If the column is loaded with whey of a neutral pH (6.5), lactoferrin (LF) and lactoperoxidase (LP) will be selectively adsorbed (Burling, 1994; Kussendrager *et al.*, 1997). In the elution stage the protein is desorbed by using specific solutions (NaOH, NaCl, phosphate). Under the various elution circumstances different whey protein fractions will be released. In case of LF and LP the elution process needs to be adjusted in such a way that overlap is avoided and reproducible ingredients are manufactured. Subsequently, the ion-exchange resin is regenerated and reloaded with negatively charged ions for a new process cycle. The eluate is further treated with UF in order to remove the elution salts and concentrate the protein component (Figure 7.3). Finally, the UF concentrate is spray dried. In the whole LF process scheme the heat stability of the protein needs to be watched closely. LF can easily be inactivated, which starts already at normal pasteurization

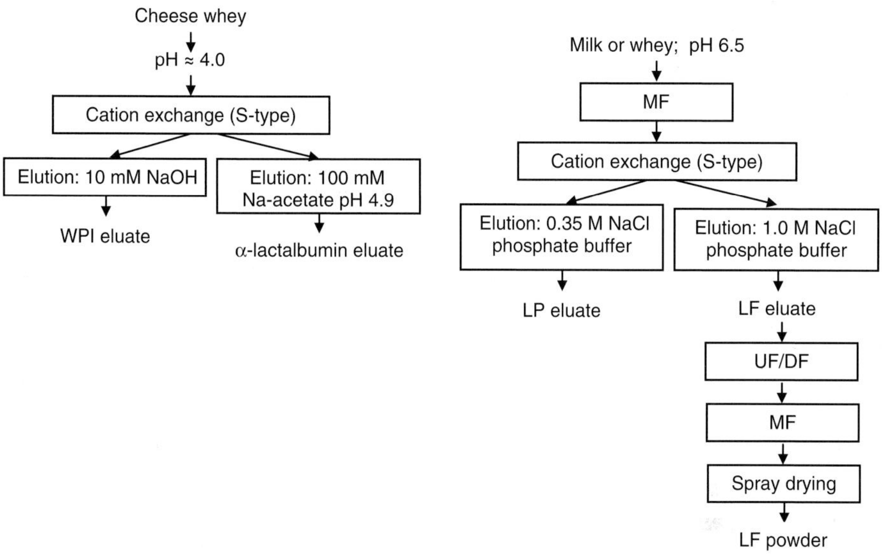

Figure 7.3 Some whey protein fractions. DF, diafiltrate; LF, lactoferrin; LP, lactoperoxidase; MF, microfiltration; UF, ultrafiltration; WPI, whey protein isolate.

temperatures of 72°C/pH 6.5 (Kussendrager, 1994). For this reason MF is used to remove bacterial contamination from the starting material. The same process step is also carried out on the UF/DF (diafiltrate) concentrate. The combination of five unit operations makes LF an expensive ingredient.

In this part the focus will be on enzymatic hydrolysis of lactose into **galacto-oligosaccharides** (GOS) and their further purification. GOS consist of a chain of galactose molecules with the glucose molecule at the reducing terminus. The degree of polymerization varies from two to eight. In Figure 7.4 a general process scheme for the production of GOS is given. Additional information regarding the production of GOS and its effect on the health of infants and adults can be found in numerous references (Ekhart & Timmermans, 1996; Angus *et al.*, 2005; Torres *et al.*, 2010; Tzortis, 2011; Lamsal, 2012). The enzyme β-galactosidase is applied. It catalyses both the hydrolysis of lactose and the transgalactosylation reaction. The ratio hydrolysis/transgalactosylation varies depending on the reaction conditions and the enzyme source. The transferase activity is favoured by a high lactose concentration, an elevated temperature and a low water activity. In industrial practice β-galactosidases are, for instance, derived from specific strains of the microorganisms *Bacillus circulans*, *Bifidobacterium bifidum* and *Aspergillus oryzea*. The glycosidic bonds between the galactose units can differ, with β1-4 bonds (4′-galactosyllactose), β1-3 bonds (3′-galactosyllactose) and β1-6 bonds (6′-galactosyllactose). Each enzyme produces a mixture of galacto-oligosaccharides with differences in glycosidic bonds and chain length. This might result in a different functionality, for instance a greater selectivity toward the growth of bifidobacteria in the large intestine (Depeint *et al.*, 2008).

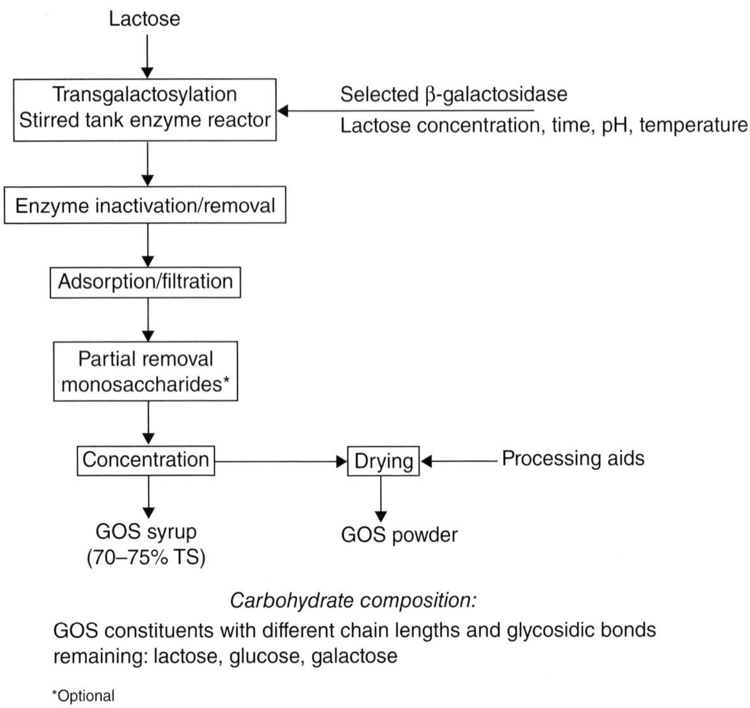

Figure 7.4 Enzymatic modification of lactose in galacto-oligosaccharides (GOS).

Batch reactors are commonly used for ease of production, as is shown in Figure 7.4. Important aspects are the yield of galacto-oligosaccharides and the thermotolerance of the enzyme applied. As the reaction might take several hours at 37°C, for example, the microbiological quality of the lactose source and the design of the equipment will need particular attention. Heat inactivation of the enzyme or its removal by ultrafiltration is an important next part of the process. Subsequently, a treatment with active carbon is carried out, which is followed by filtration. In this way the liquid is decolorized and purified. After a concentration step in an evaporator a GOS syrup is obtained. An example of the carbohydrate composition of GOS is given in the information sheet 'Galacto-oligosaccharides syrup'. It reflects the type of reaction with a majority of GOS and equal amounts of unreacted lactose and monosaccharides. As galactose is incorporated in GOS it is expected that glucose is abundantly left.

As is shown in the same information sheet, the galacto-oligosaccharide content is in the case of 'GOS-Vivinal' 59% on DM. This level can be increased by an additional process step. One of the options is the application of nanofiltration to remove part of the monosaccharides (Tzortis, 2011). Another possibility is the fractionation of GOS by applying cation-exchange chromatography (Sinclair *et al.*, 2007). The separation is based on the fact that low molecular weight carbohydrates are retained due to their higher affinity to the cation-exchange column. The elution is carried out step-wise using an aqueous mobile phase. A purer GOS ingredient, having a polymerization degree of five or higher, can be obtained. This fraction is claimed to inhibit cholera toxins (Sinclair *et al.*, 2007, 2009), and could be an alternative to antibiotic treatment to combat diarrhoeal diseases such as cholera.

Galacto-oligosaccharide is available as syrup (75% TS) as well as a powder. In dried form it can be more easily introduced to some nutritional formulations. Spray drying of GOS is made possible through the addition of processing aids such as maltodextrin or WPC.

7.1.2 Functionality

Functionality describes the usefulness/performance of an ingredient in the finished product. And again, lactose and protein are quite different in this respect. Some properties of lactose are indicated in Figure 7.2. It has a low solubility and low sweetness compared to sucrose. Both aspects play a role in its application in confectionery, resulting in a more balanced flavour and a better mouth feel. In bakery products the lactose is not fermented and remains available. Partial replacement of sucrose with lactose improves, for instance, water absorption and delays 'aging' (EWPA, 2000). As lactose is a reducing sugar, it promotes the controlled browning caused by the Maillard reaction and caramel formation. Besides the properties indicated in Table 7.1, lactose is also beneficial as a carrier in pharmaceutical applications (see Section 7.4).

In case of milk protein, more aspects of functionality need to be considered than for lactose. In Figure 7.2 six functional properties are mentioned. All relate to various food products. Solubility is, for instance, important for application in beverages. Water binding of proteins is of interest in sausage manufacture. In soups and sauces protein viscosity is a key aspect. Gelation assists in building a gel network. It is beneficial in the manufacture of processed meat products as it improves the yield. Proper emulsification is important in creamers, for instance, and prevents 'oiling off' in hot coffee. In products such as whipped toppings the foaming properties of proteins play a vital role.

As far as milk proteins are concerned, the functional properties of the two main milk proteins can be distinguished. Casein/caseinate and whey proteins differ in their solubility, heat stability and gel formation, as is indicated in Table 2.8. The

Table 7.1 Differences between lactose and sucrose.

	Lactose	Sucrose
Fermentable by baker's yeast	No	Yes
Relative sweetness (%)	30	100
Solubility (g/100 g solution, 20°C)	15	85
Reducing sugar	Yes	No

whey proteins are, for example, sensitive to heat treatment, and the degree of denaturation affects ingredient functionality (see Figure 2.5). Undenatured whey proteins are needed for optimal solubility in acid beverage applications, for instance. In this state they show a constant solubility over a wide pH range, as opposed to sodium caseinate. Both undenatured whey proteins and caseinates can be applied in desserts and toppings, for example, due to their foaming properties. Sodium caseinate is a well-known emulsifier and usually preferred in coffee whiteners. Heat stability is another important advantage of caseinate in finished products that need sterilization, for instance. Whey proteins turn into a gel after intensive heat treatment at higher concentrations; at lower protein levels these proteins result in enhanced water binding and increased viscosity.

Functional properties are only an indication of an ingredient's usefulness in food applications. Foegeding and Davis (2011) reviewed in particular the foaming and gel properties of proteins and emphasized a more comprehensive approach. The authors stated that simple tests often fail to predict the performance in a food product. These tests seldom account for the addition of non-protein solutes and all of the processing operations involved in making a food product. The use of 'model recipes' as mentioned earlier, gives a better prediction of the performance of an ingredient in a food product.

The mentioned functionality tests are still valuable. They are a tool in the manufacture of a reproducible protein ingredient. The functionality tests are often a part of a contract between the producer and the end user in which certain limits are set.

The **ingredient purity** is also a factor in the selection of an ingredient. In Figure 7.5 an example is shown of lactose; its content ranges from 75% in whey powder to more than 99.8% for pharmaceutical lactose. This difference can easily increase the cost a few times. The end users usually select the optimum price/performance ratio. Maximum lactose purity is not needed in food applications, therefore edible lactose can be an easy choice. Permeate powder can replace whey powder if the function of protein is not required and its application is focused on the browning properties. The permeate powders are part of a family of ingredients which are labelled 'dairy product solids' in the USA (Burrington, 2011). According to US law 'dairy product solids' need to have a minimum of 59% lactose and a maximum of 10% protein and 27% ash. This composition covers permeate as well as delactosed permeate. The relative high ash content and the modest sodium content of the

Whey powder 75%
↓
Permeate powder 85%
↓
Edible lactose >99%
↓
Pharmaceutical lactose >99.8%

Figure 7.5 The purity of various lactose ingredients.

delactosed ingredient is additionally recommended for its salt-enhancing properties. The mechanism is uncertain; it is possible that the remaining non-protein compounds in the delactosed permeate may serve as flavour potentiators. The ingredient can be applied as a salt replacer and play a role in dietary sodium reduction while the flavour of the end-product is still acceptable.

As described earlier and shown in the information sheets, **pharmaceutical lactose** has the highest purity level of the lactose ingredients. Considerable additional efforts have to be made to achieve this quality level. Processing and various modifications to improve functionality will be discussed. Crystallization is applied to produce pure lactose ingredients (see Section 2.5.1) from whey or permeate.

A flow scheme for the manufacture of various types of lactose is set out in Figure 7.6. The starting material in this example is cheese whey. The first part of the process is identical for edible and pharmaceutical lactose. The whey is concentrated by evaporation or a combination of RO and evaporation to approximately 60% TS in order to obtain a high yield. The concentrate is then transferred to a batch-wise crystallization process. Subsequently, the liquid is gradually cooled in the crystallization tanks to about 20°C over a period up to 24 hours. Cooling below this

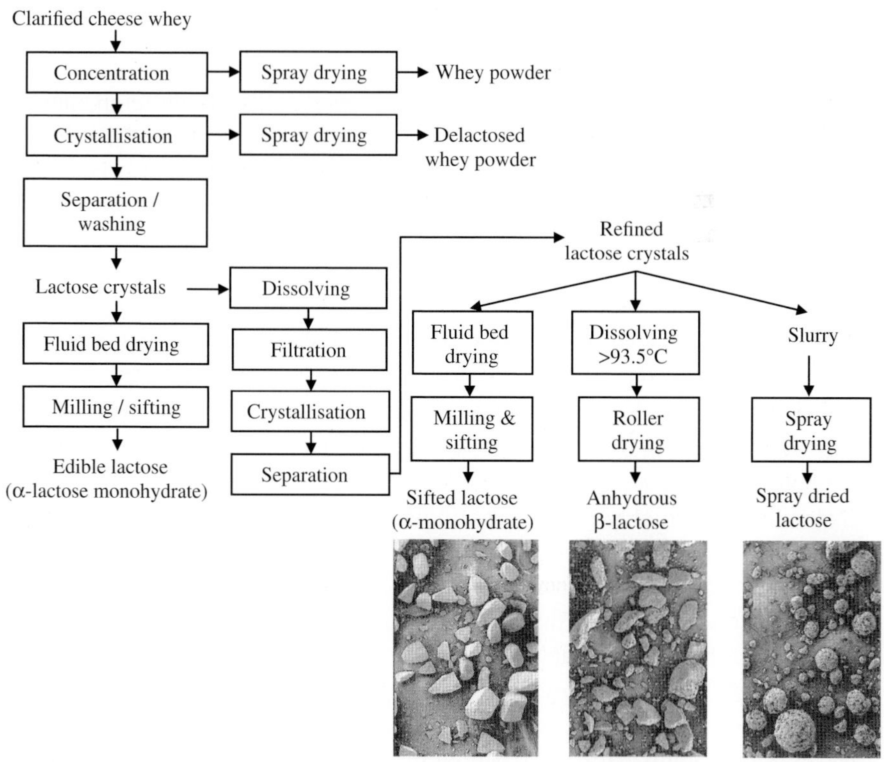

Figure 7.6 Flowcharts for the manufacture of edible lactose and three types of pharmaceutical lactose. Photos courtesy of DFE Pharma (DMV-Fonterra Excipients).

temperature is usually less profitable. It results only in a slight increase in yield at the expense of a considerable increase in process time. The α-lactose monohydrate crystals, with their typical tomahawk shape, are washed with a saturated lactose solution to remove impurities (ash, protein) from the surface of the crystals. After separation, mostly with the help of decanter centrifuges, the crystals are dried, for instance in a fluid-bed drier, to a free moisture content of 0.1–0.2%. This prevents mould growth during storage as well as caking due to recrystallization. After milling and sifting the edible lactose, α-lactose monohydrate, is available in various particle sizes (60–200 mesh).

In case of pharmaceutical lactose the crystals are not dried after the first separation step. They are dissolved again in hot water and the lactose content of the solution is adjusted to about 60%. Active carbon and a filter aid are added to the hot liquid to eliminate the residual riboflavin (yellow colour) and lower the levels of protein and ash to approximately 0.01%. This procedure also lowers the bacterial count to below 100/g. Subsequently, the liquid passes a filter press and the active carbon and filter aid are removed. The refined liquid is cooled stepwise in a second crystallization process. After separation of the crystals, three options are available, as shown in the refined lactose part of Figure 7.6. The first option is similar to that of edible lactose and results in the usual pharmaceutical α-lactose monohydrate. The crystals are dried in a fluid-bed drier and subsequently milled and/or sifted. An electron micrograph of a sifted variant is shown in Figure 7.6 in which the tomahawk crystal form can be recognized. The second option requires a dissolving temperature of the crystals greater than 93.5°C. An anhydrous lactose is obtained if the liquid is dried with the help of roller driers and the drying temperature is also kept above 93.5°C. Due to its production process the anhydrous product contains typically 80% β-lactose; the remaining part is anhydrous α-lactose. The picture shows the lactose as irregular-shaped fine and coarse particles, resulting in moderate flow properties. This disadvantage can be eliminated in the production of spray-dried lactose, which is shown in the third picture in Figure 7.6. It contains spherical agglomerates of α-lactose monohydrate (90%) bound together by amorphous lactose (10%). This can be achieved by suspending a milled α-lactose monohydrate fraction in a saturated lactose solution, which is subsequently spray dried. The presence of the hygroscopic amorphous lactose requires specially adapted packing.

Examples of the highly purified casein part of the milk proteins are given in the information sheets on 'Rennet casein – Unmatured cheese non-fat', 'Acid casein', 'Sodium caseinate' and 'Calcium caseinate'. The focus of this section is on the most versatile of these ingredients, the **caseinates**. The common variants are sodium and calcium caseinate. End users apply sodium caseinate as a functional ingredient (e.g. emulsifier), while calcium caseinate is used for nutritional as well as functional purposes. These caseinates are produced in an accurately controlled process (Muller & Hayes, 1962; Southward, 1985; Mulvihill & Ennis, 2003). From the perspective of minimizing costs and improving reproducibility, a continuous process is preferred over a batch process. The process scheme is shown in Figure 7.7.

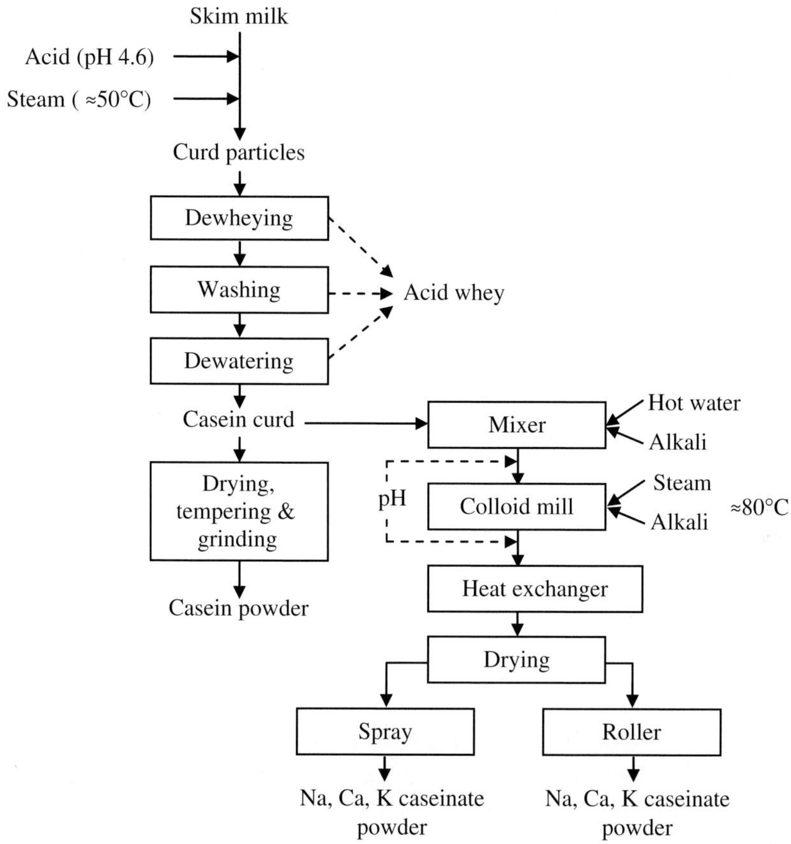

Figure 7.7 From casein micelle to caseinate.

Casein is precipitated from pasteurized skim milk to produce acid casein, which is subsequently converted into caseinates (see Section 2.3.4). Isoelectric precipitation (pH 4.6) is carried out in a continuous process in which a mineral acid solution – food-grade hydrochloric acid or sulphuric acid – is sprayed into skim milk in a countercurrent manner. In order to complete the reaction between casein micelles and acid this part of the continuous process is performed at a low temperature. The precipitation of casein starts as the temperature is increased rapidly, for instance by steam injection, and combined with a brief period of residence. Repulsion of whey takes place and curd with the required firmness is produced. A product of high quality is obtained in the subsequent multi-stage washing process in which the lactose, ash and free acid contents are substantially reduced. The amount of wash water is minimized by using a countercurrent system and by applying efficient separation of curd and (diluted) whey, for instance by means of decanter centrifuges (dewatering step). The refined, fresh casein curd is converted into caseinate. The best colour, flavour and nutritional value is produced if the lactose content is less than 0.2% of TS (Australian Society of Dairy Technology, 1972).

In Figure 7.7 a continuous dissolving technique is shown and the pH is carefully adjusted to 6.8–7.0 in two steps (Lippe *et al.*, 1980). The size of the curd is minimized by means of a colloid mill, the temperature adjusted by steam and the mode of alkali addition carefully controlled in order to handle this type of highly viscous material. The ingredient quality is a critical issue. The browning of the caseinate needs to be minimized while microorganisms/viruses are inactivated (e.g. 115°C/2 min to eliminate the foot-and-mouth virus). Subsequently, the caseinate solution is dried. Roller driers are more economical than spray driers as they can handle a higher viscosity and thereby create a feed with a higher solids content. In the information sheets spray-dried calcium caseinate with a very low bacterial count (typically 100/g) is reported. In case of nutritional applications additional hygiene measures are required.

7.1.3 Flavour

Flavour is a key factor for consumer acceptance of food products. This aspect is not limited to human foods, but is also vital for animal feed products. In case of animal feed the word 'palatability' is used. An important requirement for ingredients is often a neutral flavour and a bland odour. In this respect dairy ingredients are advantageous; they make product development less complicated. However, there are also dairy flavour ingredients available that have a characteristic flavour.

Butter is famous for its delicate flavour, which is the basis for many applications as mentioned in the information sheets. Butter's aroma is primarily derived from the volatile compounds present in the milk fat fraction. Mallia *et al.* (2008) mentioned in a review that more than 230 volatile compounds have been identified. The desirable and unique aroma depends on a variety of compounds with a low threshold level. The aroma compounds can be categorized as free fatty acids, aldehydes, ketones, lactones and sulphur-containing components. Sweet cream butter can be characterized by lactones with fruity and creamy notes and by sulphur-type compounds with corn-like and garlicky odours. The flavour compounds of sour cream butter show a wider diversity. Due to bacterial starter cultures, this product is particularly rich in volatile flavour compounds. Because of its richness in flavour, 'lactic butter' is standard in Europe. Key components are ketones such as diacetyl, with its buttery flavour, free fatty acids, for instance butyric acid with its cheesy note, and lactones with a peach/coconut-like contribution. Anhydrous milk fat, or butter oil, is produced from sweet cream. The process is completed with a vacuum treatment to obtain a water content not exceeding 0.2%. The flavour profile is less pronounced and is characterized by aldehydes conferring green and oily notes. The vacuum treatment is also applied to reduce the possibility of oxidation of unsaturated fat components and consequently the development of off flavours during storage under ambient conditions. In several applications the butter is heated. Heat often provides the driver for flavour components such as ketones and lactones, which are converted to pleasant and more flavoursome components. This can be observed in the typical flavour development in, for example, the frying of butter or the production of cakes and confectionery containing butter (Burgess, 2001).

Some whey ingredients have a salty flavour. This characteristic can be a limitation for sweet products, such as ice cream, as well as an opportunity for savoury products. Their application in savoury products is, however, limited because of their salt-rich nature. Cheese flavours are more popular in savoury snacks as they are able to provide a greater variety of flavours; they can also be applied in, for instance, soups, dips and crackers.

In this part we will focus on the preparation of **cheese flavours** and how they meet the requirements of the end users. Cheese powders are a rich source of a variety of flavours. They are made by spray-drying cheese slurries with the help of emulsifiers. As cheese requires a lengthy ripening period, efforts were made to produce alternatives that require less time and save costs. The alternatives to cheese powders are called enzyme modified cheeses (EMC). A wide range of flavours are known; two of them, Emmentaler and blue cheese, are described in the information sheets. In EMC production the aim is to mimic the profile of various natural cheeses, but at a flavour intensity that is 15–30 times greater. Generally speaking the dosage level is 0.1–2%. Compared to natural cheese, EMCs do not influence the texture of the end-product. An example of an EMC process is set out in Figure 7.8 in which the production time is typically measured in hours or days rather than in months. The EMC process is carefully controlled, and process parameters are pH, temperature, agitation, aeration and incubation time. Essentially, it starts with the incubation of a pasteurized, flavourless substrate consisting of fresh rennet curd, milk fat, emulsifiers, water and additional specific exogenous enzymes and/or microorganisms. The reproducibility of the flavour production is enhanced by the use of identified specific key components (Kilcawley *et al.*, 1998). In case of blue cheese specific key components are the methyl ketones, whereas for Emmentaler/Swiss cheese they are propionic acid and proline. The process is completed by pasteurization. The final product is standardized to the desired flavour intensity and composition. EMCs are available as pastes or spray-dried powders. During the spray-drying process the oil-in-water emulsion is encapsulated with maltodextrin and whey. Figure 7.8 shows a two-step approach in which proteolysis and lipolysis are carried out separately in order to mimic the desired complex cheese flavour in a more optimal way (Kilcawley, 2002).

7.1.4 Convenience

This requirement relates to the convenience with which an ingredient can be used in various production facilities. This aspect is also important in vending machines for coffee, and automatic calf feeding apparatus, for example. Ease of powder handling can be achieved by crystallizing lactose, clustering dry particles (agglomeration) and using a wetting agent. It will influence powder characteristics such as flowability, solubility, bulk density, caking and dust formation.

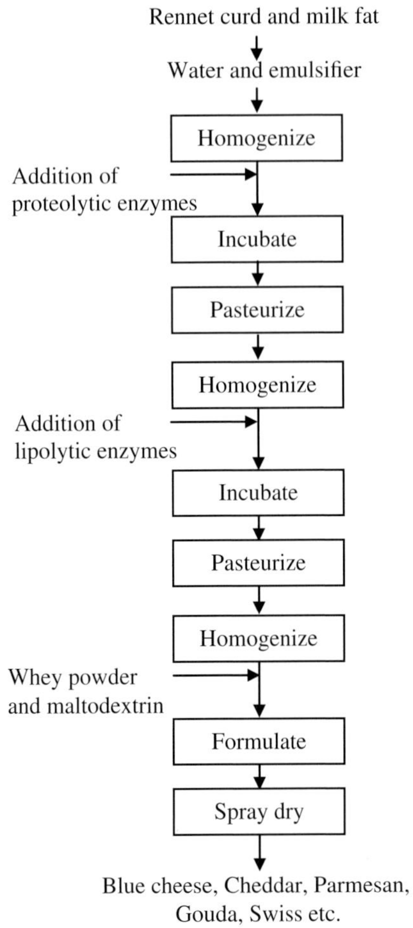

Figure 7.8 An example of enzyme-modified cheese manufacture (incubation 25–45°C/1–4 days and pasteurization 72–80°C/10–20 min). *Source:* Adapted from Kilcawley, 2002.

Caking of whey powder can be restricted by crystallizing lactose prior to spray drying (Figure 7.9). The minimal degree of crystallization of 70% is achieved within 2 hours at 20°C by concentrating whey to about 60% TS (Mol & Roetman, 1974). At lower solid levels, the buffer time has to be increased. During crystallization, α-lactose monohydrate is formed and the occurrence of amorphous lactose is reduced to such an extent that caking of the whey powder is limited. This procedure is also applied in other high-lactose ingredients such as permeate powder.

In the information sheet 'Calcium caseinate' an example of a spray-dried calcium caseinate is provided. The ease of handling by the end user can be improved by agglomerating and coating the powder agglomerates with a wetting agent (lecithin). Due to the lack of lactose in calcium caseinate, agglomeration is difficult to accomplish in the conventional driers. Westergaard (2010) describes a tall form drier in

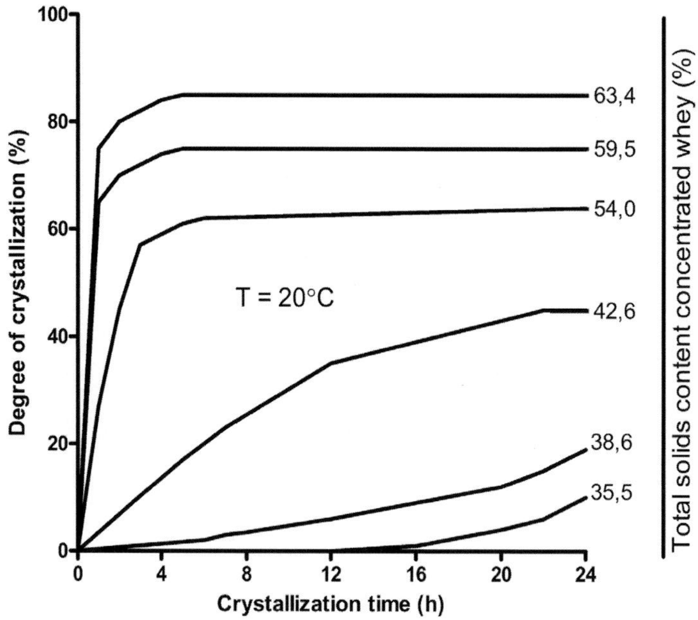

Figure 7.9 Prevention of caking in whey powder. Adapted from Mol and Roetman (1974).

which use is made of the stickiness of the 'wet' powder particles with a moisture content of 6–15%. Under special process conditions, recycled, already dried powder particles are entrained into this cloud of sticky particles and a so-called 'spontaneous secondary agglomeration' is achieved. The agglomerated powder is dried further and coated with a spray of lecithin in the final stage of the three-step drying process. In this way the 'instant' properties of the powder are improved. The same approach can be applied to other high-protein ingredients such as WPC 80.

7.1.5 Price

The prices of ingredients from milk may differ by a factor of one hundred or more – for example when comparing the commodity whey powder and the high-value-added ingredient lactoferrin. Even within the same ingredient there is variation in price. For instance a high-gel WPC 80 variant will attract a premium compared to a regular WPC 80. Price is one of the key factors in the acceptance of an ingredient by an end user. Some factors that affect ingredient prices will be discussed in this section.

The concentration of a constituent in the starting material plays an important role and has consequences for the cost of the ingredient. Figure 7.10 illustrates the concentration of numerous ingredients in different sectors of the biotechnology industry. They range from therapeutic and diagnostic enzymes, antibodies, bulk enzymes, antibiotics and amino acids to ethanol. A dominant cost element in the

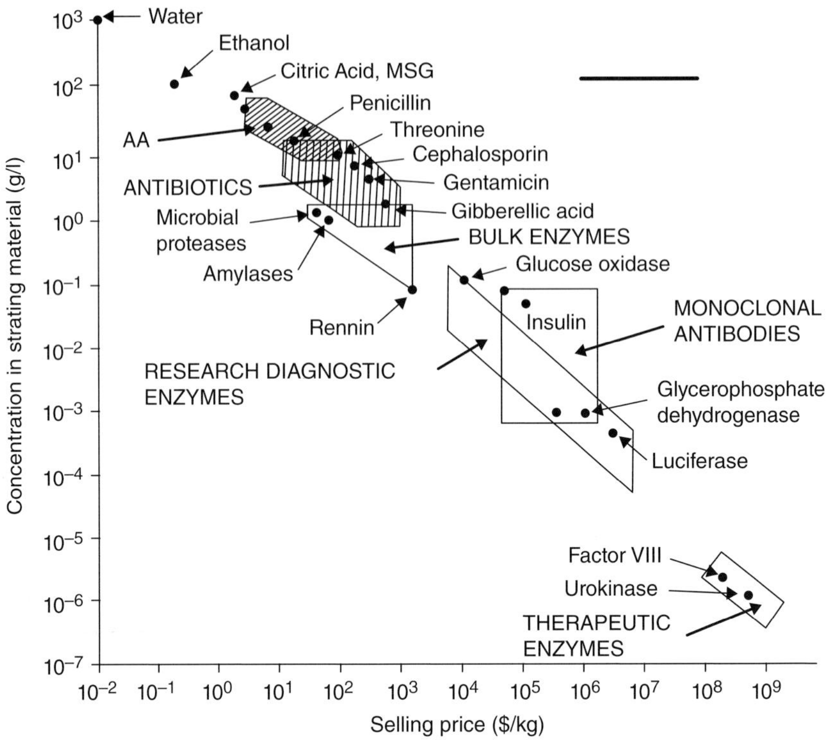

Figure 7.10 Relation between starting material and selling price. MSG, monosodium glutamate. *Source:* Dwyer, 1984. Reproduced with permission from Nature.

manufacture of these ingredients is the downstream purification process. This is particularly the case if a very high purity is required. Dwyer (1984) found in these cases a linear relationship on a logarithmic scale between the concentration in the crude starting material and the selling price. Some of the ingredients indicated in Figure 7.10 are in use in the dairy industry: for instance, rennin or rennet is used for the coagulation of cheese milk, and citric acid is a preservative. Dairy commodities like skim milk powder (SMP), WPC 80 and sodium caseinate, which are not the result of bioprocessing, can also be located in the linear relationship. They are situated in the upper left-hand part of the relationship, while lactoferrin is more in the middle. The relationship shown in Figure 7.10 underlines the importance of the concentration of the constituent in the crude starting material. Although the linear relationship should be regarded as no more than a rough indication, we can apply it to various constituents in milk.

In Table 7.2 the concentrations of bioactive constituents in whey are expressed in mg/kg, showing differences of about a factor of 1000. Lysozyme is one of them; it exhibits a strong antibacterial activity and is applied as a natural preservative (Cegielska-Radziejewska *et al.*, 2008). Hence, to produce 1 kg of lysozyme at, for

Table 7.2 Approximate content of minor bioactive components in cheese whey

Constituent	Concentration (mg/kg)
Lysozyme	0.3
Lactoperoxidase	14
Lactoferrin	60
Glycomacropeptide	1050

Source: Smithers *et al.*, 2008.

instance a yield of 66%, requires the processing of about 5 million kg of whey. Lysozyme is not uniquely sourced from dairy liquids. Larger amounts can be found in, for instance, hen's egg white, in which the lysozyme accounts for 3.5% of the total egg white protein. It is present in a concentration of approximately 3500 mg/kg. Compared to egg white, whey is not a competitive source for the isolation of lysozyme. According to Figure 7.10 the cost of lysozyme from whey is estimated to be about 10,000 times higher than that of lysozyme from egg white. Glycomacropeptide, lactoferrin and lactoperoxidase are present in substantially higher concentrations in cheese whey. All three are described in the information sheets.

The concentrations of promising constituents show considerable variation within the liquids available to the dairy industry itself. In Table 7.3 an example is given of polar lipids, which comprise both phospholipids and sphingolipids (Dewettinck *et al.*, 2008). For the isolation of the polar lipids whey and butter serum are used. From a concentration point of view butter serum is the most attractive source. However, a disadvantage is that butter serum also contains casein; this complicates purification, as indicated in Section 2.6.2 in more detail. The milk fat globule membrane (MFGM) ingredients, as illustrated in the information sheets, are not delivered in a pure form. In this way the price is kept at a lower level. Another example is calcium phosphate, which can be processed from permeate or delactosed permeate, for instance. The differences in calcium concentration of solids can vary by a factor of nine (Burrington, 2011). From a yield point of view delactosed permeate is a preferred source of calcium. However, permeate has been used for this purpose until now, as the optimization of the lactose process has to be considered as well (see Figure 6.11).

Table 7.3 Polar lipid content (g/100 g) of various dairy raw materials.

Raw material	Product basis	Total solids basis
Raw milk	0.03–0.04	0.23–0.32
Cheddar whey	0.02	0.26
Buttermilk	0.16	2.03
Butter serum	1.25	11.54

Source: Dewettinck *et al.*, 2008.

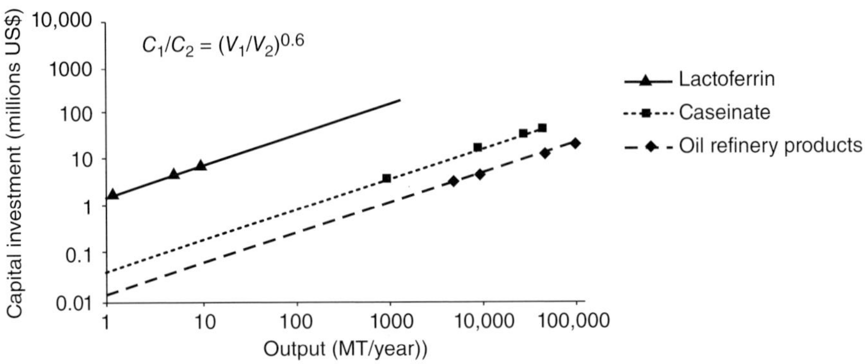

Figure 7.11 Relationship between increase in equipment cost (C_1/C_2) and increase in capacity (V_1/V_2).

The price of an ingredient is also influenced by economies of scale. From engineering it is known that an increase in the capacity of equipment is generally greater than the corresponding increase in equipment cost. This can be calculated according to the so-called '0.6 rule'[1], where the scale coefficient, n, is 0.6. The basic version of the rule originally focused on tanks and pipes. Tribe and Alpine (1986) investigated the scale coefficient for cane sugar manufacture (2500 to 10,000 tons of cane per day), and extended the basic version of the rule to other types of equipment. On average the '0.6 rule' can be more generally applied and used as a rough indication. This equation was applied to the manufacture of caseinate and lactoferrin and compared to oil refinery products produced from large plants.

Figure 7.11 uses logarithm scales to show the relationship between capital investment (millions of dollars) and output (tonnes per year), and can be used for comparison purposes. It is demonstrated that bigger caseinate plants are economically favourable: for example, an increase in capacity from 1000 to 10,000 Mt requires only a twofold greater investment. In case of lactoferrin, the economies of scale are limited as only small amounts are available in the starting material (skim milk). Furthermore, the process is more complicated and it is more difficult to achieve a favourable yield. As a consequence, lactoferrin operations are relatively expensive.

Besides the price, the functionality of an ingredient is of crucial importance too. In Figure 7.12 the functionality is expressed by the 'dosage level', which is shown on the horizontal axis. On the vertical axis, the 'acceptable price' of an ingredient is provided. A logarithm scale is applied in both cases. At a 1% dosage level, a price of a few dollars/pound is acceptable. At an addition of 0.01% the possibilities range from \$100 to \$1000/pound. The lower the required dosage level, the higher the price the end user is willing to pay. The type of functionality is also important – there is a

[1] The '0.6 rule', expressed as $C_1/C_2 = (V_1/V_2)^n$, gives the relationship between the increase in equipment cost (C) and the increase in capacity (V), where n denotes the scale coefficient (= 0.6).

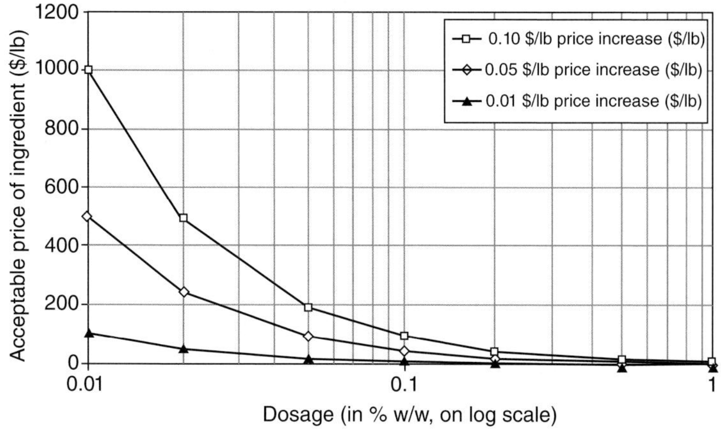

Figure 7.12 Relationship between dosage percentage and ingredient price level.

big difference between combating flu and cancer. In both cases it is important that research can substantiate the benefits.

7.2 Feed products

In this section the focus will be on the young animal. Human milk is mentioned as a reference, its replacement is discussed in Section 7.3.1 (Baby food). Young ruminants – calf and goat kid – will be considered as examples, as well as the piglet. The latter is born from the single-stomached pig, which follows a different feeding regime than ruminants.

Table 7.4 gives examples of the milk composition of humans, cows (Holstein breed, US), goats and pigs (Hurley, 2009).The whey protein/casein ratio is derived from Jenness and Sloan (1970). Due to the forces of evolution it is obvious that the milk is tailor-made for the species concerned. Piglets show the highest growth rate;

Table 7.4 Rate of growth of young animal, mature milk composition of various species and protein composition of milk replacers.

		Milk**						Milk replacer-model recipes
Mammal	Days required to double birth weight*	Total solids	Fat	Lactose	Ash	Protein total	Whey protein/ casein***	Protein composition
Human	180	12.6	4.5	6.8	0.2	1.1	60/40	60/40 whey protein/casein
Cow	47	12.2	3.5	4.9	0.7	3.1	18/82	89/11 whey protein/vegetable protein
Goat	22	12.0	3.5	4.6	0.8	3.1	14/86	100/0 whey protein/vegetable protein
Pig	14	19.9	8.2	4.8	0.6	5.8	42/58	15/85 whey protein/vegetable protein

Sources: *Webb et al. (1983), **Hurley (2009), ***Jenness & Sloan (1970),

their birth weight is doubled in about 14 days, while the human species requires about 180 days for the same increase (Webb *et al.*, 1983). Pig's milk has by far the highest total solids, protein and fat content. Cow's and goat's milk occupy an intermediate position with a lower lactose and a higher ash content than human milk. These compounds affect the osmolarity. The osmotic pressure of the milk of the animal concerned is equal to that of the blood, a factor that has to be considered in the formulation of feed for the animal.

Due to the fact that the mature milk of ruminants is sold after the neonates have received their mother's colostrum for a period of 3 days following their birth, the young calf and goat are subsequently fed with milk replacers. As a consequence of the increased cost of traditional ingredients and the availability of new by-products, milk formulations have changed considerably. Nowadays, both 'all-milk protein milk replacers' share the market with 'alternative-protein milk replacers'. The 'all-milk protein milk replacers' contain for instance skim milk powder, WPC, dried whey and delactosed whey as protein sources (Krishnamoorthy & Moran, 2011). The whey protein/vegetable protein ratio of an 'alternative-protein milk replacer' is given in Table 7.4. The 'model recipes' of the ingredients for calf (veal), goat kid and piglet milk replacers are provided in Table 7.5.

For economic reasons, in many cases the milk replacers are based on vegetable fat (palm oil, coconut oil) and do not contain casein. The casein is replaced by whey protein/vegetable protein as shown in Table 7.4. Whey protein is the most important protein source and the contribution of vegetable materials such as wheat or soya protein is limited or even avoided (goat kid). The presence of growth inhibitors and less favourable digestibility play a role in the restricted use of vegetable proteins. In case of piglets, weaning takes place after 3 weeks. This practice is also driven by economic pressures to increase the productivity of the sow and the farrowing facilities. In the early weaning period, milk-source ingredients are a part of the milk replacers. Their main advantages are high digestibility and a favourable palatability.

Table 7.5 Examples of milk replacers for early weaning of young animals.

	Calf veal	Goat kid	Piglet
Age of weaning (days)	3	3	21–28
Dairy ingredients (%)			
WPC 34	35	53	–
Delactosed whey	5	–	15
Dried whey	39	29	–
Vegetable ingredients (%)			
Fat	17	17	2
Protein concentrates	3	–	9
Grain	–	–	71
Premix			
Vitamins, minerals and amino acids	1	1	3

WPC, whey protein concentrate.

As indicated in Table 7.4, piglets can tolerate the highest share of vegetable protein. Furthermore, whey provides lactose and minor milk constituents that are important contributions to the feed (EWPA, 2007). In this way early weaning is still possible while the growth rate of the young animal is maximized and digestive disorders are minimized. Manipulation of the carbohydrate composition of the diet is a promising way to improve gut health in weaned pigs (Lallès *et al.*, 2007). In this respect galacto-oligosaccharides (GOS) could be helpful. Growth was enhanced in piglets fed a diet free of antimicrobials and supplemented with 0.03% GOS (Tsukahara *et al.*, 2010). For calves a distinction can be made in formulas for breeding and veal production. In the latter case, the amount of roughage is limited during the fattening period. As piglets are weaned at a later stage than calves and goat kids, their digestive system tolerates a higher share of non-dairy ingredients. The ingredients WPC 34, whey powder, delactosed whey or permeate form an important outlet for the dairy industry and are also extremely useful in closing the cycle.

7.3 Food products

The field of human food applications is broad and complex. Discussion can be simplified by considering a limited number of categories. Ten food categories are selected based on the application part of the information sheets. A survey of the 10 categories is given in Table 7.6. Each category presents several products in which ingredients from milk are used. Some of these categories have a rather long history, such as confectionery, bakery products, soups and sauces and meat products. Baby foods are an important driver for nutritional applications, and have benefited from the use of lactose, for instance. 'Nutrition/supplements' is a later development.

Table 7.6 Product diversification in various food segments based on information sheets.

Food segment	Products
Baby food	Infant formulas, follow-on formulas
Nutrition/supplements	Dietary, sport, slimming and geriatric supplements; clinical/sip feeding; health foods/bars
Beverages	Clear beverages, fruit juices, cream liqueurs
Confections	Milk chocolate, toffees, caramels, chocolate bars
Bakery products	Crackers, bread, puffed pastry products, sandwiches, burgers, breakfast cereals, croissants, pound cake, pastries
Meat products	Fish products, processed meats
Soups and sauces	Soups, sauces, dressings, spreads
Savoury snacks	Crisps/chips, coatings, soufflés, fried snacks
Dairy (type) products	Reconstituted milk, concentrated milk, flavoured milk drinks, yoghurt, spreadable butter, ice cream, standardization of milk powder, natural cheese, processed and imitation cheese, coffee whiteners, whipped toppings, fat blends, desserts and frozen desserts
Convenience meals	Meal toppings, pizzas, pasta, salads, gratin dishes

Table 7.7 Some important fat-free components (g/L) in cow's and human milk.

	Cow's milk	Human milk
Protein	34	10
Casein	28	4
Whey	6	5.5
α-Lactalbumin	1.2	1.5
β-Lactoglobulin	3.0	–
Lactoferrin	0.01–0.05	2.0
Lactose	48	71
Oligosaccharides	0.03–0.06	3–8
Minerals	7	2

Source: Adapted from Playne *et al.*, 2003; Thompkinson *et al.*, 2007.

The category 'Dairy (type) products' is more complicated. It includes regular dairy products as well as their spin-offs. Examples are coffee cream and whiteners, whipped cream and toppings, natural cheese and processed cheese. Beverages, convenience meals and savoury snacks are more recent. Each of the ten categories will be discussed in the following paragraphs.

7.3.1 Baby food

As human milk is the only source of nutrients during the neonatal period it is generally assumed that it is the 'gold standard' for the development of baby food formulas. Due to its availability, cow's milk is used as a starting material. Cow's milk has been 'humanized' to make its composition resemble that of human milk (Thompkinson & Kharb, 2007).

Table 7.7 shows differences in the fat-free part of both milks (Playne *et al.*, 2003). Human milk has a lower protein content, a higher whey protein/casein ratio (60/40 for human milk cf. 20/80 for cow's milk), a higher lactose content, a lower mineral percentage and 50–100 times the amount of lactoferrin and oligosaccharides found in cow's milk. The smaller amount of casein in human milk helps to produce a softer and more flocculent curd, which remains for a shorter time in the stomach leading to faster digestion in the gastrointestinal tract. As far as the composition of whey proteins is concerned, α-lactalbumin has a more dominant position in human milk compared to cow's milk. Due to its favourable levels of tryptophan and sulphur-containing amino acids (cyst(e)ine, methionine) and its modest threonine content, α-lactalbumin can be applied to improve the amino acid pattern of baby food formulas (Table 7.8). In this respect 'reference proteins' for human milk protein are an important tool (FAO, 2011; EC, 2003). As a result the protein content of the infant food formula can be reduced. Research has shown that the safe protein level in these formulas is 1.8 g/100 kcal (Bernstorf Schröder & Hogh Petersen, 2009).

Another significant protein in human milk is lactoferrin, with its specific biological activity. It prevents the growth of pathogenic organisms in the intestine, while

Table 7.8 Some key amino acids in human milk and four bovine protein sources (g/100 g protein).

	Tryptophan	Methionine	Cyst(e)ine	Threonine
Human milk protein (reference protein*)	1.7	3.3		4.4
Bovine milk protein (milk powder)	1.3	0.8	2.7	4.4
Bovine caseine (caseinate)	1.4	0.4	3.0	4.6
Bovine whey protein (WPC 80sw)	2.2	2.5	2.3	7.9
Bovine α-lactabumin	4.8	4.8	1.0	4.3

WPC, whey protein concentrate.
**Source:* Reference protein for infants (FAO, 2011).

stimulating the immune system and regulating iron metabolism. Lactoferrin is used in several baby food formulas in Asian countries (see information sheet 'Lactoferrin'). In recent studies more than one hundred host defence-related milk proteins were revealed in cow's milk. Besides lactoferrin, lactoperoxidase and lysozyme a range of additional antimicrobial proteins and peptides were found. Amongst them are the cathelicidins. These peptides have an extremely potent and broad-spectrum antimicrobial activity, and are significantly more potent than lactoferrin (Wheeler *et al.*, 2012). Hettinga *et al.* (2011) further studied the differences in the host-defence proteomes of human and bovine milk. Hundreds of proteins were identified, of which several were related to the host defence system. Thirty-three proteins were found in both species. Better understanding of the function of milk in the development of the immune system might result in new ingredients that could be beneficial for 'humanized' baby food.

The carbohydrate make-up of human milk is also quite different, when considering the oligosaccharide fraction. The use of bovine GOS (see Figure 7.4) is an approach to mimic the functionality of the oligosaccharide fraction of human milk. GOS have a prebiotic effect that stimulates the development of beneficial bifidobacteria and lactobacilli present in the intestine. The composition of bovine GOS is less complex than its human counterpart, which contains about 130 different oligosaccharides. Sialic acid is a major part of human oligosaccharides. A concentrate of a sialic acid-containing oligosaccharide of bovine origin has been isolated (Holst *et al.*, 2007). The dried ingredient (7% sialyllactose) is a next step in mimicking the composition of human milk, for instance.

Allergy may be one of the problems encountered in applying bovine ingredients to baby foods (Monaci *et al.*, 2006). Food allergy is an abnormal immunological response due to sensitization to a food or food component. The most common foods causing an allergic reaction are cow's milk, egg, soya, wheat, peanut, tree nuts, sesame and fish. For this reason, milk has to be declared on the label of the product. The incidence of cow's milk allergy (CMA) is 2–3% in the first year. It is a complex disorder that usually disappears by the age of 3 years. The prevalence of the allergic reaction varies with dietary habits of various countries. CMA ranges from second to fourth in published allergy lists (Dalal *et al.*, 2002). Milk allergy is an adverse reaction to the proteins present in milk, including casein and whey proteins. The adverse reaction can be reduced by hydrolysis of the protein (see information sheet 'Casein hydrolysate'). Extensive enzymatic hydrolysis of casein combined with adequate removal of the residual high-molecular-weight substances is undertaken to reduce the risk of an allergic reaction. Its effectiveness is strictly regulated in the European nutritional standards for infant formulas (EC, 1996). In hypersensitive infants the allergenicity of the proteins, from which the hydrolysate is made, needs to be reduced by at least 90%.

7.3.2 Nutrition/supplements

According to EU regulations, food supplements are concentrated sources of nutrients or other substances with a nutritional or physiological effect whose purpose is to supplement normal diets. Dairy supplements are used for various purposes, whether in sport, 'senior' or clinical nutrition or weight management.

Milk proteins are an important source of essential amino acids needed for growth and tissue repair. These proteins can be easily digested. Some milk proteins are even available in a predigested form, for instance, as whey protein hydrolysates (see information sheet 'Whey protein hydrolysate: WPH 80'). The essential amino acids form about half of the total amino acids. Overall the whey protein concentrates have a slightly more favourable amino acid profile than caseinates (see Section 2.3.2). Calcium caseinate combines an enhanced calcium content with a good amino acid profile and is used in clinical nutrition. The branched chain amino acids – valine, leucine and isoleucine – play a pivotal role in muscle metabolism. In this respect whey proteins are favourable; they contain about 30% more of these amino acids than caseinate or soya protein isolate. Whey proteins are therefore applied in sport and 'senior' nutrition (Cribb, 2005; Kreider, 2005). In the latter case, they are helpful in compensating for age-related loss of muscle mass. Furthermore, whey proteins are a rare source of cysteine and therefore also of interest to sport and senior nutrition. The cysteine concentration is about 2–5 times higher than in caseinate and soya protein isolate. Cysteine boosts production of glutathione which is the centrepiece of the body's antioxidant defence system. In the case of weight management, satiety is an important aspect in which dietary protein plays a role. A specific effect of glycomacropeptide in appetite suppression might be helpful (Thomä-Worringer *et al.*, 2006). The peptide is split off by rennet during cheese making and can be isolated from cheese whey (see Section 2.3.3).

Milk fat globule membrane ingredients are a new source for nutrition/ supplements. The concentrated MFGMs are rich in potential ingredients. They differ in composition from other sources such as soyabean lecithins and egg lecithins, as shown in Table 7.9 (Burling & Graverholt, 2008).The milk source has a higher sphingomyelin (SM) and phosphatidylserine (PS) content. SM participates in the regulation of the uptake of cholesterol and PS is claimed to be an active health-promoting compound for improved cognitive performance.

Table 7.9 Composition of phospholipids, expressed in molar %, from milk, egg and soya.

Phospholipid constituent	Soya	Egg	Milk
Phosphatidylcholine (PC)	34	75	27
Phosphatidylethanolamine (PA)	21	15	25
Phosphatidylinositol (PI)	18	0.4	8
Sphingomyelin (SM)	0	1.5	24
Phosphatidylserine (PS)	0.5	0	12
Phosphatidic acid (PA)	9	0	0
Others	17.5	8.1	4

Source: Burling & Graverholt, 2008.

7.3.3 Beverages

A great variety of shelf-stable beverages for protein fortification are available today. Chocolate beverages that have a neutral pH need to be sterilized. Consequently, caseinate or MPC are usually preferred. In acidified beverages, which are subjected to a mild pasteurization, whey proteins are often the ingredient of choice. These proteins are stable over a wide pH range. For acidified drinks with a high clarity usually the low-fat whey protein isolate (WPI) is selected. In the most expensive drink segment, the alcohol-containing cream liqueurs are present. In these dairy-based oil-in-water emulsions, sodium caseinate is used as an emulsifier. The caseinate enhances the long-term stability of the cream compared to skim milk. The sodium caseinate production requires accurate process control, particularly during heat treatment of the protein suspension (Lynch & Mulvihill, 1997).

Besides the proteins, 'milk calcium' is applied to enrich beverages. This whey fraction has a high bioavailability and is used, for instance, to build up sufficient bone density in the young and to mitigate against osteoporosis in seniors.

7.3.4 Confections

'Confections' is the collective term applied to products usually compounded of sugar. Some examples of this category are provided in Table 7.6. Functional dairy ingredients are an important part of these recipes, and a product such as milk chocolate has a long history. A survey of 12 dairy ingredient possibilities for milk chocolate is set out in Figure 1.2 and discussed in Section 1.1. A combination of skim milk powder and anhydrous milk fat (AMF) is attractive from the point of view of process flexibility. In these cases usually a high-heat skim milk powder is applied. This ingredient has a low bacterial count, which is important for the shelf life of confections. The heat treatment is also beneficial for the development of a cooked flavour. The percentage of milk-solids non-fat (MSNF) in confections varies from 10 to 25% (Bouzas, 1999). Cost savings can be obtained and a high quality level maintained by using whey-based ingredients up to 5% of the total cocoa mass. Examples of such ingredients are WPC 34 and demineralized whey powder. Lactose can feasibly be applied for its reduced sweetness and can accentuate the chocolate flavour. Because lactose is also a reducing sugar it has a positive effect on colouring products like candy, gum, milk chocolate and coatings. This is a result of the Maillard reaction, which improves taste and mouth feel as well. As far as the fat component is concerned, fat fractions with higher melting points are more effective in preventing loss of gloss in dark chocolate (Haylock & Dodds, 2009). This effect is called bloom inhibition. A new application is nutritional bars with a maximum of 50% protein. Combinations of WPC 80/WPI and hydrolysed whey proteins are used to maximize protein content and minimize bar hardening (Burrington, 2007). Hydrolysed whey proteins do not bind water and are useful in maintaining a softer texture. Furthermore, whey proteins provide an attractive amino acid profile.

7.3.5 Bakery products

Dairy ingredients can be applied in a number of bakery products (see Table 7.6). Casein/caseinates are used to increase the lysine content in breakfast cereals and biscuits, for instance. Replacement of wheat gluten is, however, difficult due to the fact that its properties are not similar enough to those of milk protein. WPCs are used in, for instance, biscuits, cookies, sponges and cakes to improve texture and crust colour (Burrington, 1999). In cake recipes, part of the egg products can be substituted by WPC 80/WPI. They are better suited for the replacement of egg white than whole eggs. When whole eggs are used, the results can be improved if fat and WPC are pre-emulsified (de Wit, 2001). In pastry recipes part of the sucrose can be replaced with lactose, and in that case a reduction in fat of up to 50% is possible without detracting from the taste (EWPA, 2000). The lactose can absorb more water than sucrose, consequently the texture remains tender in spite of the lower fat content.

7.3.6 Meat products

Quality improvement and cost-effectiveness are key drivers for the use of dairy ingredients in meat products. Minced, processed meat products like sausages, frankfurters and luncheon meats are their main applications. Sodium caseinate is used as an emulsifier. Due to its heat stability the emulsion can be maintained at higher temperatures. As the caseinates are preferentially adsorbed by the fat globules, meat proteins are more available to assist in binding of the water. With respect to WPC no single type of WPC can be used for all meat applications. It is necessary to match the application requirement with the specific functionality (Prabhu, 2006). The gel-forming properties of WPC can be attractive. WPC supports the network formed during the cooking of the product and prevents moisture loss. An example is the use of a high-gel WPC in the seafood surimi (Nopianti *et al.*, 2011), in which it replaces the traditional egg white. The high-gel WPC is preferably derived from casein whey. The gel-forming properties are improved due to the absence of glycomacropeptide (see information sheet 'WPC 80_{aw}'). Another application of milk ingredients is whole-muscle meat products (e.g. cooked hams, poultry breast). WPC, soluble over a wide pH range, is used in 'injection brines' to add weight and tenderness.

Lactose, being a reducing sugar, is applied in sausages to enhance browning during frying. Furthermore, lactose is able to mask salty and bitter flavours without causing too much of a sweet taste.

7.3.7 Soups and sauces

Milk proteins are widely used in 'convenience foods'. In this category gravies and dressings can also be included. The heat-stable caseinates are preferably used as emulsifying agents and viscosity controllers in canned cream soups and sauces; they improve creaminess and texture. Whey protein-caseinate blends are used as whitening agents in gravies. They are a cheap replacement for skim milk powders. Whey proteins are particularly suitable in a low-pH environment such as salad dressings. Another possibility is the use of specially treated WPC's that form gels simply by adding salt (see Figure 2.5). These cold-set gel WPC's can, for instance, be used in low-fat salad dressings and mayonnaise type products (Johnson, 2000b).

7.3.8 Savoury snacks

Savoury snacks are produced by topical coating of the 'product' with flavoured seasonings. Coating can be carried out by dusting the surface of a fried snack, spraying of a flavour suspended in oil or oil spraying of the baked product followed by dusting its exterior. For the flavouring of savoury snacks, various dairy ingredients can be applied such as whey powders (acid, sweet), WPC, skimmed milk or buttermilk powder, cheese powders and EMC (Johnson, 2000a). Whey powder provides uniform coating with excellent visual appeal. Due to the presence of salts it enhances the savoury character. Cheese powder and EMC provide the specific cheesy notes, such as Cheddar or Parmesan. Both whey powder and EMC are also cost efficient as they can partially replace the more expensive cheese powders.

7.3.9 Dairy (type) products

This category includes a range of dairy products as well as some of their non-dairy replacements. The latter are based on vegetable fat.

One of the alternatives for creamers is a powdered coffee whitener in which sodium caseinate acts as an emulsifier. Small fat droplets are created, which play an important role in whitening of coffee. Topping powders can be a shelf stable alternative for whipped cream. Sodium caseinate is one of the emulsifiers used in toppings. The caseinate is particularly useful during the manufacture itself and stabilizes the emulsion prior to drying.

Ice cream and other frozen dairy desserts are an important field of applications. The development of processes and recipes for ice cream is described in Section 1.1. The recipe started with three constituents and became a complex multi-ingredient-system over the course of time (see Figure 1.1). One of the forerunners in partially replacing the milk solids non-fat (MSNF) was whey powder. Due to the risk of 'sandiness' – caused by lactose crystallization – the replacement of MSNF was limited to 25% in the USA (CFR, 1997). Nowadays, WPC 34 is more commonly applied and the substitution of the naturally occurring casein in MSNF can be increased to about 50% on a protein-to-protein basis. At this level the unique functional properties of casein – body, smoothness and resistance to heat shock – are retained (Young, 2007). In case of frozen dairy desserts there is no 'standard of identity', so freedom of formulation is greater. By keeping the lactose content of the total mix below 7.5%, 'sandiness' in the end-product can be avoided.

In case of yoghurt, industry sometimes utilizes caseinate/WPC blends, which subsequently changes the casein/whey protein ratio. The blend is added to the starting milk, which is then heated to 90°C/5 min. In this way an optimum with respect to firmness, smoothness and reduction of whey separation can be obtained. The structural and sensory properties are even more critical in low-fat yoghurts. Evidence is available that stirred low-fat yoghurt with 4.5% protein and a casein-to-whey protein ratio of 60/40 provides attractive textural properties comparable to the characteristics of full-fat yoghurt (Krzeminski *et al.*, 2011). However, the addition of whey protein did not enhance the fat-related sensory properties of the stirred low-fat yoghurt (Tomaschunas *et al.*, 2012). The important role of milk fat could not be completely compensated.

Figure 7.13 provides an example of how an ingredient can be efficiently used in the process of the end user. In the butter-making process cooling, heating and ripening

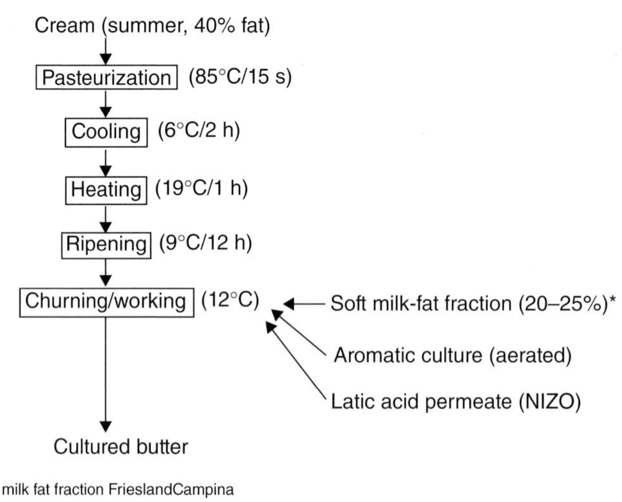

Figure 7.13 Application of soft milk-fat fraction in spreadable butter.

are applied in such a way that the hardness of the butter is minimized (Walstra *et al.*, 2006). A butter with improved spreadability and suitable for use direct from the refrigerator can be produced by adding a soft, or low-melting-point, anhydrous milk fat fraction. This fraction is 'worked' into the fresh butter immediately after churning. The percentage added can be adjusted according to the wishes of the end user and the properties of the cream available.

In conventional cheese production as well as in the manufacture of recombined cheese, MPC is applied in order to enhance the protein content of the milk and standardize the fat/protein ratio. A constant quantity of cheese can be produced from the existing equipment in this way. Furthermore, the output is independent of the seasonal variations in the protein content of the milk.

7.3.10 Convenience meals

Convenience meals are the most complicated of the 10 food categories considered here. These products are intended to serve as snacks or the main course of a meal, but are produced outside the household. Costa *et al.* (2001) distinguished four convenience classes:

1 ready to eat (e.g. chilled sandwiches and salads);

2 ready to heat (e.g. chilled and frozen pizzas);

3 ready to end-cook (e.g. chilled and frozen lasagne);

4 ready to cook, a lesser known class, that has been minimally prepared for cooking.

Olsen *et al.* (2012) investigated the health aspects of the ready-to-heat meals. Their results indicate that the likelihood of buying healthy convenience food is primarily affected by the sensory product attributes such as appearance, flavour and odour.

Ingredients from milk can contribute in various ways to the convenience and preference of a meal. Sometimes the contribution is complex, for instance in case of a ham or cheese burger. As set out in Table 7.6, the bread may contain a dairy ingredient, as can the meat and sauce. The cheese slice is a more visible component of the burger and certainly contributes to the overall appeal of these products. In the following part, special attention will be paid to the cheese ingredients.

Cheese is available in various forms, such as natural cheese, processed cheese, imitation cheese and cheese powder/EMC. They are useful ingredients for different applications (Figure 7.14). The choice of a cheese ingredient is based on various considerations. Natural and processed cheeses are chosen for texture and flavour, whereas cheese powder/EMC is predominantly selected for flavour reasons. In case of imitation cheese, costs are the driving force.

An impression of the composition of an imitation cheese is given in the ingredient declaration of Table 1.4, while its process is reviewed in Figure 1.3. Natural

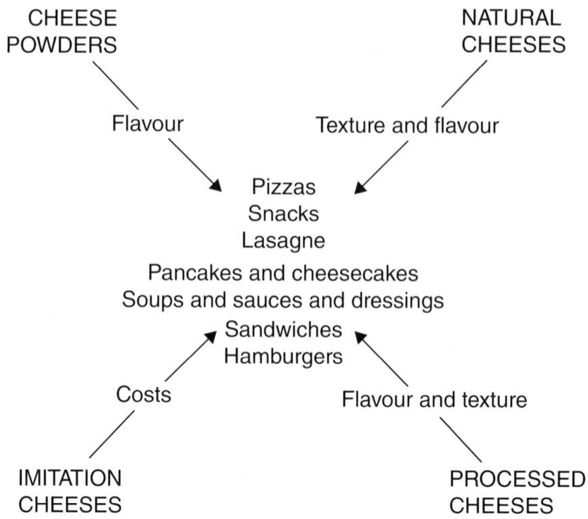

Figure 7.14 Examples of applications of various cheese ingredients.

cheese is highly preferable but there are a few disadvantages. The melt of natural cheese is, for instance, less predictable during heating, while oiling-off can also take place. Furthermore, its shelf life is limited. For these reasons, processed cheese is sometimes selected. Four main variants of processed cheese are described in the information sheets. The non-melt hard variant is suitable for gratin dishes. In sandwiches and cheeseburgers the semi-melt hard processed cheese is used. The high-melt hard variant is usually utilized for the manufacture of pizzas, while high-melt soft processed cheese is used in sauces, for instance. A variety of recipes are described in the literature (Berger *et al.*, 2002). The composition (moisture and fat content), the emulsifying agents and the selected raw materials play a crucial role in obtaining the desired characteristics. The dosage level of emulsifying agents is adjusted to the calcium content of the cheese applied (Table 7.10).

The texture and flavour are a compromise between the use of a young and medium-ripe cheese. They have respectively an 'intact casein' level of 90–95% and 70–90%. It is necessary to have a minimum quantity of 'intact casein' otherwise separation

Table 7.10 Some natural cheeses: typical composition, function and flavour.

Cheese type	Moisture (%)	Flat/TS (%)	Protein (%)	Calcium (%)	Function	Favour
Parmesan	18	0.35	42	1.3	Grated	Rich/sharp
Swiss	37	0.45	28	0.9	Melted	Sweet/mellow
Cheddar	37	0.5	25	0.7	Shredded processed	Mild-sharp
Gouda	38	0.5	24	0.8	Sliced	Mild-rich
Mozzarella (LMPS)	48	0.3	28	0.7	Stretched	Mild/bland
Feta	55	0.5	14	0.5	Crumbled	Salty/strong
Cream cheese	55	0.7	9	0.1	Spread	Cream/acid

Source: Data from MILKingredients, ca; Nevo table (Nederlands Voedingsstoffenbestand) 2000; Harbutt 2009.

will occur during the melting process. The minimum value of 'intact casein' in processed cheese needs to be at least 12%.

Natural cheese with its balanced, subtle taste is a preferred source; its applications in meals are very specific. The flavour and function of seven popular ingredient cheeses as well as their composition are indicated in Table 7.10 (MILKingredients.ca; NEVO, 2000; Harbutt, 2009). The differences in texture are vast. The hard Parmesan cheese can be grated and is applied as a topping. The crumbly texture of feta cheese is particularly suitable for salads. On the other hand the excellent stretch properties of mozzarella cheese are very useful in high quality pizzas. Cream cheese is an excellent source if a spreadable ingredient is required. The differences in cheese structure are the result of the process chosen. Moisture and fat content are useful in influencing the hardness of a cheese. Another essential parameter in determining cheese texture is the calcium content (Lucey, 2008). By manipulating the calcium level the cheese texture can be made suitable for stretching or crumbling, for instance.

7.4 Pharmaceutical products

Pharmaceutical lactose is a high value-added application of whey solids. Its share of the total lactose market is modest, amounting to approximately 10%. This ingredient has to meet the high purity standards of pharmaceutical excipients as described in the various pharmacopoeias. Pharmaceutical lactose is particularly applied for the preparation of tablets. It is the most widely used ingredient for tablets, capsules, etc. in medicine, which may contain up to 80% lactose. Advantages are a bland taste, low hygroscopic properties and good compatibility with other ingredients. This form of lactose is, for instance, the most common constituent in sleeping pills (see Table 1.8). Its manufacture is mentioned in Section 7.1.2. Four types of pharmaceutical lactose are described (they are also included in the information sheets) as: monohydrate, anhydrous, spray dried and inhalation lactose.

In Figure 7.15 the two main processes for the production of tablets are demonstrated (DMV-Fonterra, 2005). The traditional approach is wet granulation. Granulation can be defined as the production of aggregates from smaller particles by the use of mechanical energy and/or binding agents. Liquid, usually water, is added. The binder solution is mixed with various ingredients, dried and finally compressed into tablets. In this multi-stage process, α-lactose monohydrate is applied as a filler. For this purpose, the finer milled grades of α-lactose are used. The number of process steps can be reduced considerably by applying the direct compression process, which is also shown in Figure 7.15. In this case two other types of lactose are used as filler-binder. The roller-dried variant is more efficient in the production of tablets than the common α-lactose monohydrate. This anhydrous lactose has a good compressibility as well as a high solubility. The enhanced solubility is caused by the presence of a high percentage of β-lactose. The roller-dried variant is commonly utilized in formulations in which moisture-sensitive drugs are used. Spray-dried lactose combines good flowability with a favourable tablet compressibility. The latter is based on the presence of a small proportion of amorphous lactose.

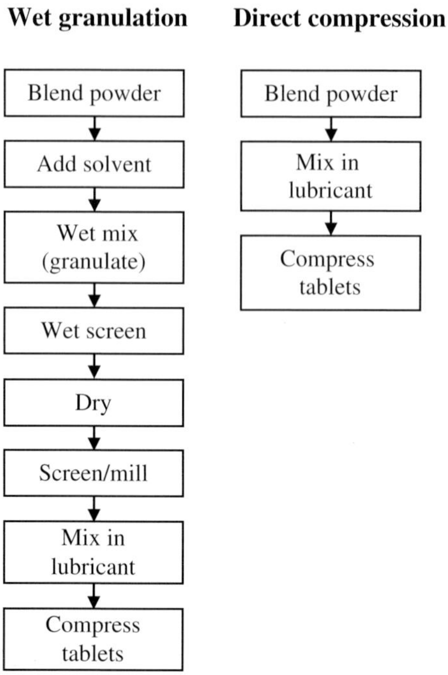

Figure 7.15 Schemes for two pharmaceutical tablet production processes. *Source:* DMV-Fonterra, 2005. Courtesy of K. Kussendrager.

The use of lactose is not limited to tablets; it can also be applied in capsules and sachets. Good flow properties are required and therefore the coarser variants of α-lactose monohydrate are used. The finer grades of lactose can be used in inhalation delivery systems, for instance as used by asthma patients (te Wierink *et al.*, 2002). The powder – i.e. the drug and its lactose carrier – is administered via the lungs. The therapeutic drug, which is a few micrometres in diameter, is attached to α-lactose monohydrate particles that are considerably bigger. These coarse lactose particles increase powder flow, facilitate drug aerosol formation and act as a bulking agent for uniform filling. Optimal delivery is generally obtained by mixing a relatively coarse and very finely ground lactose, as indicated in the information sheet 'Lactose inhalation pharma'. According to theory the very fine lactose particles force the drug to weaker binding sites on the surface of the coarse carrier particles. As a result the drug particles are more easily liberated from the surface of the coarse carrier particles after actuation (Xu *et al.*, 2011). Inhalation lactose requires special sieving and/or milling techniques and is the most expensive custom-made lactose variant.

References

Angus F, Smart S & Short C (2005) Prebiotic ingredients with emphasis on galacto-oligosaccharides and fructo-oligosaccharides. In: Tamine AY (ed.), *Probiotic Dairy Products*. Blackwell Publishing, pp. 120–37.

Australian Society of Dairy Technology (1972) *Casein Manual*. ASDT, pp. 25–7.

Bennett M (2012) On the management of innovation in sugar technology: A personal tribute to Tate & Lyle's ventures into R&D. *International Sugar Journal* **114**: 235–41.

Berger W, Klostermeyer H, Merkenich K & Uhlmann G (2002) In: *Processed Cheese Manufacture*. BK Giulini Chemie, pp. 200–30.

Bernstorf Schröder J & Hogh Petersen S (2009) What is the optimum protein level in infant formulas? *AgroFOOD industry hi-tech* **20**: 22–5.

Bouzas J (1999) Whey products and lactose in confectionery applications. *Applications Monograph–Confectionery*. US Dairy Export Council (http://innovatewithdairy.com).

Burgess KJ (2001) Milk fat as ingredients. *International Journal of Dairy Technology* **54**: 56–60.

Burling H (1979) Method and apparatus for desalination of whey. US 4,159,350 [patent].

Burling H (1994) Isolation of bioactive components from cheese whey. *Scandinavian Dairy Information* **8**: 54–6.

Burling H & Graverholt G (2008) Milk – a new source for bioactive phospholipids for use in food formulations. *Lipid Technology* **20**: 229–31.

Burrington K (1999) Whey products in baked goods. *Applications Monograph–Bakery*. US Dairy Export Council (http://innovatewithdairy.com).

Burrington KJ (2007) US whey ingredients in nutrition bars and gels. *Applications Monograph–Bars and Gels*. US Dairy Export Council.

Burrington KJ (2011) Permeate for sodium reduction. *Applications Monograph–Permeate*. US Dairy Export Council (http://www.innovatewithdairy.com).

Business Consulting 3A (2008) The world market for whey and lactose products 2006 to 2010. *European Dairy Magazine* **1**: 26–8.

Cegielska-Radziejewska R, Leśierowski G & Kijowski J (2008) Properties and application of egg white lysozyme and its modified preparations – a review. *Polish Journal of Food and Nutrition Sciences* **58**: 5–10.

CFR (Code of Federal Regulations) (1997) 7CFR 58.2825 – United States Standard for ice cream.

Costa AIA, Dekker M, Beumer RR, Rombouts FMN & Jongen WMF (2001) A consumer-oriented classification system for home meal replacements. *Food Quality and Preference* **12**: 229–42.

Cribb P (2005) Whey proteins and body composition. *Applications Monograph–Body Composition*. US Dairy Export Council.

Dalal I, Binson I, Reifen R *et al.* (2002) Food allergy is a matter of geography after all: sesame as a major cause of severe IgE-mediated food allergic reactions among infants and young children in Israel. *Allergy* **57**: 362–5.

Depeint F, Tzortis G, Vulevic J, I'Anson K & Gibson GR (2008) Prebiotic evaluation of a novel galactooligosaccharide mixture produced by the enzymatic activity of *Bifidobacterium bifidum* NCIMB 41171, in healthy humans: a randomized, double-blind, crossover, placebo-controlled intervention study. *American Journal of Clinical Nutrition* **87**: 785–91.

Dewettinck K, Rombaut R, Thienpont N, Le TT, Messens K & Van Camp J (2008) Review: Nutritional and technological aspects of milk fat globule membrane material. *International Dairy Journal* **18**: 436–57.

DMI (1996) *Dairy Ingredient Application Guides: Milk fat and butter ingredients/Concentrated and dry milk ingredients/Whey ingredients*. Dairy Management, Inc.

DMV-Fonterra (2005) *Introduction to Lactose New Zealand – Pharmaceutical Lactose*. Lactose New Zealand.

Doultani S, Turhan KN & Etzel MR (2004) Fractionation of proteins from whey using cation exchange chromatography. *Process Biochemistry* **19**: 1737–43.

Dwyer JL (1984) Scaling up bio-product separation with high performance liquid chromatography. *Bio /Technology* November, pp. 957–64.

EC (1996) Commission directive 96/4 EC of 16 February 1996 amending directive 91/321/EC on infant formulae and follow-on formulae. *Official Journal of the European Commission* **49**: 12–16.

Ekhart PF & Timmermans E (1996) Techniques for the production of transgalactosylated oligosaccharides (TOS). *International Dairy Federation Bulletin 313*, pp. 59–64.

EWPA (2000) *Milk Sugar, a Natural Multi-purpose Ingredient, the Benefits of Using Milk Sugar in Pastries and Plain Cakes*. European Whey Products Association.

EWPA (2007) *Whey in Animal Nutrition – A Valuable Ingredient*. European Whey Products Association.

Foegeding EA & Davis JP (2011) Food protein functionality: A comprehensive approach. *Food Hydrocolloid* **25**: 1853–64.

González-Chávez SA, Arévalo-Gallegos S & Rascón-Cruz Q (2009) Review, Lactoferrin: structure, function and applications. *International Journal of Antimicrobial Agents* **33**: 301.e1–301.e8.

Harbutt J (ed.) (2009) *World Cheese Book*. Dorling Kindersley Ltd.

Haylock SJ and Dodds TM (2009) Ingredients from milk. In: Beckett ST (ed.), *Industrial Chocolate Manufacture and Use*. Blackwell Publishing, pp. 76–100.

Hettinga K, Valenberg van H, de Vries S, van Hooijdonk T, van Arendonk J & Vervoort J (2011) The host defense proteome of human and bovine milk. *PLoS ONE* **6**(4): e19433, pp. 1–8.

Holst HH, Gunther WS, Mogensen MT & Jorgensen AS (2007) Concentrate derived from a milk product enriched in naturally occurring sialyllactose and a process for preparation thereof. US 2007/0104843A1 [patent].

Houldsworth DW (1980) Demineralization of whey by means of ion exchange and electrodialysis. *Journal of Society of Dairy Technology* **33**: 45–51.

Hurley WL (2009) *Milk Composition and Synthesis*. Resource Library – University of Illinois Urbana, Champaign.

Jenness R & Sloan RE (1970) The composition of milks of various species: a review. *Dairy Science Abstracts* **31**: 599–612.

Johnson AH (1980) The composition of milk. In: Webb BR, Johnson AH & Alford JA (eds), *Fundamentals of Dairy Chemistry*. AVI Publishing Co., p. 7.

Johnson B (2000a) US whey products in snacks and seasonings. *Applications Monograph–Snacks and Seasoning*. US Dairy Export Council; available at: http://www.usdec.org/files/pdfs/10snacks.pdf

Johnson BR (2000b) Whey protein concentrates in low-fat applications. *Applications Monograph–Low Fat Applic*ations. US Dairy Export Council (http://www.usdec.org).

Kilcawley KN (2002) Studies on the biochemistry and technology of enzyme-modified Cheddar cheese. PhD thesis, University of Cork.

Kilcawley KN, Wilkinson MG & Fox PF (1998) Review: Enzyme-Modified Cheese. *International Dairy Journal* **8**: 1–10.

Kramer F (1991) *De fascinerende wereld van melk*. DMV, pp.75–82.

Kreider RB (2005) Whey proteins and senior nutrition. *Application Sheet – Seniors Nutrition*. US Dairy Export Council (http://www.usdec.org).

Krishnamoorthy U & Moran J (2011) *Rearing Young Ruminants on Milk Replacers and Starter Feeds*, pp. 27–31.

Krzeminski A, Groszhable K & Hinrichs J (2011) Structural properties of stirred yoghurt as influenced by whey proteins. *LWT – Food Science and Technology* **44**: 2134–40.

Kussendrager KD (1994) Effects of heat treatment on structure and iron-binding of bovine lactoferrin. *Bulletin of the International Dairy Federation –Indigenous Antimicrobial Agents of Milk*, pp. 133–46.

Kussendrager KD, Kivits MGC & Verver AB (1997) Process for isolating lactoferrin and lactoperoxidase from milk and milk products obtained by such a process. US 5596082 [patent].

Lallès JP, Bosi P, Smidt H & Stokes CR (2007) Weaning – A challenge to gut physiologists. *Livestock Science* **108**: 82–93.

Lamsal BP (2012) Review: Production, health aspects and potential food uses of dairy prebiotic galactooligosaccharides. *Journal of the Science of Food and Agriculture* **92**: 2020–8.

Lippe F, Ottenhof HAWE & de Boer R (1980) Continuous method for dissolving milk proteins containing casein and an apparatus for performing the method as well as water soluble milk protein product of a novel type. US 4407747 [patent].

Lucey JA (2008) Some perspectives on the use of cheese as a food ingredient. *Dairy Science and Technology* **88**: 573–94.

Lynch AG & Mulvihill DM (1997) Effect of sodium caseinate on the stability of cream liqueurs. *International Journal of Dairy Technology* **50**: 1–7.

Macfarlane GT, Steed H & Macfarlane S (2008) Review article: Bacterial metabolism and health-related effects of galacto-oligosaccharides and other prebiotics. *Journal of Applied Microbiology* **104**: 305–44.

Mallia S, Escher F & Schlichtherle-Cerny H (2008) Aroma-active compounds of butter: a review. *European Food Research and Technology* **226**: 315–25.

McSweeney SL, Mulvihill DM & Callaghan DM (2004) The influence of pH on the heat-induced aggregation of model milk protein ingredient systems and model infant formula emulsions stabilized by milk protein ingredients. *Food Hydrocolloids* **18**: 109–25.

MILKingredients.ca. Ingredient profiles: Cream cheese, Feta, Mozzarella, Cheddar cheese, Swiss-style cheese. URL: www.MILKingredients.ca

Mol JJ & Roetman R (1974) Manufacture of non-caking whey powder. *Officieel Orgaan FNZ*, pp. 458–61.

Monaci L, Tregoat V, van Hengel A & Anklam E (2006) Milk allergens, their characteristics and their detection in food: A review. *European Food Research and Technology* **223**: 149–79.

Muller LL & Hayes JF (1962) Improved equipment for continuous precipitation of acid casein. *Australian Journal of Dairy Technology* **17**: 189–92.

Mulvihill DM & Ennis MP (2003) Functional milk proteins: production and utilization. In: Fox PF & McSweeney PLH (eds), *Advanced Dairy Chemistry, Vol. 1: Proteins*. Kluwer Academic/Plenum, pp. 1175–228.

NEVO (Nederlands Voedingsstoffenbestand) (2000) Gouda cheese. RIVM (Rijksinstituut voor Volksgezondheid en Milieu), p. 74.

Nopianti R, Huda H & Ismail N (2011) A review on the loss of the functional properties of proteins during frozen storage and the improvement of gel-forming properties of surimi. *American Journal of Food Technology* **6**: 19–30.

Olsen NV, Menishelli E, Sørheim O & Næs T (2012) Likelihood of buying healthy convenience food: An at-home testing procedure for ready-to-heat meals. *Food Quality and Preference* **24**: 171–8.

Outinen M, Rantamäki P & Heino A (2010) Effect of milk treatment on the whey composition and whey powder functionality. *Journal of Food Science* **75**: E1–E10.

Playne MJ, Bennett LE & Smithers GW (2003) Functional dairy foods and ingredients. *Australian Journal of Dairy Technology* **58**: 242–64.

Prabhu G (2006) US whey proteins in processed meats. *Applications Monograph–Processed Meats*. US Dairy Export Council (http://www.innovatewithdairy.com).

Schaafsma G (2008) Lactose and lactose derivatives as bioactive ingredients in human nutrition. *International Dairy Journal* **18**: 458–65.

Sinclair HR, de Slegte J & Klarenbeek G (2007) Inhibition of cholera toxins by galacto-oligosaccharides., US 8,202,842B2 [patent].

Sinclair HR, de Slegte J, Gibson GR & Rastall RA (2009) Galactooligosaccharides (GOS) inhibit *Vibrio cholerae* toxin binding to its GM1 receptor. *Journal of Agricultural Food Chemistry* **57**: 3113–19.

Smithers GW (2008) Whey and whey proteins – from gutter-to-gold. *International Dairy Journal* **18**: 695–704.

Southward CR (1985) Manufacture and applications of edible casein products I. Manufacture and properties. *New Zealand Journal of Dairy Science and Technology* **20**: 79–101.

Thomä-Worringer C, Sørensen J & López-Fandiño R (2006) Health effects and technological features of caseinomacropeptide. *International Dairy Journal* **16**: 1324–33.

Thompkinson DK & Kharb S (2007) Aspects of infant food formulation. *Comprehensive Reviews in Food Science and Safety* **6**: 79–102.

Tomaschunas M, Hinrichs J, Köhn E & Busch-Stockfisch M (2012) Effects of casein-to-whey protein ratio, fat and protein content on sensory properties of stirred yoghurt. *International Dairy Journal* **26**: 31–5.

Torres DPM, Gonçalves M, do Pilar F, Teixeira JA & Rodrigues LR (2010) Galacto-oligosaccharides: Production, properties, applications, and significance as prebiotics. *Comprehensive Reviews in Food Science and Food Safety* **9**: 438–54.

Tribe MA & Alpine RLW (1986) Scale economies and the '0.6 rule'. *Engineering Costs and Production Economics* **10**: 271–8.

Tsukahara T, Inoue R, Shimojo N *et al.* (2010) Alpha-linked galactooligosaccharide suppresses small intestinal eosinophil infiltration and improves growth performance in weaning pigs. *Journal of Veterinary Medical Science* **72**: 547–53.

Tzortzis G (2011) Development and evaluation Bimuno, a novel second-generation prebiotic galactooligosaccharide mixture. In: Paeschke TM & Aimutis WR (eds), *Nondigestible Carbohydrates and Digestive Health*. Wiley-Blackwell, pp. 295–311.

Wakabayashi H, Yamauchi K & Takase M (2006) Review: Lactoferrin research, technology and applications. *International Dairy Journal* **16**: 1241–51.

Walstra P, Wouters JTM & Geurts TJ (2006) Dairy Science and Technology. CRC/Taylor & Francis, pp. 467–95.

Webb BH, Johnson AH & Alford JA (1983) *Fundamentals of Dairy Chemistry*. AVI Publishing Co., p. 7.

Westergaard V (2010) *Milk Powder Technology – Evaporation and Spray Drying*. GEA Niro, pp.147–59, 233–9, 269–79.

Wheeler TT, Smolenski GA, Harris DP *et al.* (2012) Host-defence-related proteins in cows' milk. *Animal* **6**: 415–22.

de Wit JN (2001) *Lecturer's handbook on whey*. EWPA, p. 54.

te Wierink H, Diepenmaat P & Damhuis R (2002) Formulation of lactose for inhaled delivery systems. *Pharmaceutical Technology Europe* November, pp. 1–4.

Xu Z, Mansour HM & Hickey AJ (2011) Particle interactions in dry powder inhaler unit processes: a review. *Journal of Adhesion Science and Technology* **25**: 451–82.

Young S (2007) Whey products in ice cream and frozen desserts. *Applications Monograph–Ice Cream*. US Dairy Export Council (http://www.innovatewithdairy.com).

Further reading

DMI (1996) *Dairy Ingredient Application Guides: Milk fat and butter ingredients/Concentrated and dry milk ingredients/Whey ingredients*. Dairy Management, Inc.

USDEC (1999 to present) Ingredients and applications. US Dairy Export Council (http://www.innovatewithdairy.com).

Information Sheets

– Milk
– Lipids
– Cheese
– Whey

From Milk By-Products to Milk Ingredients: Upgrading the Cycle, First Edition. Ruud de Boer.
© 2014 John Wiley & Sons, Ltd. Published 2014 by John Wiley & Sons, Ltd.

Whole milk powder

Description

Whole milk powder (Codex Alimentarius) or dry whole milk (USDA/FDA) are identical terms, which can be used interchangeably. Whole milk powder (WMP) is produced from fat-standardized, pasteurized milk. Prior to the water removal through evaporation, an additional heat treatment and homogenization step is carried out. Subsequently, the concentrate is dried. The vast majority of WMP is produced by using the spray-drying process.

Product analysis

Chemical[a]

Protein	26.3%
Fat	26.7%
Lactose	38.4%
Ash	6.1%
Calcium	0.91%
Phosphorus	0.78%
Potassium	1.33%
Sodium	0.37%
Moisture	2.5%
pH	6.7
Protein/lactose ratio	0.7

[a]National nutrient database for standard release 18 (2009).

Physical[b]

Titratable acidity	≤0.15%
Solubility index	≤1.0 mL
Scorched particles	≤15.0 mg

[b]USDA Dairy standards extra grade of dry whole milk (spray).

Microbiological[c]

Standard plate count	≤10,000/g
Coliforms	neg./g
E. coli	neg./g
Staphylococcus CP	neg./g
Salmonella	neg./750 g
Yeast and mould	≤50/g

[c]Fonterra.

Applications

WMP is a dairy commodity with a long tradition. A main outlet is the dairy industry. It is used in large quantities in regions that lack the necessary milk production. In these regions it is applied for reconstitution purposes. The reconstituted milk can be used itself or applied in milk products such as ice cream. Another important application is the confectionery industry. A well-known use is milk chocolate. Here it is applied for its creamier taste and soft texture. Bakery products are another segment. WMP is used here for its gold-brown colour, improved flavour and texture. Besides, WMP is applied in miscellaneous products such as soups and sauces.

Properties

The properties of WMP depend to a great extent on the applied drying process. There are differences in powder properties, which partially rely on the atomizers used (wheel or nozzle). The rate of dissolvability is mainly influenced by the increase of the particle size and the coating with lecithin. The latter is added after the first step of a usually multi-stage drying process. The fat component can be a source of off-flavours. Nitrogen flushing is applied during packaging to increase the shelf life. Usually the shelf life is 6–9 months. An alternative to spray drying is roller drying. Its use is very limited nowadays. Roller-dried WMP is attractive for chocolate production due to its high amount of free fat.

Literature

Beckett ST (2008) *The Science of Chocolate.* RSC Publishing, pp. 30–8.

Codex Stan 207-1999, *Codex Standard for milk powders and cream powder.*

Smith K (2008) *Dried dairy ingredients.* Wisconsin Center for Dairy Research.

Skimmed milk powder

Description

Skimmed milk powder (Codex Alimentarius) is also known as non-fat dry milk (USDA/FDA). There are only minor differences between the two. The abbreviations are respectively SMP or NFDM. SMP is produced from pasteurized milk from which the cream is removed. The milk receives additional heat treatments (low, medium and high heat). Subsequently, it is concentrated by using evaporation. In the vast majority the concentrate is spray dried.

Product analysis

Chemical[a]

Protein	35.1%
Fat	0.7%
Lactose	52.2%
Ash	8.0%
Calcium	1.23%
Phosphorus	0.99%
Potassium	1.71%
Sodium	0.55%
Moisture	4.0%
pH	6.7
Protein/lactose ratio	0.7

[a]National nutrient database for standard release 18 (2009).

Physical[b]

Whey Protein Nitrogen Index (WPNI):	
Low-heat	>6.0 mg/g
Medium-heat	1.51–5.99 mg/g
High-heat	<1.5 mg/g
Titratable acidity	≤0.15%
Solubility index	≤1.2; high-heat ≤2.0 mL
Scorched particles	≤15.0 mg

[b]USDA Dairy standards extra grade of NFDM (spray).

Microbiological[c]

Standard plate count	≤10,000/g
Coliforms	neg./g
E. coli	neg./g
Staphylococcus CP	neg./g
Salmonella	neg./750 g
Yeast and mould	≤50/g

[c]Fonterra regular low-heat.

Applications

SMP is a dairy commodity with a long tradition for various kinds of applications. It can be characterized by a bland flavour. The ingredient can be used in combination with other fat sources. Also, it is combined with anhydrous milk fat for the manufacture of recombined milk products in high-temperature areas. The applications are mostly based on the heat treatment employed. Low-heat powders are used for most dairy purposes, for instance ice cream, drinks, standardization of cheese milk. Other dairy applications are infant food and calf milk replacers, albeit with a different set of bacteriological requirements. Medium-heat powders are used in confectionery/chocolate, meat products and concentrated milks. An application for high-heat SMP is, for instance, bakery products, in which water binding and a gold-brown colour is desirable.

Properties

SMP's most valuable component is its protein. The level is standardized ≥34% by means of milk permeate or lactose. SMP is produced in numerous varieties. Dissolvability and the degree of denatured whey protein – usually designated as Whey Protein Nitrogen Index (WPNI) – are important aspects. The rate of dissolvability can be influenced by agglomeration of the powder particles and lecithin addition. WPNI is manipulated by applying the heat treatment. Typical process conditions for low-heat, medium-heat and high-heat are respectively 72°C/15 s, 75°C/3 min and 80–90°C/20–30 min.

Literature

Codex Stan 207-1999, *Codex Standard for milk powders and cream powder*.

International Dairy Federation (1988) Special issue no. 9001: *Recombination of milk and milk products*. IDF.

Kieseker FG & Aitken B (1993) Recombined full-cream milk powder. *Australian Journal of Dairy Technology* 48: 33–37.

Pisecký J (1997) *Handbook of milk powder manufacture*. NIRO A/S, pp. 19–78, 149–219.

Skimmed milk powder colostrum

Description

Skimmed milk powder (SMP) colostrum is derived from healthy cows. The milk is obtained from the first days of milking after parturition and is called colostrum. Colostrum has its own characteristic composition and is not considered a product of normal mammary secretion. The presence of colostrum in milk is undesirable for regular dairy processing and creates problems such as reduced heat stability and poor cheese-making properties. For this reason, colostrum is not allowed to be added to the bulk milk supply. However, colostrum has certain health benefits and therefore has inherent value. The process for making this SMP colostrum includes fat removal, mild heat treatment and spray drying.

Product analysis[a]

Chemical

Protein	≥65%
Immunoglobulin (IgG)	≥25%
Fat	≤5%
Lactose	≤10%
Ash	≤5%
Moisture	≤5%

Microbiological

Standard plate count	≤10,000/g
Coliforms	≤10/g
E. coli	≤10/g
Staphylococcus CP	neg./g
Salmonella	neg./25 g
Listeria	neg./25 g
Enterobacter sakazaki	neg./g
Yeast and mould	≤10/g

[a] FrieslandCampina.

Applications

SMP colostrum is a high-value-added ingredient. The level of the valuable components can be increased by fat removal, so the dosage level of the ingredient into food products can be reduced. It is applied as a dietary supplement for the enhancement of immunity, and used, for instance, by athletes to recover faster after a bout of intensive training. Furthermore, it is applied for support of enteric and oral health.

Properties

SMP colostrum contains several bioactive proteins such as immunoglobulins, growth factors, lactoferrin, lactoperoxidase and lysozyme in substantially elevated amounts. The immunoglobulin IgG is the most abundant protein. Its concentration can be about one hundred times higher than in mature milk. Colostrum is an ideal source of nutrition for the calf as the newborn animal has virtually no immunoglobulins at birth. Immunoglobulins are important in protecting against infections. The amount of colostrum mostly exceeds the need of newborns. The surplus can be collected separately from farms and further processed into colostrum products. An alternative for the 'natural' product is the milk obtained through the immunization of cows. These products are often regarded as a pharmaceuticals.

Literature

Commission Regulation (EC) no. 1662/2006 'Specific hygiene rules for food of animal origin'.

de Koning PJ, de Jong J & Kaper J (1989) Colostrum in cows: a pilot investigation into the situation in the Netherlands. *Voedingsmiddelentechnologie* **22**: 24–6.

Mehra R, Marnila P & Korhonen H (2006) Review: Milk immunoglobulins for health promotion. *International Dairy Journal* **16**: 1262–71.

Scammell AW (2001) Production and uses of colostrum. *Australian Journal of Dairy Technology* **56**: 74–82.

Skimmed milk powder low lactose

Description

Skimmed milk powder (SMP) low lactose is produced from pasteurized milk from which the cream is removed. In a subsequent process step, lactose hydrolysis takes place by means of microbial β-galactosidase (lactase). After the hydrolysis has been completed, the enzyme is inactivated and the treated milk is concentrated by means of evaporation. Drying is usually carried out with the help of an integrated belt drier.

Product analysis (typical)[a]

Chemical

Protein	37%
Fat	1%
Lactose	5%
Glucose	24%
Galactose	24%
Ash	8%
Calcium	1.20%
Phosphorus	0.95%
Potassium	1.70%
Sodium	0.42%
Moisture	≤2.5%
Protein/lactose ratio	7.4

Physical

Solubility index	≤1.25 mL
Scorched particles	min. B (≤15.0 mg)

Microbiological

Standard plate count	<10,000/g
Enterobacteriaceae	<10/g
Salmonella	neg./25g
Yeast and mould	<10/g
Bacillus cereus	<100/g

[a]Valio.

Applications

SMP low lactose is a specialized ingredient designed to meet the demands of the lactose-intolerant population. There are no harmonized rules at EU level for 'low lactose'. In a number of EU countries, a threshold value for foodstuffs, other than infant foods, is 1 g/100 g in the final product. SMP low lactose is applied in a range of low-lactose food products such as confectionery (chocolate, toffee), ice cream, bakery products and prepared meals. Due to a high level of reducing sugars, some adaptation of process and recipe of the end-product may be necessary.

Properties

The properties of SMP low lactose are different from those of regular SMP. The ingredient has, for instance, an increased amount of reducing carbohydrates (glucose, galactose, lactose). The consequence is that the browning reaction develops faster during baking and cooking. Due to the presence of glucose and galactose, the sweetness of the product is also enhanced. The sweetness is about twice that of lactose and approximately two-thirds that of sucrose. Furthermore, the powder has a thermoplastic and sticky nature, which affects the selection of the drying process applied. When kept in proper conditions (cool, dry), the powder has a shelf life of one year.

Literature

Piseckỳ J (1997) *Handbook of Milk Powder Manufacture*. NIRO A/S, pp. 75–6.

Schaafsma G (2008) Review: Lactose and lactose derivatives as bioactive ingredients in human nutrition. *International Dairy Journal* **18**: 458–65.

Shrestha AK, Howes T, Adhikan BP & Bhandari BR (2007) Water sorption and glass transition properties of spray dried lactose hydrolysed skim milk powder. *LWT - Food Science and Technology* **40**: 1593–660.

Skimmed milk powder lactose-free

Description

Skimmed milk powder (SMP) 'lactose-free' is produced in such a way that the lactose component is eliminated as far as technically feasible. It is manufactured from pasteurized milk, from which the cream is removed. Subsequently, membrane-based processes are applied to remove part of the lactose and keep the maximum of the milk minerals in the product. Then an enzymatic hydrolysis of the residual lactose takes place. After hydrolysis the enzyme is inactivated and the treated milk is concentrated by means of evaporation. The drying is usually carried out with the help of an integrated belt drier.

Product analysis (typical)[a]

Chemical

Protein	47%
Fat	1%
Lactose (HPLC)	0.06%
Glucose	19%
Galactose	19%
Ash	9%
Calcium	1.40%
Phosphorus	1.10%
Potassium	1.90%
Sodium	0.46%
Moisture	3%
pH	6.7
Protein/lactose ratio	780

Physical

Solubility index	≤1.25 mL
Scorched particles	min. B (≤15.0 mg)

Microbiological

Standard plate count	<10,000/g
Enterobacteriaceae	<10/g
Staphylococcus CP	<10/g
Salmonella	neg./25 g
Yeast and mould	<10/g
Bacillus cereus	<100/g

[a]Valio.

Applications

SMP 'lactose-free' is a specialized ingredient. It is targeted to meet the demands of people with a severe lactose intolerance and to make the manufacturing of 'lactose-free' products possible. There are no harmonized rules at EU level for 'lactose-free' products. The threshold level varies from 10 to 100 mg/100 g in the final application and depends on the EU member state. To be more precise, it relates to foodstuffs other than those intended for infants. Examples of the application of SMP 'lactose-free' products are milk chocolate, flavoured milk drinks, ice cream and prepared foods. Due to the higher protein content compared to regular SMP, the dosage level can be reduced.

Properties

A bland flavour is an important asset of milk powder, in which minerals also play a part. Mineral losses during processing, can be compensated, for instance, by recycling or by adding some whey salts.

SMP 'lactose-free' is beneficial for lactose-intolerant people and avoids gastrointestinal symptoms like cramps, bloating, flatulence and diarrhoea. However, the ingredient also contains a high level of galactose. SMP 'lactose-free' products are not suitable for patients with galactosaemia. As opposed to the more widespread lactose intolerance, galactosaemia is a rare genetic, metabolic disorder.

Literature

Allen M, Vessanen K & Harju M (2004) Whey salt powder, process for its production and its use. EP 1061811 B1 [patent].

EFSA Panel on Dietetic Products, Nutrition and Allergies (NDA) (2010) Scientific Opinion on lactose thresholds in lactose intolerance and galactosaemia. *EFSA Journal* **8**: 1777.

Tossavainen O & Sahlstein J (2011) Lactose-free milk product and process for producing the same. US 7,829,130 B2 [patent].

Acid casein

Description

Acid casein is manufactured from pasteurized skim milk by isoelectric precipitation of the casein through acidification. The precipitation is carried out in a continuous process under strict conditions. After 'cooking' the casein curd, it is separated from the whey mostly by decanting centrifuges. Subsequently, the curd is washed in a counter-current way to remove the lactose. It is dried on a fluidized bed, tempered and ground/sieved to obtain the desired mesh size.

Product analysis

Chemical[a]

Protein (N × 6.38)	85.4%
Fat	1.3%
Lactose	0.1%
Ash	1.8%
Calcium	0.1%
Sodium	<0.1%
Moisture	11.4%
pH	4.6–5.4

[a]http://nzic.org.nz/ChemProcesses/dairy/3E.pdf

Physical[b]

Solubility in water (%)	0
Scorched particles	≤15 mg/25 g(B)
Particle size:	
% <30 mesh (600 μm)	70
% <90 mesh (160 μm)	60

Microbiological[b]

Standard plate count	≤30,000/g
Coliform	≤10/g
Salmonella	neg./25 g
Yeast and mould	≤30/g

[b]FrieslandCampina.

Applications

Casein is available in different varieties, which refer to the acidification agent used. Lactic acid casein, sulphuric acid casein and hydrochloric acid casein can be distinguished. They are interchangeable in most cases. Also the particle size can be chosen; for instance, 30 as well as 90 mesh sizes are available. In its fine-powdered form the insoluble casein can be applied in imitation cheese, non-dairy creamers and protein supplements, for instance.

Properties

Casein is not soluble in water. Its functional properties, such as its emulsifying capacity, can be clearly improved during the processing of the final product. Under optimal conditions (time, temperature, processing aids) casein is converted into caseinates.

The acids needed for casein precipitation are released into the whey. The acid whey or its residual streams limit casein(ate) production. New techniques are currently being developed, in which the acid used to achieve precipitation can be recovered and at the same time a sweet whey is produced.

Literature

Balster J, Pünt I, Stamatialis DF, Lammers H, Verver AB & Wesseling M (2007) Electrochemical acidification of milk by whey desalination. *Journal of Membrane Science* **303**: 213–20.

Codex Standard for edible casein products, Codex Stan 290-1995.

Muller LL & Hayes (1962) Improved equipment for continuous precipitation of acid casein. *Australian Journal of Dairy Technology* **17**: 189–93.

Sodium caseinate

Description

Sodium caseinate is produced from acid casein. It is manufactured from pasteurized skim milk and the casein is precipitated by means of an acid. After separation of the curd, it is washed to remove the lactose. In a continuous process the curd is subsequently transformed into sodium caseinate. Dissolving of casein is followed by an intensive heat treatment. Due to the high viscosity of the liquid, this process is carried out under very strict conditions. Particularly, roller driers are suitable to handle suspensions with a high solids content. The example given here relates to a sodium caseinate dried with a roller dryer.

Product analysis (typical)[a]

Chemical

Protein (N × 6.38)	91.4%
Fat	1.1%
Lactose	0.1%
Ash	3.6%
Calcium	0.1%
Sodium	1.2–1.4%
Moisture	3.8%
pH	6.8–7.0
Protein/lactose ratio	≤910

[a] http://nzic.org.nz/ChemProcesses/dairy/3E.pdf

Physical[b]

Solubility in water (%)	>99
Scorched particles	≤7.5 mg/25 g (A)
Particle size:	
% <800 μm (20 mesh)	99–100
Bulk density (tapped)	520 g/L

Microbiological[b]

Standard plate count	500/g
Enterobacteriaceae	<1/g
Staphylococcus CP	neg./g
Salmonella	neg./50 g
Yeast and mould	<10/g

[b] FrieslandCampina.

Applications

Sodium caseinate can be used in various applications such as coffee whiteners, whipped toppings, cheese products, yoghurts, cream liqueurs, meat products, high-fat powders, soups and sauces. It acts as an emulsifier, a texture improver and an air stabilizer.

Properties

As opposed to casein micelles sodium caseinate has a random coil structure. It is a protein-based emulsifier. Sodium caseinate provides body/viscosity to the final application. Furthermore, it is very stable and can withstand high temperatures as well as elevated ethanol percentages. The blandest and most stable flavours are obtained by using fresh curd casein.

Literature

Lippe F, Ottenhof AWEM & de Boer R (1983) Continuous method for dissolving milk proteins containing casein and an apparatus for performing the method as well as water soluble milk protein products of a novel type. US 4,407,747 [patent].

Southward CR (1985) Review: Manufacture and applications of edible casein products I. Manufacture and properties. *New Zealand Journal of Dairy Science and Technology* **20**: 79–101.

Calcium caseinate

Description

Calcium caseinate is produced from acid casein. The casein is precipitated from pasteurized skimmed milk using acid. After separation of the casein curd, it is washed to remove the lactose. Subsequently, the curd is dissolved by adding a calcium hydroxide solution followed by an intensive heat treatment. In the example set out here, the suspension of calcium caseinate is converted into powder by means of a spray drier. As the drier can handle only suspensions of a limited viscosity, the powder is rather bulky.

Product analysis (typical)[a]

Chemical

Protein (N × 6.38)	91.2%
Fat	1.1%
Lactose	0.1%
Ash	3.8%
Calcium	1.3–1.6%
Sodium	<0.1%
Moisture	3.8%
pH	6.8–7.0
Protein/lactose ratio	910

[a]http://nzic.org.nz/ChemProcesses/dairy/3E.pdf

Nutritional (g per 100 g of protein)[b]

Alanine	3.1
Arginine	3.8
Aspartic acid	7.3
Cystine	0.4
Glutamic acid	22.3
Glycine	1.9
Histidine*	3.2
Isoleucine*°	5.8
Leucine*°	10.1
Lysine*	8.3
Methionine*	3.0
Phenylalanine*	5.4
Proline	10.5
Serine	6.3
Threonine*	4.6
Tryptophan*	1.4
Tyrosine	5.8
Valine*°	7.4

*Essential amino acid.
°Branched-chain amino acid.
Total amino acids >100 due to water-containing molecules.

Microbiological[b]

Standard plate count	100/g
Enterobacteriaceae	<1/g
E. coli	neg./10g
Staphylococcus CP	neg./g
Salmonella	neg./50 g
Yeast and mould	<10/g

[b]FrieslandCampina.

Applications

Calcium caseinate is used in numerous nutritional applications due to its high protein level, high protein/lactose ratio and favourable calcium content. Examples of its applications are dietetics, sport nutrition, health foods, slimming products, and clinical, infant and geriatric formulas.

Properties

Calcium caseinate has a favourable amino acid profile, although compared to whey protein it has less sulphur-containing amino acids. As opposed to sodium caseinate with its random coil casein structure, calcium caseinates exist in water as a colloidal dispersion (small 'submicelles'). The viscosity of a calcium caseinate solution is less than sodium caseinate at a comparable protein level. Therefore, it can be used at a higher dosage rate in its final application. The solubility of the spray-dried powder can be improved by agglomeration (cluster of powder particles) and coating with lecithin. A powder characteristic of calcium caseinate is its bulk density (345 g/L – tapped). The spray-dried calcium caseinate has a lower value for bulk density than the roller-dried variant.

Literature

Augustin MA, Oliver CM & Hemar Y (2011) Casein, caseinates and Milk Protein Concentrates. In: Chandan RC & Kilara A (eds), *Dairy Ingredients for Food Processing*. Wiley-Blackwell, pp. 161–78.

European Dairy Association (EDA) (1997) *Nutritional quality of proteins*.

Zwijgers A (1992) Outline of milk protein concentrate. *International Ingredients* no. 3, pp. 18–23.

Casein hydrolysate

Description

Casein hydrolysate is produced from casein. The enzymatic hydrolysis of the dissolved protein is carried out in a carefully controlled batch process (combination of proteases, substrate concentration, temperature, pH, time). The post-hydrolysis processes include, for instance, inactivation of enzyme, filtration for removal of residual high-molecular-weight substances and clarification. Subsequently, further concentration and spray drying takes place. The example given here is of an extensively hydrolysed casein.

Product analysis (typical)[a]

Chemical

Protein (N × 6.72)	86.9%
Fat	<0.1%
Lactose	<0.1%
Ash	6.0%
Moisture	≤5.0%
pH	6.8
Degree of hydrolysis	33%
Average MW (Da)	267
Free amino acids	13%

Nutritional (g/100 g of protein)

Alanine	3.1
Arginine	3.7
Aspartic acid	7.4
Cysteine	0.3
Glutamic acid	23.3
Glycine	1.7
Histidine*	1.6
Isoleucine*°	4.9
Leucine*°	7.9
Lysine*	7.7
Methionine*	2.5
Phenylalanine*	3.7
Proline	10.2
Serine	4.6
Threonine*	3.9
Tryptophan*	0.7
Tyrosine	4.4
Valine*°	6.1

*Essential amino acids.
°Branched-chain amino acids.

Microbiological

Standard plate count	≤1000/g
Coliform	≤10/g
E. coli	neg./g
Salmonella	neg./100g
Yeast and mould	≤100/g

[a] FrieslandCampina.

Applications

This casein hydrolysate is intended for infants with existing cow's milk allergy (CMA). The prevalence of CMA in early childhood is estimated to be 2–6%. These infants are allergic to the intact cow's milk proteins. Extensive hydrolysis of the protein is required to reduce the allergic reaction. This application is strictly regulated; for instance, the effectiveness needs to be at least 90%.

Properties

As CMA is dependent on the peptide size, the protein component is hydrolysed to a high degree. Peptides with a molecular weight exceeding 1500 Da are undesirable. As casein contains a high content of hydrophobic amino acids, the risk of bitterness is reduced by a second hydrolysis. The content of sulphur-containing amino acids is lower than that of whey protein hydrolysates. The amino acid profile of casein hydrolysate can be slightly different from that of regular casein due to post-hydrolysis processes.

Literature

Clemente A (2000) Enzymatic protein hydrolysates in human nutrition. *Trends in Food Science & Technology* **11**: 254–62.

Pasupuleti VK & Braun S (2010) State of the art manufacturing of protein hydrolysates. In: Pasupuleti VK & Demain AL (eds), *Protein Hydrolysates in Biotechnology*. Springer, pp. 11–32.

Saha BC & Hayashi K (2001) Review: Debittering of protein hydrolysates. *Biotechnology Advances* **19**: 355–70.

Casein phosphopeptides (CPP)

Description

Casein phosphopeptides (CPP) are bioactive peptides. They can be derived from enzymatic hydrolysis, in which casein is used as a starting material. The hydrolysis is carried out in a carefully controlled batch process (proteases, substrate concentration, temperature, pH, time). The post-hydrolysis processes include, for instance, enzyme inactivation by heat treatment and are followed by concentration and spray drying. The example given here illustrates an ingredient with an elevated CPP level.

Product analysis (typical)[a]

Chemical

Protein (N × 6.55)	92.4%
Casein phosphopeptides	26%
Fat	0.1%
Lactose	NA
Ash	7.3%
Moisture	4.2%
pH (6% solution)	7.3
Degree of hydrolysis	14%
Average MW (Da)	655
Free amino acids	2.1%

Microbiological

Standard plate count	≤5000/g
Coliform	≤10/g
E. coli	neg./g
Salmonella	neg./100g
Yeast and mould	≤100/g

[a]FrieslandCampina.

Applications

CPP are mineral carrier peptides. Based on this characteristic, they are used in infant, sport and medical nutrition, for instance. They are also used in non-dairy application such as oral care products, where the CPP prevent enamel demineralization. Products such as toothpaste, mouthwash, chewing gums and additives for the prevention of dental caries are examples.

Properties

CPP, being tasteless peptides, have elevated concentrations of various phosphoserine-rich peptides with divalent metal sequestering activity. They are able to increase the mineral bioavailability of calcium, and that of trace minerals such as zinc, copper and manganese. In this way CPP enhance uptake in the digestive tract. Results in the final application can be influenced by the hydrolysis process and the enzymes applied, the food matrix and the dosage level used. An indication of the release of phosphopeptides in the process is obtained by the degree of hydrolysis. Less known is the ability of CPP to prevent oxidation stress by sequestering ferrous ions.

Literature

Adamson NJ & Reynolds EC (1997) Relationship between degree of casein hydrolysis and phosphopeptide release. *Journal of Dairy Research* **64**: 505–14.

Aimutis WR (2004) Bioactive properties of milk proteins with particular focus on anticariogenesis. *Journal of Nutrtion* **134**(4S): 989S–995S.

Kitts DD (2005) Antioxidant properties of casein phosphopeptides: review. *Trends in Food Science & Technology* **16**: 549–54.

Miquel E & Farré R (2007) Effects and future trends of casein phosphopeptides on zinc bioavailability; review. *Trends in Food Science & Technology* 18: 139–43.

Milk protein concentrate: MPC 80

Description

Milk protein concentrate (MPC) is produced in various protein contents ranging from 42 to 85%. These variants are all manufactured from pasteurized skimmed milk followed by ultrafiltration. Both milk proteins, casein and whey protein, are concentrated at the same time, while most of the lactose and part of the minerals are removed. The higher protein levels are obtained through diafiltration (washing process). After another pasteurization treatment, further concentration by evaporation takes place. Subsequently, the concentrate is spray dried. Here an example is given of MPC 80.

Product analysis (typical)[a]

Chemical

Protein	81.3%
Fat	1.6%
Lactose	4.6%
Ash	6.8%
Calcium	2.2%
Sodium	0.1%
Moisture	5.7%
pH (5% solution at 20°C)	7.1
Protein/lactose ratio	17.7

Microbiological[a]

Standard plate count	<10,000/g
E. coli	neg./g
Staphylococcus CP	neg./g
Salmonella	neg./750 g
Listeria	neg./25 g
Yeast and mould	<10/g

[a] Fonterra.

Applications

MPCs are used in a variety of applications, which range from replacement of skimmed milk powder to a more significant protein fortification. An important market is the dairy industry. Examples are yoghurt, ice cream, protein standardization of cheese milk, recombined cheese, imitation and processed cheeses. MPCs are also present in the nutritional food segment – amongst others in infant formulas, sport and dietetic drinks, breakfast cereals and chocolate bars.

Properties

MPCs are characterized by the natural structure of milk proteins. They exist as casein micelles and, in cases of mild heat treatment, in undenatured whey protein. The casein/whey protein ratio is the same as in milk. MPC has a favourable amino acid profile and is rich in calcium. The protein/lactose ratio of MPC 80 is favourable for the formulation of low-lactose-containing end-products. The ingredient has an excellent heat stability, which is important for the sterilization of liquid formulas. Due to their compact molecular structure, MPCs have a lower viscosity at high protein levels compared to, for instance, caseinates. A limitation of the ingredient is its insolubility at around pH 4.5, which has an impact on its functionality.

Literature

Novak A (1991) Milk protein concentrate. In: *New applications of membrane processes.* International Dairy Federation special issue no. 9201, pp. 51–66.

Singh H (2009) Protein interactions and functionality of milk protein products. In: Corredig M (ed.), *Dairy-derived Ingredients - Food and Nutraceutical Uses.* Woodhead, pp. 664–74.

Zwijgers A (1992) Outline of milk protein concentrate. *International Food Ingredients* no. 3, pp. 18–23.

Milk protein concentrate: MPC 80 low calcium

Description

MPC 80 with a low calcium content is produced from pasteurized skimmed milk. Subsequently, ultrafiltration takes place followed by diafiltration (washing process) to increase the protein content and further reduce lactose and part of the minerals. After pH adjustment, the liquid passes through a strong cation-exchange column loaded with sodium. During this process, the calcium attached to the casein micelle is largely replaced. After an additional pasteurization step, further concentration by evaporation is carried out. Subsequently, the concentrate is spray dried.

Product analysis (typical)[a]

Chemical

Protein	81.7%
Fat	1.7%
Lactose	3.9%
Ash	7.1%
Calcium	0.3%
Sodium	2.3%
Moisture	5.6%
pH (5% solution at 20°c)	7.0
Protein/lactose ratio	21

Microbiological

Standard plate count	<10,000/g
Coliform	<1/g
E. coli	neg./g
Staphylococcus CP	neg./g
Salmonella	neg./750 g
Listeria	neg./125 g
Yeast and mould	<10/g

[a]Fonterra.

Applications

This new variant has been developed to improve the use of MPC as a cheese milk extender. The formation of so-called 'nuggets' – insoluble matter manifest as protein gels of a different colour – is reduced substantially. A second important field of application is its use as an alternative to sodium caseinate. The latter is mostly applied as a milk-protein-based emulsifier. A wide range of applications of this calcium-depleted MPC is emerging such as whipped toppings, processed cheese, coffee whiteners, soups, sausages, yoghurts, ice cream and caramels.

Properties

During MPC processing, both milk proteins – casein and whey protein – are incorporated. This means that the protein yield compared to the production of casein(ates) is increased close to 20%. However, due to the remaining lactose in the MPC, its shelf life is a more critical issue than is the case with casein(ates).

MPC 80 low calcium has an improved solubility at ambient temperature in comparison with regular MPC 80. This results from the disintegration of the casein micelle structure arising from the exchange of calcium by sodium. In this process about 85% of the calcium is removed. The treatment improves the 'cold solubility' as well as the emulsifying properties of this modified MPC 80.

Literature

Bhaskar GV, Singh H & Blazey ND (2007) Milk protein products and processes. US 7,157,108 B2 [patent].

Bhaskar GV (2010) Milk product and process. US 2010/0021595 A1 [patent].

Milk protein isolate: MPI 85

Description

MPIs are not described by a standard. In industrial practice their protein content ranges from 85 to 90%. They are produced from pasteurized skimmed milk followed by ultrafiltration and an intensive diafiltration step (washing process). Both milk proteins, casein and whey protein, are concentrated and nearly all the lactose and part of the minerals are removed. An ultra-high-temperature (UHT) treatment can be applied to obtain an ingredient with a low bacterial count. Furthermore, concentration by means of evaporation takes place, which is followed by spray drying.

Product analysis (typical)[a]

Chemical

Protein (N × 6.38)	85.5%
Fat	1.5%
Lactose	0.7%
Ash	6.3%
Moisture	4.8%
pH (10% solution, 20°C)	7.0
Protein/lactose ratio	122

Nutritional (g per 100 g of protein)

Alanine	3.4
Arginine	4.3
Aspartic acid	7.9
Cysteine	0.7
Glutamic acid	22.8
Glycine	1.9
Histidine*	2.9
Isoleucine*°	5.5
Leucine*°	10.3
Lysine*	8.8
Methionine*	3.1
Phenylalanine*	5.1
Proline	11.4
Serine	6.0
Threonine*	4.6
Tryptophan*	1.3
Tyrosine	5.8
Valine*°	7.4

*Essential amino acid.

°Branched-chain amino acid.

Total amino acids >100 due to water-containing molecules.

Microbiological

Standard plate count	250/g
Enterobacteriaceae	<1/g
E. coli	neg./g
Staphylococcus aureus (CP)	neg./g
Salmonella	neg./50g
Yeast and mould	<20/g

[a] FrieslandCampina.

Applications

MPIs are applied mainly for nutritional purposes such as protein enrichment of beverages, which endure intensive processing, ice cream, bars and breakfast cereals. If they are used for so-called 'lactose-free' products, the dosage percentage has to be considered. The threshold level varies from 10 to 100 mg lactose/100 g in the final food application. Within the EU, the level depends on the member state.

Properties

MPIs have a favourable nutritional value. They are rich in essential and branched-chain amino acids. If necessary, the ash content can be decreased by lowering the pH during part of the filtration process. In that case the calcium is released from the casein micelle. As a consequence, calcium and other minerals are further reduced and the protein content of the MPI is increased to about 90% 'as is'. Other important properties of MPI are its heat stability and, compared to sodium caseinate, a relatively low viscosity at elevated protein levels.

Literature

EFSA Panel on Dietetic Products, Nutrition and Allergies (NDA) (2010) Scientific Opinion on lactose thresholds in lactose intolerance and galactosaemia. *EFSA Journal* **8**: 1777.

Hiddink J, de Boer R & Romijn DJ (1978) Removal of milk salts during ultrafiltration of whey and buttermilk. *Netherlands Milk & Dairy Journal* **32**: 80–93.

Rajagopalan N & Cheryan M (1991) Total Protein Isolate from milk by ultrafiltration: factors affecting product composition. *Journal of Dairy Science* **74**: 2435–9.

Micellar casein concentrate: MCC 80

Description

Micellar casein concentrate is produced by a combination of membrane processes. Skimmed milk is used as a starting material. The milk is treated by microfiltration (MF) to reduce the bacterial count at low temperature. Subsequently, microfiltration fractionation (MFF) combined with an intensive diafiltration step (washing process) is carried out. The lactose and a part of the minerals pass through this membrane as well as a substantial amount of the whey protein. The concentrate, which contains an elevated amount of casein micelles, is subsequently spray dried. Here, MCC 80 is provided as an example.

Product analysis (typical)[a]

Chemical

Protein	83%
Casein micelles/proteins	92%
Fat	1.5%
Lactose	2%
Ash	8.5%
Calcium	3%
Moisture	5.0%
Acidity	0.15%
Protein/lactose ratio	42

Microbiological

Standard plate count	10.000/g
Enterobacteriaceae	neg./0.1g
E. coli	neg./g
Salmonella	neg./25g
Staphylococcus CP	neg./0.1g
Yeast and mould	<10/g

[a]Ingredia.

Applications

MCCs are a new type of ingredient and their applications are still in infancy. An example of this is the use of liquid MCC in cheese manufacture in order to achieve a constant capacity of an existing production line. Specific nutritional applications are also mentioned, for instance, in sterilized liquid formulas (tube feeding) and sugar-free chocolates. An advantage of MCC production is its 'by-product'. This permeate, or 'native whey', is an excellent source for the production of serum protein concentrates.

Properties

The current state of the art in industry does not make a complete separation of casein and whey protein possible. During processing, the casein/whey protein ratio changes from about 80/20 to 92/8. However, this partial removal of whey protein by MFF significantly reduces the detrimental effects of heating on rennet coagulation during cheese-making. Its suitability for tube feeding is also based on a high stability to sterilization and a reduced viscosity compared to, for instance, sodium caseinate solutions.

Literature

Jeantet R, Schuck P, Thierry S, Andre C & Delaplace G (2010) The influence of stirring speed, temperature and solid concentration on the rehydration time of micellar casein powder. *Dairy Science and Technology* **90**: 225–36.

Nestec (1997) Enteral tube feeding composition with a native micellar protein component. US 5,683,984 [patent].

Pierre A, Fauquant J, Le Graet Y, Piot M & Maubois JL (1992) Preparation de phosphocaseinate natif par microfiltration sur membrane. *Lait* **72**: 461–74.

Saboya LV & Maubois JL (2000) Review: Current developments of microfiltration technology in the dairy industry. *Lait* **80**: 541–53.

Milk serum protein concentrate: MSPC 60

Description

MSPCs are made from milk serum. Serum is the liquid that is left after the removal of fat globules and casein micelles. By means of centrifugation, the fat is separated from the milk and subsequently the casein micelles are concentrated through microfiltration fractionation (MFF). The MFF concentrate can, for instance, be mixed with part of the cream and used for cheese-making. The protein of the serum phase is concentrated by means of ultrafiltration. In addition, the solids content of the concentrate is increased by using an evaporator and afterwards it is spray dried. Here an example is given of MSPC 60.

Product analysis (typical)[a]

Chemical

Protein (N × 6.38)	60%
Fat	0.2%
Lactose	31.3%
Ash	5%
Moisture	3.5%
pH	6.8
Protein/lactose ratio	1.9

Microbiological[a]

Standard plate count	≤30,000/g
Enterobacteriaceae	neg./g
E. coli	neg./10 g
Salmonella	neg./125 g
Yeast and moulds	≤50/g
Staphylococcus CP	≤100/g

[a] FrieslandCampina.

Applications

MSPCs are a new type of dairy ingredient. As they are directly derived from milk, their consistency is improved compared to WPCs. They have potential in beverage applications, where a clear appearance is desirable. For the bakery and confectionery industry, their foaming qualities are attractive. The infant food sector is another promising possibility.

Properties

MSPCs are separated from milk before it is used to make cheese. The starting material for the production of MSPCs is sometimes called 'ideal whey' or 'native whey'. It does not contain residuals of cheese manufacture (e.g. metabolites, cultures, coagulants, calcium chloride, colour agents) and can be sterile. The fat content is extremely low; therefore, its foaming properties and its clarity in solution are better compared to WPCs. The amino acid profile of its protein is modified due to the absence of glycomacropeptide. The latter, which is rich in threonine, is released into cheese whey after coagulation. If MFF is carried out at approximately 10°C, MSPC can be enriched with some β-casein (rich in proline).

Literature

Evans J, Zulewska J, Newbold M, Drake MA & Barbano DM (2010) Comparison of composition and sensory properties of 80% whey protein and milk serum protein concentrates. *Journal of Dairy Science* **92**: 1824–43.

Glas C (2010) Functional serum protein product for use in infant food and therapeutic compositions and methods for the preparation thereof. US 2010/0168017 A1 [patent].

Nelson BK & Barbano DM (2005) A microfiltration process to maximize removal of serum proteins from skim milk before cheese making. *Journal of Dairy Science* **88**: 1891–900.

Milk fat: butter

Description

Butter is one of the oldest forms of preserving milk fat. Nowadays, it is produced on an industrial scale. The cream is separated from the milk, afterwards it is intensively pasteurized to inactivate the milk lipase and de-aerated. The melting behaviour of the final product is improved by a special cooling treatment of the cream. Subsequently churning takes place, in which a water-in-oil emulsion is formed by an intensive beating with air. Usually butter-making is carried out in a continuous process. Various variants of butter are available; here an unsalted sweet butter is given as an example.

Product analysis

Chemical

Total fat (wt.%)	82
Unsaturated fatty acids (wt.%)	29.4
Saturated fatty acids (wt.%)	70.6
SCFA[a] (wt.%)	9.2
SCFA[a] (mol.%)	18.2
Milk solids non-fat (wt.%)	<2
Moisture (wt.%)	≤16
pH	>6.5

[a]SCFA, short-chain fatty acids (C_4–C_{10}).

Physical

Melting/dropping point	32–34°C
Free fatty acids	<0.3%
Peroxide (meq oxygen/kg)	≤0.3

Microbiological

Standard plate count	≤1000/g
Lipolytic count	≤50/g
Proteolytic count[b]	≤100/g
Coliform[b]	≤10/g
Yeast and mould[b]	≤20/g

[b]USDA.

Applications

Butter is a well-known consumer product, but it is also applied for industrial purposes. Due to its flavour, it is a preferred ingredient in the bakery industry (e.g. pastries, croissants). It can be delivered in sheet form for quicker conditioning prior to use. Other applications are processed cheese, ice cream, and confectionery.

Properties

Butter is composed of a complex mixture of triglycerides. Several hundred fatty acids are present of which 15 account for over 95% of those present in the triglycerides. Unique are the short-chain fatty acids, which are important for the typical butter flavour. Usually the fatty acid composition is expressed in weight %. However, there is a large variation in molecular weight. For this reason, the mol. % demonstrates more clearly the large contribution of the short-chain fatty acids to the total fatty acid profile.

Butter is available in unsalted and salted variants as well as in sweet and lactic flavours. Lactic butter is standard in Europe. The lactic flavour is obtained through the use of dairy cultures. The milk solids non-fat contributes to the typical flavour of lactic butter. The shelf life of butter is 6 months at 6°C. Freezing butter at −20°C can extend shelf life to up to 2 years.

Literature

Hettinga D (2005) Butter. In: Shahidi F (ed.), *Bailey's Industrial Oil and Fat Products*, vol. 6. John Wiley & Sons, Inc., pp. 1–59.

International Dairy Federation (1991) *Utilization of milk fat*. IDF Bulletin no. 260.

Walstra P, Wouters JTM & Geurts TJ (2006) *Dairy Science and Technology*. CRC Press, pp. 37–63, 467–488.

Milk fat anhydrous

Description

Anhydrous milk fat (AMF) is the purest form of milk fat. Its shelf life is an important advantage. AMF can be manufactured from butter or cream. The latter is preferred as a raw material as it results in a more consistent quality. The cream of the milk is concentrated in a two-step process to approximately 75% fat. By using a homogenizer a phase inversion is carried out. The final product is obtained through another centrifugation step and vacuum treatment. An example of the major fatty acids of AMF is given here.

Product analysis[a]

Chemical/nutritional

Fat	≥99.8%
Fatty acid profile (wt.%):	
C_4 butyric	3
C_6 caproic	2
C_8 caprylic	1
C_{10} capric	3
C_{12} lauric	4
C_{14} myristic	12
C_{16} palmitic	29
C_{18} stearic	10
$C_{16\text{-}1}$ palmitoleic	2
$C_{18\text{-}1}$ oleic	23
$C_{18\text{-}2}$ linoleic	1
$C_{18\text{-}3}$ linolenic	1
Moisture	≤0.2%

Physical

Melting/dropping point	32–34°C
Free fatty acids	≤0.3%
Peroxide (meq oxygen/kg)	≤0.2

Microbiological

Standard plate count	≤100/g
Coliform	neg./g
Salmonella	neg./25 g
Listeria	neg./10 g
Yeast and mould	≤10/g

[a]Royal VIV Buisman.

Applications

Anhydrous milk fat is convenient to use in a liquid form, because it is easy to mix and meter into other products. It is applied in recombined dairy products, confectionery, bakery goods, ice cream and fat blends. Due to its extremely low moisture content, it is more suitable for cooking, such as shallow frying, than butter. As far as chocolate is concerned, milk fat is one of the few fats that is compatible with cocoa butter. Currently, a fresh frozen milk fat is available as an alternative for unsalted butter.

Properties

AMF is a natural product, which is reflected in the variation in the content of the numerous fatty acids. Its composition varies with the season, the breed of cow and the animal feed used. To some extent, its composition can be standardized by mixing summer and winter fat. The complex fatty acid composition is reflected in its melting behaviour. The melting range of milk fat varies from −30°C to +37°C. Similar ingredients with a slightly lower purity are butter oil and ghee.

As opposed to butter, AMF is stored in cans and drums at ambient temperature. Under these conditions the fat is solid. It has to be tempered in a hot room before use.

The maximum shelf life is 1 year. Oxidation of the fat can be avoided by using a nitrogen top layer.

Literature

Codex Standard for milk fat products, Codex. Stan 28-1973.

Hettinga D (2005) Butter. In: Shahidi F (ed.), *Bailey's Industrial Oil and Fat Products*, vol. 6. John Wiley & Sons, Inc., pp. 1–59.

International Dairy Federation (1991) *Utilization of milk fat*. IDF Bulletin no. 260.

Tetra Pak Processing Systems AB (2003) *Dairy Processing Handbook*, 2nd edn, pp. 293–9.

Milk fat anhydrous hard fraction

Description

The milk fat hard fraction is produced from anhydrous milk fat (AMF) by a so-called dry fractionation process – a process in which the triglycerides are separated according to their melting point. The use of solvents or additives is avoided. The AMF is first melted and then cooled down slowly to 25–30°C, for instance. The hard fraction, which is also called stearin, is separated from the soft fraction (olein) by a filtration system. An example of the characteristics and the composition of the major fatty acids of a hard fraction is given here. The composition of AMF is also given for reference.

Product analysis[b]

Chemical/nutritional

Fat		≥99.6%
Fatty acid profile (wt.%):	AMF	Hard fraction
C_4 butyric	3	3
C_6 caproic	2	2
C_8 caprylic	1	1
C_{10} capric	3	3
C_{12} lauric	4	4
C_{14} myristic	12	12
C_{16} palmitic	29	33
C_{18} stearic	10	12
C_{16-1} palmitoleic	2	1
C_{18-1} oleic	23	19
C_{18-2} linoleic	1	1
C_{18-3} linolenic	1	1
Unsaturated	31%	25%
Saturated	69%	75%
Moisture		≤0.4%

Physical

Melting/dropping point	40°C
Free fatty acids	≤0.4%
Peroxide (meq oxygen/kg)	≤0.6

Microbiological

Standard plate count	≤100/g
Coliform	neg./g
Salmonella	neg./25 g
Yeast and mould	≤10/g

[a]Royal VIV Buisman.

Applications

The milk fat anhydrous hard fraction has a high melting point, which is favourable for improving the physical functionality of milk fat in puff pastry products. In confectionery, milk fat behaves as a bloom inhibitor. Bloom is the appearance of white spots on the surface of the chocolate. The hard fraction has a particularly beneficial effect. The same fraction is also applied as a coating, for instance, on chocolate bars.

Properties

Fractionation causes only a limited change in the fatty acid profile. The change is only obvious in the longer chain saturated fatty acids. The amounts of palmitic and stearic acid are slightly higher in the hard fraction, while the level of oleic acid is lower. These modest differences in the triglyceride molecule still have an important effect on the melting properties of the fat fraction. The manner in which the fractionation and the separation process is carried out is often a trade secret.

Literature

Boer F, Royal VIV Buisman (2008) Fractioneren maakt melkvet meer dan een fractie beter. *Genootschap Melkkunde* November 5.

Fouad FM, van de Voort FR, Marshall WD & Farrell PG (1990) A critical evaluation of thermal fractionation of butter oil. *Journal of American Oil Chemists' Society* **67**: 981–7.

Kaylegian KE, Hartel RW & Lindsay RC (1993) Applications of modified milk fat in food products. *Journal of Dairy Science* **76**: 1782–96.

Milk fat anhydrous soft fraction

Description

The milk fat soft fraction is produced from anhydrous milk fat (AMF) by applying a two-step dry fractionation process. Firstly, a hard and a soft fraction is obtained from AMF. The soft fraction is also called olein. Secondly, the olein is fractionated again to produce a second or third soft fraction with a lower melting point. This fraction is separated by a filtration system. An example of the characteristics and composition of the major fatty acids of such a soft fraction is set out here. The composition of AMF is given as a reference.

Product analysis[a]

Chemical/nutritional

Fat		≥99.6%
Fatty acid profile (wt.%):	AMF	soft fraction
C_4 butyric	3	4
C_6 caproic	2	3
C_8 caprylic	1	2
C_{10} capric	3	3
C_{12} lauric	4	5
C_{14} myristic	12	11
C_{16} palmitic	29	22
C_{18} stearic	10	7
$C_{16\text{-}1}$ palmitoleic	2	2
$C_{18\text{-}1}$ oleic	23	27
$C_{18\text{-}2}$ linoleic	1	2
$C_{18\text{-}3}$ linolenic	1	1
Unsaturated	31%	45%
Saturated	69%	55%
Moisture		≤0.4%

Physical

Melting/dropping point	10–12°C
Free fatty acids	≤0.4%
Peroxide (meq oxygen/kg)	≤0.6

Microbiological

Standard plate count	≤100/g
Coliform	neg./g
Salmonella	neg./25 g
Yeast and mould	≤10/g

[a] Royal VIV Buisman.

Applications

The milk fat anhydrous soft fraction has a low melting point, which is useful for the production of a (low-fat) butter spreadable at refrigerator temperature. Other special outlets are flavour applications, chocolate for ice cream bars and some bakery products (e.g. pound cakes).

Properties

The milk fat soft fraction is obtained through a complicated process. The background is the complex composition of milk fat with a large variety of fatty acids and a huge number of triglycerides. Dry fractionation is preferred over solvent fractionation as the use of chemicals is avoided and there is no loss of flavours. The selected process – e.g. residence time, fractionation temperature, agitation, separation efficiency – is decisive for the physical properties of the soft fraction. This fraction also contains the various flavour compounds in a concentrated form.

Literature

Deffense E (1993) Milk fat fractionation today: a review. *Journal of the American Oil Chemists' Society* **70**: 1193–201.

Vanhoutte B, Dewettinck K, Vanlerberghe B & Huyghebaert A (2002) Monitoring milk fat fractionation: Effect of agitation, temperature, and residence time on physical properties. *Journal of the American Oil Chemists' Society* **79**: 1169–76.

Vanhoutte B, Dewettinck K, Vanlerberghe B & Huyghebaert A (2003) Monitoring milk fat fractionation: Filtration properties and crystallization kinetics. *Journal of the American Oil Chemists' Society* **80**: 213–18.

MFGM buttermilk

Description

MFGM is an abbreviation of milk fat globule membrane. Buttermilk is used as a raw material. During butter-making the cream is turned into butter and buttermilk. In this 'churning' process the milk fat globules are destabilized through mechanical agitation in the presence of air. The destroyed MFGM migrates partially to the buttermilk. Subsequently, the buttermilk is concentrated by evaporation and spray dried. An example of a MFGM buttermilk or sweet buttermilk powder is given here.

Product analysis

Chemical/nutritional

Protein (N × 6.38)	≥30%
Lactose	± 50%
Ash	≤9%
Total lipids[a]	7–13%
Phospholipids[a]	1.6%
% of total phospholipids:	
Phosphatidyl ethanolamine (PE)	39
Phosphatidyl choline (PC)	24
Phosphatidyl serine (PS)	8
Phosphatidyl inositol (PI)	9
Sphingomyelin (SM)	19
Moisture	≤4%
pH	6.0–6.5

[a]Britten *et al.* (2008).

Microbiological[b]

Standard plate count	≤20,000/g
Aerobic thermophilic spores	≤1000/g
Coliforms	neg./g
E. coli	neg./g
Salmonella	neg./750 g
Staphylococcus CP	neg./g
Yeast and mould	≤50/g

[b]Fonterra: for ultra-heat-treated milk.

Applications

MFGM buttermilk has a pH value similar to that of milk. This is of importance for many applications. Also by changing the process lactic butter production releases sweet buttermilk. The ingredient can to some extent replace skim milk powder. It is used in various dairy products as (recombined) evaporated milk and cream. Stability and flavour can be improved. Limited percentages can also be used in the production of cheese (moisture retention) and yoghurt (texture). Outside the dairy industry, it is applied in bakery products due to its flavour and softening effect on crumb.

Properties

MFGM buttermilk powder has a higher fat content than skim milk powder. It contains increased levels of the highly surface active phospholipids, of which PE, PC and SM are the main ones. Also the protein composition of MFGM buttermilk is different: besides casein micelles and whey proteins, it contains an enhanced level of specific MFGM proteins. Both membrane proteins and phospholipids have an amphiphilic nature, which contributes to the emulsifying properties. Compared to milk fat the phospholipids have on average a higher level of unsaturated fatty acids. This fatty acid composition makes MFGM buttermilk more susceptible to oxidation. The shelf life is 6–9 months.

Literature

Britten M, Lamothe S & Robitaille G (2008) Effect of cream treatment on phospholipids and protein recovery in butter-making process. *International Journal of Food Science and Technology* **43**: 651–7.

Gallier S, Gragson D, Cabral C, Jiménez-Flores R & Everett DW (2010) Composition and fatty acid distribution of bovine milk phospholipids from processed milk products. *Journal of Agricultural Food Chemistry* **58**: 10503–11.

Vanderghem C, Bodson P, Danthine S, Paquot M, Deroanne C & Blecker C (2010) Milk fat globule membrane and buttermilks: from composition to valorization. *Biotechnology, Agronomy, Society and Environment* **14**: 485–500.

MFGM whey

Description

MFGM is an abbreviation of milk fat globule membrane. This MFGM variant refers to the cheese whey from which it is produced. The ingredient is manufactured from a side stream of whey protein isolate (WPI). It contains MFGM components and a part of the whey proteins. Three different membrane processes are applied in order to concentrate whey proteins, remove bacteria and separate the induced aggregation of a phospholipoprotein complex. Subsequently, the concentrate is spray dried.

Product analysis[a]

Chemical/nutritional

Protein	73%
Lactose	≤3%
Ash	≤3%
Sialic acid	± 2%
Total lipids	16%
Phospholipids	7.5%
% of total phospholipids:	
Phosphatidyl ethanolamine (PE)	25
Phosphatdyl choline (PC)	25
Phosphatidyl serine (PS)	12
Phosphatidyl inositol (PI)	7
Sphingomyelin (SM)	24
Others	7
Moisture	≤6%
pH	6.4

Microbiological

Standard plate count	<10,000/g
Enterobacteriaceae	<10/g
Salmonella	neg./125 g
Staphylococcus aureus	neg./g
Bacillus cereus	<50/g
Yeast and mould	<10/g

[a] Arla Foods Ingredients.

Applications

This whey protein concentrate is enriched with various phospholipids and contains bioactive compounds. It is used for nutritional reasons, for instance in clinical nutrition and/or infant food. In infant food it can be useful to simulate the natural composition and performance of human milk.

Properties

MFGM whey is derived from cheese-making. The fat is concentrated in the cheese curd and the MFGM layer is damaged during processing. Fragments of the 10–20 nm thick membrane migrate into the cheese whey. Compared to other dairy streams, cheese whey is a modest source of phospholipids (0.26% on whey solids); thus concentration thereof is desirable. After aggregation, the residual lipid-containing material is large in size relative to the major whey proteins. It can be concentrated by utilizing a combination of membrane processes. The ingredient contains five phospholipids. Besides this, it includes sialic acid, immunoglobulins, lactoferrin and a small amount of gangliosides (reference is made to 'others'). MFGM whey stimulates intestinal maturation in infants. The gangliosides and sialic acid are also of interest as they might play a role as 'brain nutrients'.

If kept cool, the shelf life is estimated to be a minimum of 12 months.

Literature

Jarrow P, Schnell N, Jumpsen J & Clandinin T (2009) Influence of dietary gangliosides on neonatal brain. *Nutrition Reviews* **67**: 451–63.

Maubois JL & Olliver G (1997) Extraction of milk proteins. In: Damodaran S & Paraf A (ed.), *Food Proteins and their Application*. Marcel Dekker, p. 558.

Rueda R, Barranco A, Ramirez M *et al.* (2008) Enriched infant formulas. US2008/0003329 [patent].

MFGM cream serum

Description

MFGM is an abbreviation of milk fat globule membrane. MFGM cream serum is a by-product of the production of anhydrous milk fat (AMF). During processing, the MFGM material, comprising closely associated phospholipids and proteins, migrates to the serum phase. For AMF production, cream is separated from milk in two steps and concentrated to approximately 75% fat. Subsequently, homogenization takes place, breaking the emulsion and releasing the cream serum containing the MFGM material. A similar product is butter serum. In this case butter is used as a raw material. For quality purposes cream serum is given as an example. It is concentrated by evaporation followed by spray drying.

Product analysis[a]

Chemical/nutritional

Protein	29%
Lactose	47%
Ash	7%
Total lipids	14%
Phospholipids	7%
% of total phospholipids:[b]	
Phosphatidyl ethanolamine (PE)	29
Phosphatidyl choline (PC)	25
Phosphatidyl serine (PS)	10
Phosphatidyl inositol (PI)	9
Sphingomyelin (SM)	27
Moisture	≤5%
pH	6.5

Microbiological

Standard plate count	≤20,000/g
Coliforms	≤10/g
Salmonella	neg./25g
Staphylococcus aureus	≤100/g
Yeast and mould	≤100/g

[a]Corman.
[b]Britten *et al.* (2008).

Applications

From the MFGM-containing streams, butter serum is often used in animal feed in a liquid form. It is less consistent in quality, as often use is made of various butter operations with differences in process history (e.g. heat treatment of the cream and storage time of the butter). As AMF and cream serum are derived from one process line, quality control is stricter. A consistent quality is important for value-added applications. Examples of use are bakery goods (anti-staling), margarines (prevention of spattering) and confectionery (crystallization behaviour).

Properties

MFGM cream serum contains nearly five times the amount of phospholipids as buttermilk. Due to the increased phospholipid content it is suitable for the manufacture of more tailor-made ingredients. As the complex structure of MFGM is prone to modification during isolation and processing, strict standardization is needed to avoid variation in composition and functionality. The technological properties of MFGM cream serum are focused on its use as emulsifying agent, as well as the promising nutritional properties of the various MFGM compounds.

Literature

Britten M, Lamothe S & Robitaille G (2008) Effect of cream treatment on phospholipids and protein recovery in butter-making process. *International Journal of Food Science and Technology* **43**: 651–7.

Dewettinck K, Rombaut R, Thienpont N, Le TT, Messens K & Van Camp J (2008) Review: Nutritional and technological aspects of milk fat globule membrane material. *International Dairy Journal* **18**: 436–57.

El-Loly MM (2011) Composition, properties and nutritional aspects of milk fat globule membrane – a review'. *Polish Journal of Food and Nutrition Sciences* **61**: 7–32.

MFGM cream serum concentrate

Description

MFGM is an abbreviation of milk fat globule membrane. In this MFGM cream serum concentrate the phospholipids – also called polar lipids – are increased twice. As raw material the cream serum of the anhydrous milk fat (AMF) process is used. For this purpose, cream is separated from milk and in a two-step process concentrated to approximately 75% fat. Homogenization is a next step, in which the emulsion is broken and the MFGM material released. Subsequently, the cream serum is treated with membrane processing to reduce the lactose content and enhance the level of phospholipids. An example of a spray-dried product is set out here.

Product analysis[a]

Chemical/nutritional

Protein	52%
Lactose	≤10%
Ash	≤6%
Total lipids	27%
Phospholipids	≥16%
% of total phospholipids:	
Phosphatidyl ethanolamine (PE)	22
Phosphatidyl choline (PC)	27
Phosphatidyl serine (PS)	12
Phosphatidyl inositol (PI)	8
Sphingomyelin (SM)	27
Others	4
Moisture	≤5%
pH	6.2–0.8

Microbiological

Standard plate count	<10,000/g
Enterobacteriaceae	<10/g
Salmonella	neg./125 g
Bacillus cereus	<100/g
Yeast and mould	<200/g

[a] Arla Foods Ingredients.

Applications

MFGM cream serum concentrate is an emulsifying agent as well as an ingredient for nutritional applications. Particularly in the latter case, a high percentage of phospholipids is desirable to restrict the dosage level in the final product. Applications thereof are, for instance, infant food and health drinks. It is also used in cosmetic products (e.g. hair and skin care).

Properties

In this MFGM ingredient, different fat compounds as well as various proteins are concentrated concurrently. In this example the lipid fraction contains neutral lipids and phospholipids as well as other complex lipids such as the carbohydrate-containing gangliosides. Milk phospholipids differ in composition from soybean and egg phospholipids with respect to SM and PS. The nutritional properties of milk phospholipids are promising. However, they are subject to ongoing debate with respect to amongst others brain function, anticancer properties, antimicrobial activity and cholesterol reduction.

The phospholipid level can be further increased by 'deproteinization', which includes use of $CaCl_2$, pH adjustment and heating at elevated temperatures.

Literature

Burling H & Graverholt G (2008) Milk – A new source for bioactive phospholipids for use in food formulations. *Lipid Technology* **20**: 229–31.

Burling H, Andersson I & Schneider M (2006) Phosphatidyl serine enriched milk fractions for the formulation of functional foods. WO 2006/128465 [patent].

Dalemans D, Blecker C, Bodson P, Danthine S, Deroanne C & Paquot M (2010) Milk ingredient enriched in polar lipids and uses thereof. US2010/0068293 [patent].

Unmatured cheese non-fat

Description

Unmatured cheese non-fat is available in a powder form (rennet casein) as well as in a moist form (natural cheese 0^+). Both are manufactured from pasteurized skim milk by proteolytic coagulation of the casein. For this purpose calf rennet or similar enzymes such as enzymes from microbial sources are applied. The coagulum is further processed and the sweet whey is separated. In case of natural cheese 0^+, the normal cheese-making procedures are followed. For rennet casein the remaining curd is thoroughly washed to remove all lactose, subsequently dried with the help of a vibratory type drier and milled to the required particle size. The product analysis of rennet casein is demonstrated here.

Analysis of rennet casein

Chemical

Protein (N × 6.38)	80%
Fat	1%
Lactose	0.1%
Ash	8%
Calcium	2.8%
Phosphorus	1.4%
Moisture	11%
pH (5% TS - 20°C)	7

Microbiological[a]

Standard plate count	<10,000/g
Coliform	neg./0.1g
Salmonella	neg./375g
Staphylococcus CP	neg./g
Thermophiles	<2000/g
Yeast and mould	<10/g

[a]Glanbia Ingredients.

Applications

The industrial applications of rennet casein are limited. In the past it was used for the production of plastic materials. Nowadays it is utilized as a food ingredient, particularly in analogues or imitation cheeses. It offers the possibility of using vegetable fat in formulas. The ingredient is available in a coarse variant (30 mesh – 600 μm) or a finer ground alternative (90 mesh – 160 μm). An alternative for rennet casein is natural cheese 0^+ in, for instance, a block form.

Properties

By splitting off the hydrophilic glycomacropeptide during the rennet treatment, the casein is converted into a hydrophobic substance. For this reason rennet casein is insoluble in water. It can be transformed into a soluble form again by adding emulsifying agents, which are used in the production of processed or imitation cheeses. This results in excellent emulsifying and melting characteristics. The ingredient has a clean/bland flavour. Non-fat cheese is sometimes preferred over rennet casein as it may be processed more conveniently. The ingredient has a minimum moisture content of 50% and may or may not be brine-treated. A non-fat, unsalted cheese has a maximum shelf life of 1 month when kept at 2°C. A frozen variant is less critical in this respect.

Literature

Lucey JA (2008) Perspectives on the use of cheese as a food ingredient. *Dairy Science and Technology* **88**: 573–94.

Mulvihill DM (1989) Casein and caseinates manufacture. In: Fox PF (ed.), *Developments in Dairy Chemistry - 4*. Elsevier Applied Science, pp. 104–6.

Southward CR, Casein products, available at http://nzic.org.nz/ChemProcesses/dairy/ 3E.pdf.

Unmatured cheese low salt

Description

This unmatured cheese is a cheese curd with a low salt content. It is produced from cow's milk. The cheese product is used for industrial purposes and not aimed for direct consumption. The ingredient is mostly referred to as cagliata cheese. It is produced by following normal cheese-making procedures. The shelf life of the product is improved by using a steam treatment. A product specification of a cagliata cheese is set out here.

Product analysis[a]

Chemical

Dry matter	>53%
Fat on dry matter	47%
Salt on dry matter	0.5–0.8%
Protein	±23%
pH	5.3–5.6

Microbiological

Coliform	<1000/g
E. coli	<10/g
Staphylococcus CP	<100/g
Salmonella	neg./25 g
Listeria monocytogenes	neg./25 g
Yeast and moulds	<100/g
Shelf life at 0–4°C	4 months

[a]Google: product specifications cagliata cheese; see also Milchwerk Jäger GmbH.

Applications

The unmatured low-salt cheese is specifically intended for use in the manufacture of processed mozzarella or mozzarella-type cheeses with a long shelf life. The final product goes through a dry stretching process instead of the traditional stretching method of using a brine solution. In the stretching procedure special equipment such as extruders may be utilized.

Properties

The unmatured low-salt cheese has a mild lactic flavour and a firm/elastic texture. The salt content of the starting product is kept low. In the following 'mozzarella process' the required fibrous structure of the final product is induced at ±70°C. Standardization on the required salt content is carried out afterwards. Then part of the calcium of the casein micelles is solubilized and replaced by sodium.

Literature

Gaucheron F, Le Graët Y & Briard V (2000) Effect of NaCl addition on the mineral equilibrium of concentrated and acidified casein micelles. *Milchwissenschaft* **55**: 82–6.

Kempter K & Wolfschoon-Pombo A (2005) Processed Mozzarella cheese and method for its production. US 6,933,000 [patent].

Meyer M, Hofacker D & Henning M (2008) Verfahren zum Herstellen von fermentierten Milchprodukten, insbesondere Mozarella. EP2168429A1 [patent].

Smith RM, Sorley AG & Kirkby MW (2009) Method and apparatus to produce a pasta filata cheese. US 2009/0117248A1 [patent].

Processed cheese non-melt hard

Description

The non-melt hard processed cheese retains its shape when reheated. It is called thermally form-stable. Processed cheese is not made directly from milk. The product is manufactured from shredded natural cheese using a special emulsifying salt. The process is followed by heating and continuous blending to form a product with an extended shelf life and the required firm structure. An illustration of a non-melt hard version with 45% fat in dry matter is given. The ingredient is based on young (1.5 month) Cheddar cheese.

Product analysis

Chemical

Dry matter	53%
Fat on dry matter	45%
pH	5.6–5.7
Intact casein	80–90%

Microbiological[a]

Standard plate count	≤10,000/g
Enterobacteriaceae	≤100/g
E. coli	≤100/g
Staphylococcus CP	≤100/g
Listeria monocytogenes	Absent
Shelf life at 20°C	max. 1 year

[a]Kaasfabriek Eyssen: data at the end of the shelf life period.

Applications

The processed cheese non-melt hard can be grated, cut and shaped into blocks. It can be used in fried snacks and still does not show any leakage. It is also used in, for instance, bread products and gratin dishes. Under these conditions the processed cheese particles stay visible and attractive.

Properties

This processed cheese variant is particularly suitable in frying situations. It maintains its structure and shows no flow and stretch. It has similar characteristics and salt content (>4.5% on moisture) as the Hispanic fresh cheese called queso fresco. However, the shelf life of the processed cheese non-melt is counted in months instead of days and it does not expel serum.

Literature

Berger W, Klostermeyer H, Merkenich K & Uhlmann G (2002) *Processed Cheese Manufacture – a Joha Guide*. BK Giulini Chemie GmbH, p. 222.

Glass K & Doyler ME (2005) Safety of processed cheese – a review. Available at: http://fri.wisc.edu/docs/pdf/ProcCheese_May2005_v2.pdf

Johnson ME & Lucey JA (2006) Calcium: a key factor in controlling cheese functionality. *Australian Journal of Dairy Technology* **61**: 147–53.

Processed cheese semi-melt hard

Description

An important attribute of processed cheese is its consistency. This variant melts slightly during heating. Processed cheese is not made directly from milk. It is produced from natural cheese using emulsifying salts. The process is carried out by heating and continuous blending to form a product with an extended shelf life. A semi-melt hard variant of 45% fat on dry matter is rendered as an example. The basis of this ingredient is young (1 month) Cheddar and/or Gouda cheese.

Product analysis

Chemical

Dry matter	57%
Fat on dry matter	45%
pH	5.7
Intact casein	80–90%

Microbiological[a]

Standard plate count	≤10,000/g
Enterobacteriaceae	≤100/g
E. coli	≤100/g
Staphylococcus CP	≤100/g
Listeria monocytogenes	Absent
Shelf life at 20°C	max. 1 year

[a]Kaasfabriek Eyssen: data at the end of the shelf life period.

Applications

This processed cheese variant melts slightly, when heated for use in its final application. The cheese has good toasting properties and can, for instance, be applied in sandwiches and cheeseburgers.

Properties

The processed cheese semi-melt needs to have good remelting properties. Its texture is a compromise between being firm enough to be sliced and still suitable for toasting. A high intact casein content, the type of emulsifying agents and the processing are important. The product is pasteurized and hot-packed under vacuum. Subsequently, the cheese blocks are cooled rapidly. The shelf life is determined by the pasteurization process, the design of the process line and the formula used. In particular, the formula needs to avoid the outgrowth of surviving spores of clostridia bacteria. They can be pathogenic or cause defects such as blowing. Factors such as moisture content, salt content, pH, water activity and use of preservatives have an impact.

Literature

Berger W, Klostermeyer H, Merkenich K & Uhlmann G (2002) *Processed Cheese Manufacture – a Joha Guide*. BK Giulini Chemie GmbH, p. 221.

Glass KA, Kaufman KM & Johnson EA (1998) Survival of bacterial pathogens in pasteurized process cheese slices stored at 30°C. *Journal of Food Protection* **61**: 290–4.

Processed cheese high-melt hard

Description

High-melt hard processed cheese flows when heated. It is called stretch behaviour. Processed cheese is not derived directly from milk. In the example given here a major part of the natural cheese is replaced by rennet casein. Rennet casein, which may be considered a very young cheese, has been converted into a dry form. The high-melt hard variant is manufactured from rennet casein, shredded young Cheddar cheese (1 month), butter and a selected emulsifying salt. Subsequently, the mix of ingredients is heated and continuously blended to form a product with an extended shelf life. The ingredient contains 40% fat on dry matter in this example.

Product analysis

Chemical

Dry matter	51%
Fat on dry matter	40%
pH	±5.7
Intact casein	>90%

Microbiological[a]

Standard plate count	≤10,000/g
Enterobacteriaceae	≤100/g
E. coli	≤100/g
Staphylococcus CP	≤100/g
Listeria monocytogenes	Absent
Shelf life at 20°C	max. 1 year

[a]Kaasfabriek Eyssen: data at the end of the shelf life period.

Applications

Processed cheese high-melt hard can be sliced or grated into pieces of uniform size. Mostly, it is applied as a ready-made meal topping. The ingredient is used, for instance, on industrially produced pizzas, pastas and salads. This processed cheese is a component of so-called convenience food formulas.

Properties

This processed cheese variant is a cheese-like product. As it is composed of dairy constituents, the name processed cheese is used instead of imitation cheese or cheese analogue. The product has a bland flavour. Due to its stretching properties, namely its capacity to form fibrous strands, the ingredient is applied as a topping. It is a less expensive replacer for a low-moisture, part-skim mozzarella. The shelf life is also considerably longer.

Literature

Berger W, Klostermeyer H, Merkenich K & Uhlmann G (2002) *Processed Cheese Manufacture – a Joha Guide.* BK Giulini Chemie GmbH, p. 224.

Jana AH & Upadhyay KG (2003) Mozzarella cheese analogue – a review. *Journal of Food Science Technology* **40**: 1–10.

Processed cheese high-melt soft

Description

This high-melt soft processed cheese variant is characterized by a low viscosity. The product is also referred to as 'pumpable' or liquid cheese. Processed cheese is not derived directly from milk. It is produced from shredded natural cheese using emulsifying salts. The process is followed by continuous blending and an intensive heat treatment to form a product with the required viscosity and an extended shelf life. An example of a product suitable for filling into tubes is given here. The processed cheese concerned is based on a medium-ripe (2-month-old) Gouda cheese.

Product analysis

Chemical

Dry matter	38%
Fat on dry matter	45%
pH	5.6
Intact casein	60–75%

Microbiological[a]

Standard plate count	≤10,000/g
Enterobacteriaceae	≤100/g
E. coli	≤100/g
Staphylococcus CP	≤100/g
Listeria monocytogenes	Absent
Shelf life at 7°C	max. 9 months

[a]Kaasfabriek Eyssen: data at the end of the shelf life period.

Applications

Of the nearly unlimited number of processed cheeses – varying in fat, salt, moisture and flavour – the high-melt soft variant is just one example. This variant is further processed into food products. It is used in snacks, soufflés, sauces and soups. The high-melt soft processed cheese is also suitable as a filling for tubes.

Properties

The low viscosity of this type of processed cheese, even at low temperatures is a key aspect. This feature makes the product easy for dosing. A controlled breakdown of the protein matrix during manufacturing and storage is therefore essential. A reduced percentage of intact casein is a valuable indication of protein breakdown. Another aspect is the substantial increase in moisture content. Both factors will improve the flow properties. A disadvantage is that the product is more vulnerable from a microbiological point of view. High hygienic standards and the use of preservatives are also critical for the desired shelf life.

Literature

Berger W, Klostermeyer H, Merkenich K & Uhlmann G (2002)) *Processed Cheese Manufacture – a Joha Guide.* BK Giulini Chemie GmbH, p. 213.

Lee SK, Buwalda RJ, Euston SR, Foegeding EA & McKenna AB (2003) Changes in the rheology and microstructure of processed cheese during cooking. *Lebensmittel-Wissenschaft & Technologie* **36**: 339–45.

Otto F & Dubielzyk A (1993) Process for producing pumpable foodstuffs, in particular processed cheese. US 5,240,724 [patent].

Enzyme-modified cheese: Emmentaler type

Description

This enzyme-modified cheese (EMC) has an Emmentaler or 'Swiss' flavour. In this type of ingredient the bulk of the cheese mass is not used and the emphasis is on a highly intense and stable flavour. The ingredient originates not directly from milk but from an immature cheese, which is submitted to an accelerated ripening. Cheese, water and emulsifying agents are converted into a paste. The mixture is homogenized and pasteurized. Exogenous enzymes and/or microorganisms are added and incubation is carried out in strictly controlled conditions (pH, temperature, time, agitation). The process, which is usually a closely guarded trade secret, is followed by a heating step to inactivate the enzymatic activity and achieve long-term stabilization. An example of an EMC Emmentaler type (in a paste form) is given here.

Product analysis[a]

Chemical

Moisture	44–48%
Fat	19–23%
Salt	2–4%
Titratable acidity (mL 0.05N NaOH)	50–54/g
pH	4.0–4.4

Microbiological

Standard plate count	≤5000/g
Coliform	≤10/g
E. coli	neg.
Yeast and mould	≤100/g
Shelf life at 3–8°C	12 months

[a]Jeneil Biotech.

Applications

EMC Emmentaler type is only one ingredient in a family of dairy flavours. These ingredients are used in a wide range of convenience foods such as processed/imitation cheeses, cheese spreads, snack coatings, crisps, crackers, soups, sauces, salad dressings and ready meals. Due to its intense flavour, the dosage level is considerable lower than that for natural cheese. As a result thereof a desired flavour can be achieved without causing an increase in fat. A percentage of 1 to 5% is applied, depending on the finished product.

Properties

In natural Emmentaler cheese specific cultures are used that result in a certain flavour profile. It is characterized, particularly, by propionic acid, free proline and acetic acid. The amino acid proline contributes to the typical sweetish taste. In the manufacture of an EMC Emmentaler variant, proline may be used as a specific key component to determine the completion of the fermentation process. In spite of many efforts, the flavour profile of EMCs is not exactly identical to that of natural cheese. Sometimes 'top notes' – natural flavours – are added to obtain a more balanced overall flavour. When diluted in a suitable blend, a desirable flavour effect may be achieved.

Literature

Guinee TP & Kilcawley KN (2004) Cheese as an ingredient. In: Fox PF, McSweeney PLH, Cogan TM & Guinee TP (eds), *Cheese: Chemistry, Physics and Microbiology, vol. 2: Major Cheese Groups*. Elsevier, pp. 395–428.

Kilcawley KN, Wilkinson MG & Fox PF (1998) Review: enzyme-modified cheese. *International Dairy Journal* **8**: 1–10.

Enzyme-modified cheese: blue cheese type

Description

This enzyme-modified cheese (EMC) has the 'sharpness' of a blue cheese flavour. As opposed to blue cheese itself, it consists of intensified flavour constituents without undesirable cheese mass. The ingredient does not stem directly from milk. An accelerated ripening process is applied to an aqueous cheese dispersion that contains an emulsifying agent. Lipolytic and proteolytic enzymes, as well as the spores of the highly lipolytic mould *Penicillium roquefortii*, may be added. Fermentation is carried out in a slurry system under controlled conditions. The process is completed by homogenization and a heat treatment to inactivate the enzymes and moulds. In this way a flavour ingredient with a long shelf life is created. An illustration of an EMC blue cheese type, in paste form, is given here.

Product analysis[a]

Chemical

Moisture	50–54%
Fat	22–26%
Salt	4–6%
Titratable acidity (mL 0.05N NaOH)	21–25/g
pH	4.7–5.1

Microbiological

Standard plate count	≤5000/g
Coliform	≤10/g
E. coli	neg.
Yeast and mould	≤100/g
Shelf life at 3–8°C	12 months

[a]Jeneil Biotech.

Applications

EMC blue cheese type is a member of a family of about 12 cheese flavours. The ingredient is also available in powder form, which is convenient for use in dry blends. In this case food-acceptable flavour carriers need to be added, such as milk solids or maltodextrin. The EMC ingredients are used in a wide range of convenience foods, for instance processed/imitation cheeses, cheese spreads, snack coatings, crisps, crackers, soups, sauces, salad dressings and ready meals. A dosage level of 1 to 5% is applied depending on the specific food product.

Properties

In general the intensity of the flavour components in EMCs may be enhanced 15–30 times. The EMC blue cheese variant is characterized by a strong lipolysis of fat. Particularly, the short-chain fatty acids that are typically present in milk fat play a role. In the manufacture of EMC blue cheese type, methyl ketones can be used to determine the completion of the fermentation process. The procedure can be optimized by applying a two-step process, in which the proteolysis part is carried out separately from the lipolysis step.

Literature

Groesbeck CA, Kwon SS-Y & Vadehra DV (1993) Blue cheese flavorant. EP 622.028 [patent].
Kilcawley KN (2002) Studies on the biochemistry and technology of enzyme-modified Cheddar cheese. PhD thesis, University College Cork.

Whey powder$_{sw}$

Description

Whey powder is produced from clarified, sometimes decolorized, sweet cheese whey. It is concentrated by means of an evaporator and the lactose is partially crystallized. The resulting concentrate is subsequently spray dried. The composition is to some extent dependent on the type of cheese. The annual variation in sweet whey constituents within one province of The Netherlands is shown here.

Product analysis

Chemical[a]

Protein (N × 6.4)	12–14% on TS
Fat	0.7–1.3% on TS
Lactose	78–81% on TS
Ash	7.7–8.7% on TS
Calcium	0.5–0.7%
Magnesium	0.1%
Potassium	2.2–2.7%
Sodium	0.6%
Phosphorus	0.6–0.7%
Chloride	1.5–2.3%
Citric acid	2.2–2.7% on TS
Lactic acid	0.5–2.0% on TS
Moisture	4.0%
pH	6.3–6.5
Protein/ash ratio	1.6

[a]Olling (1963).

Microbiological[b]

Standard plate count	≤30,000
Coliform	≤10/g
E. coli	neg./g
Salmonella	neg./100 g
Staphylococci	neg.

[b]US Dairy Export Council.

Applications

Due to its low protein/ash ratio whey powder has only a limit number of applications in food products. Examples are bakery, snacks and frozen products. However, whey powder is mostly applied in animal feed products for nutritional purposes.

Properties

Lactose is by far the most abundant component in whey. It is a reducing sugar and is useful in development of a brown colour in bakery products. The stickiness and susceptibility to caking of whey powder, which is caused by amorphous lactose, can be reduced by pre-crystallizing the lactose before drying.

Cheese whey has a pH of at least 5.7. The pH of acid whey, a by-product of casein or quark manufacture, for instance, is less than 5. For this reason acid whey has a higher ash level and therefore its calcium and phosphorus content is 2–3 times higher.

Literature

Durham RJ (2009) Modern approaches to lactose production. In: Corredig M (ed.), *Dairy-Derived Ingredients – Food and Nutraceutical Uses*. Woodhead, pp. 103–44.

Olling CCJ (1963) Composition of Friesian whey. *The Netherlands Milk and Dairy Journal* **17**: 176–84.

United States Dairy Export Council (2009) Sweet Whey Powder. Available at: http://www.usdec.org/products/content

Demineralized whey powder

Description

Demineralized whey powder is manufactured from clarified sweet cheese whey. The whey is treated with a cation- and anion-exchange resin to reduce the ash content by at least 90%. It is concentrated by an evaporator, the lactose is partially crystallized and afterwards the concentrate is spray dried.

Product analysis (typical)

Chemical

Protein (N × 6.38)	13%
Fat	1%
Lactose	82%
Ash	1%
Calcium	0.2%
Potassium	0.3%
Sodium	<0.1%
Phosphorus	0.2%
Chloride	<0.1%
Moisture	3%
pH	6.8
Protein/ash ratio	13

Microbiological[a]

Standard plate count	<10,000/g
Enterobacteriaceae	neg. in 3 × 1 g
E. coli	neg./10 g
Salmonella	neg. in 5 × 25 g
Staphylococci CP	neg./g
Yeasts and moulds	≤100/g

[a] FrieslandCampina.

Applications

Whey powder with a desalting degree of more than 90% is a well-established ingredient for infant foods. For other applications such as confectionery, bakery products and frozen desserts, a lower degree of demineralization, for instance 50%, can be sufficient. In this case, alternative processes may be considered too.

Properties

The high protein/ash ratio, the favourable amino acid profile and the elevated lactose content are of value for its application in 'humanized' baby food. Due to its salty effluent stream the ion-exchange process is limited in the choice of production locations.

Literature

Greiter M, Novalin S, Wendland M, Kulbe KD & Fischer J (2002) Desalination of whey by electrodialysis and ion exchange resins: analysis of both processes with regard to sustainability by calculating their cumulative energy demand. *Journal of Membrane Science* **210**: 91–102.

Kjærgaard Jensen G & Oxlund JK (1988) Concentration and drying of whey and permeates. In: *Trends in utilization of whey*. Bulletin of the International Dairy Federation no. 233, pp. 4–20.

Whey protein concentrate: WPC 34_{sw}

Description

This whey protein concentrate is made from clarified, sometimes decolorized, sweet cheese whey. The protein content is brought to the level of skim milk by ultrafiltration. Subsequently, the solids content of the concentrate is increased by an evaporator and spray dried into a powder.

Product analysis

Chemical

Protein (N × 6.38)	34%
True protein	30.6%
Glycomacropeptide	1.5%
Other NPN	2.3%
Fat	3%
Lactose	50%
Ash	8%
Calcium	0.6%
Phosphorus	0.6%
Moisture	5%
pH	6.5
Protein/ash ratio	4.3

Microbiological[a]

Standard plate count	<30,000/g
Coliform	<10/g
E. coli	neg./g
Salmonella	neg./100 g
Staphylococci	neg.

[a]US Dairy Export Council.

Applications

WPC 34 is used mainly to partially replace skim milk powder in a range of food products (e.g. dairy, confectionery, bakery, snacks). Furthermore, it can be applied for nutritional reasons in animal feed products (e.g. calf milk replacer). An ingredient with similar composition and application can be obtained by combining demineralization of whey with removal of part of the lactose (crystals).

Properties

These WPCs are different from skim milk powder as far as protein and mineral composition is concerned. The protein has a good solubility over a wide pH range, which is, for instance, beneficial for application in cloudy beverages. The gelling properties of the whey protein make it favourable for use in cultured dairy products such as yoghurt. Reproducibility can be improved by careful selection of the source of cheese whey and its process history (acidification, temperature, bleaching).

Literature

Evans J, Zulewska J, Newbold M, Drake MA & Barbano DM (2009) Comparison of composition, sensory, and volatile components of thirty-four percent whey protein and milk serum protein concentrates. *Journal of Dairy Science* **92**: 4773–91.

United States Dairy Export Council (2009) WPC 34. Available at: http://www.usdec.org/products/content

de Wit JN (1998) Nutritional and functional characteristics of whey proteins in food products. *Journal of Dairy Science* **81**: 597–608.

Whey protein concentrate: WPC 80_{sw}

Description

This whey protein concentrate is produced from clarified sweet cheese whey. Ultrafiltration combined with diafiltration (washing step) is performed to achieve a high protein content. Subsequently, the solids level of the concentrate is increased by an evaporator and spray dried into a powder. WPC 80_{sw} is available in various versions with approximately the same chemical composition.

Product analysis

Chemical

Protein (N × 6.38)	80%
Fat	6%
Lactose	6%
Ash	4%
Calcium	0.3%
Phosphorus	0.4%
Moisture	4%
pH	6.4

Nutritional (g per 100 g of protein)

Alanine	5.2
Arginine	2.7
Aspartic acid	11.3
Cysteine	2.5
Glutamic acid	18.2
Glycine	2.0
Histidine*	2.1
Isoleucine*°	7.2
Leucine*°	11.4
Lysine*	9.8
Methionine*	2.3
Phenylalanine*	3.8
Proline	6.4
Serine	6.1
Threonine*	7.9
Tryptophan*	2.2
Tyrosine	3.2
Valine*°	6.9

*Essential amino acids.

°Branched-chain amino acids.

Total amino acids >100 due to water-containing molecules.

Microbiological[a]

Standard plate count	<10,000/g
Coliform	<10/g
E. coli	<3/g
Salmonella	neg./25 g
Yeast and mould	<50/g

[a]Hilmar Ingredients.

Applications

WPC_{sw} is used as an ingredient, for instance, in dietary products (infant foods), sports nutrition (e.g. drinks, snacks, bars) and meat/fish products (surimi).

Properties

The whey protein concentrate can be modified by means of a heat treatment to provide good heat stability and gel characteristics in various food products. However, in beverages it forms a cloudy solution.

Literature

Evans J, Zulewska J, Newbold M, Drake MA & Barbano DM (2010) Comparison of composition and sensory properties of 80% whey protein and milk serum protein concentrates. *Journal of Dairy Science* **93**: 1824–43.

Mallee L & Steijns J (2007) Whey protein concentrates from acidic whey: benefits for use in infant formulas. *Agrofood industry hi-tech* **18**: 14–16.

Moon Bokyung & Mangino ME (2004) The effect of preheating on whey protein concentrates. *Milchwissenschaft* **59**: 165–9, 294–7.

Nopianti R, Huda H & Ismail N (2011) A review on the loss of the functional properties of proteins during frozen storage and the improvement of gel-forming properties of surimi. *American Journal of Food Technology* **6**: 19–30.

de Wit JN (2002) Lecturer's handbook on whey, EWPA.

Whey protein concentrate: WPC 80_{aw}

Description

This whey protein concentrate is made from clarified acid casein whey. A high protein content is obtained through ultrafiltration followed by diafiltration (washing step). The concentrate is standardized to neutral pH and subsequently spray dried.

Product analysis

Chemical

Protein (N × 6.38)	80%
Fat	5%
Lactose	6%
Ash	4%
Calcium	0.25%
Moisture	5%
pH	6.5

Nutritional (g per 100 g of protein)

Alanine	5.1
Arginine	2.9
Aspartic acid	12.1
Cysteine	3.2
Glutamic acid	18.6
Glycine	1.9
Histidine*	2.4
Isoleucine*°	6.0
Leucine*°	12.8
Lysine*	10.9
Methionine*	2.5
Phenylalanine*	4.0
Proline	5.5
Serine	4.8
Threonine*	5.4
Tryptophan*	2.6
Tyrosine	3.6
Valine*°	5.3

*Essential amino acids.
°Branched-chain amino acids.
Total amino acids >100 due to water-containing molecules.

Microbiological[a]

Standard plate count	<10,000/g
Enterobacteriaceae	<10/g
Yeasts and moulds	<25/g
Staphylococci CP	neg./g
Salmonella	neg./50g

[a] FrieslandCampina.

Applications

WPC_{aw} is a nutritional as well as a functional ingredient. WPC 80 is used in dietary, infant and sports nutrition. As a gelling agent, it can be used in a variety of foods, for instance in desserts and meat/fish products.

Properties

This WPC 80 does not contain glycomacropeptide (GMP), which is released during cheese manufacture. Compared to WPC 80_{sw}, it results in a significant decrease in the amino acid threonine. Together with cysteine and tryptophan, a better match with the amino acid pattern of breast milk is obtained. The absence of GMP also has a beneficial effect on the gelling properties of WPC 80_{aw}.

Literature

Mallee L & Steijns J (2007) Whey protein concentrates from acidic whey: benefits for use in infant formulas. *Agrofood industry hi-tech* **18**: 14–16.

Veith PD & Reynolds EC (2004) Production of a high gel strength whey protein concentrate from cheese whey. *Journal of Dairy Science* **87**: 831–40.

Whey protein hydrolysate: WPH 80

Description

This whey protein hydrolysate is manufactured from WPC. The enzymatic hydrolysis of the dissolved whey protein takes place in a batch process under carefully controlled conditions (proteases, substrate concentration, pH, temperature, time). The post-hydrolysis processes include, for instance, enzyme inactivation, filtration, concentration and spray drying. Here an example is given of a partially hydrolysed WPH.

Product analysis (typical)[a]

Chemical

Protein (N × 6.38)	77.2%
Fat	5.1%
Lactose	3.9%
Ash	4.0%
Moisture	<5%
pH	7.2
Degree of hydrolysis	6%
Average MW (Da)	5124
Free amino acids	1%

Nutritional (g per 100 g of protein)

Alanine	4.9
Arginine	2.6
Aspartic acid	10.0
Cysteine	3.1
Glutamic acid	16.9
Glycine	1.8
Histidine*	1.4
Isoleucine*°	6.5
Leucine*°	10.0
Lysine*	8.8
Methionine*	2.0
Phenylalanine*	3.1
Proline	6.1
Serine	4.0
Threonine*	6.1
Tryptophan*	1.3
Tyrosine	2.7
Valine*°	6.0

*Essential amino acids.
°Branched-chain amino acids.

Microbiological

Standard plate count	≤5000/g
Coliform	≤10/g
E. coli	neg./g
Salmonella	neg./g
Yeast and moulds	≤100/g

[a] FrieslandCampina.

Applications

This partially hydrolysed WPH 80, with its low residual content of allergens, is used in infant foods with the aim of reducing the risk of allergies developing. These infant formulas, in which the protein breakdown is limited, are described as 'hypoallergenic' or 'hypoantigenic'. The same type of ingredient is also applied as a 'predigested protein'. It facilitates protein absorption in the human body. In this case it is used for nutritional purposes, for instance, in sport, clinical and geriatric foods.

Properties

Whey protein, which has a very favourable amino acid pattern, is broken down by partial hydrolysis into various peptides. Their molecular weight ranges from less than 500 to more than 10,000 Da. The limited hydrolysis and the use of whey protein improve palatability and reduce the costs. WPH can also be produced from WPI as a starting material. In this case an ingredient with a higher purity is obtained. In solution this WPH gives a clear appearance.

Literature

Alexander DA, Schmitt DF, Tran NL, Barraj LM & Cushing CA (2010) Partially hydrolyzed 100% whey protein infant formula and atopic dermatitis risk reduction: a systematic review of literature. *Nutrition Reviews* **68**: 232–45.

Exl B-M (2001) A review of recent developments in the use of moderately hydrolyzed whey formulae in infant nutrition. *Nutrition Research* **21**: 355–79.

Shin HS, Kim SB, Kang SC *et al.* (2007) Production of low antigenic cheese whey protein hydrolysates using mixed proteolytic enzymes. *Journal of the Science of Food and Agriculture* **87**: 2055–60.

Whey protein isolate: WPI 90_{sw}

Description

Whey protein isolate is a highly purified form of whey proteins. It is made from clarified sweet cheese whey. A heat treatment can be eliminated by applying microfiltration. The composition, as set out here, is obtained through ion-exchange chromatography. Critical process parameters are pH, loading volume, rinsing and elution procedure. The whey is adjusted to a pH of about 4, which is below the isoelectric point of most whey proteins (4.2–5.5). The positively charged proteins are adsorbed by a selected cation exchanger. After loading and rinsing, the proteins are desorbed from the cation exchanger by an elution procedure using a high pH value. The liquid obtained through elution is concentrated by means of ultrafiltration and subsequently spray dried.

Product analysis[a]

Chemical

Protein	≥90%
Fat	≤1.0%
Lactose	≤1.0%
Ash	≤3.0%
Moisture	≤5%
pH	6.7–7.5

Microbiological

Standard plate count	≤10,000/g
Coliform	≤10/g
E. coli	neg./g
Salmonella	neg./1500 g
Staphylococcus (CP)	<10/g
Listeria	neg./25 g
Yeast and moulds	≤10/g

[a]Davisco Foods International.

Applications

Due to its complicated processing, including attention to the side streams, WPI 90 qualifies as a higher priced food ingredient. Examples of its applications are sports and clinical nutrition (sip feeding), protein-fortified clear beverages, re-formed fish products and nutrition bars. An extruded version of WPI can be used in snacks.

Properties

This bland-flavoured WPI 90 has a low lactose content, which allows it to be used in several nutritional applications. The low fat content is advantageous for its functional properties, such as foaming and clarity in solutions. The protein, comprising mainly β-lactoglobulin and α-lactalbumin, can be modified by heat treatment to provide, for instance, favourable gelling properties.

A WPI with similar overall composition can also be obtained through a combination of membrane processes. It can be distinguished from the ion-exchange product by the presence of glycomacropeptide.

Literature

Chatterton DEW, Smithers G, Roupas P & Brodkorg A (2006) Review: Bioactivity of β-lactoglobulin and α-lactalbumin – Technological implications for processing. *International Dairy Journal* **16**: 1229–40.

Doultani S, Turhan KN & Etzel MR (2003) Whey protein isolate and glycomacropeptide recovery from whey using ion exchange chromatography. *Journal of Food Science* **68**: 1389–95.

Qi PX & Onwulata CI (2011) Physical properties, molecular structures and protein quality of texturized whey protein isolates: effect of extrusion temperature. *Journal of Agricultural and Food Chemistry* **59**: 4668–75.

Alpha-lactalbumin

Description

Alpha-lactalbumin is the second of the various whey proteins. It can be produced by, for instance, ultrafiltration, precipitation and ion-exchange chromatography. The latter results in a rather pure form and is given here as an example. Sweet, microfiltered whey can be used as a starting material. After adjustment to about pH 4 the major whey proteins have a positive charge. These proteins are adsorbed on a cation-exchange column. The α-lactalbumin fraction is desorbed using a specific elution liquid. The protein solution obtained is concentrated through ultrafiltration and in addition spray dried.

Product analysis[a]

Chemical

Protein (N × 6.25)	≥95%
α-Lactalbumin (% of protein)	90%
Fat	≤0.5%
Lactose	≤0.2%
Ash	≤3.5%
Moisture	≤6.5%
pH	6.0–7.5

Nutritional (g per 100 g of protein)

Alanine	2.0
Arginine	1.3
Aspartic acid	16.4
Cysteine	4.8
Glutamic acid	12.6
Glycine	2.7
Histidine*	3.0
Isoleucine*°	6.0
Leucine*°	10.8
Lysine*	10.9
Methionine*	1.0
Phenylalanine*	4.1
Proline	3.3
Serine	3.4
Threonine*	4.3
Tryptophan*	4.8
Tyrosine	4.4
Valine*°	4.3

*Essential amino acids.
°Branched-chain amino acids.

Microbiological

Standard plate count	≤2500/g
Coliform	≤10/g
E. coli	neg./g
Salmonella	neg./375 g
Staphylococcus aureus	<10/g
Yeast and moulds	≤10/g

[a] Davisco Foods International.

Applications

Alpha-lactalbumin is the primary protein in human milk and is therefore of interest for infant nutrition. It is mostly applied in the so-called 'humanized' infant formulas. The nutrient composition of human milk is the 'gold standard' for infant food development. The ingredient can also be used in other nutritional applications.

Properties

Alpha-lactalbumin is the primary protein in human milk. Due to its favourable tryptophan and cysteine levels and its modest threonine content, bovine α-lactalbumin can improve the overall amino acid composition of infant food products. Another advantage of α-lactalbumin is that the protein content of the formula can be reduced.

Literature

Chao W (2001) Whey treatment process for achieving high concentration of α-lactalbumin. US 6,613,377 [patent].

Doultani S, Azan Turhan K & Etzel MR (2004) Fractionation of proteins from whey using cation exchange chromatography. *Process Biochemistry* **39**: 1737–43.

Kamau SM, Cheison SC, Chen W, Xiao-Ming Liu & Rong-Rong Lu (2010) Alpha-lactalbumin: Its production technologies and bioactive peptides. *Comprehensive Reviews in Food Science and Food Safety* **9**: 197–212.

Beta-lactoglobulin

Description

Beta-lactoglobulin is the most abundant whey protein. It can be manufactured from clarified cheese whey in the same process scheme as α-lactalbumin. An example of a process is ion-exchange chromatography. The pH of the whey is adjusted below the isoelectric point of the major whey proteins (4.2–5.5). Acidification gives these proteins a positive charge. The acidified whey flows to a cation-exchange column and the proteins are adsorbed. After loading, specific desorption of β-lactoglobulin can be carried out at a selected pH value. Subsequently, this liquid is concentrated by ultrafiltration and spray dried. The example given here is of a highly purified β-lactoglobulin powder.

Product analysis[a]

Chemical

Protein (N × 6.38)	≥95%
β-Lactoglobulin	≥90%
Fat	≤1.0%
Lactose	≤1.0%
Ash	≤3.5%
Moisture	≤6.5%
pH	≤7.5

Microbiological

Standard plate count	≤10,000/g
Coliform	≤10/g
E. coli	neg./g
Salmonella	neg./375 g
Staphylococcus (CP)	<10/g
Listeria	neg/25 g
Yeast and moulds	≤10/g

[a] Davisco Foods International.

Applications

The application of β-lactoglobulin is mainly based on its functional properties. It can be used in a wide range of applications such as desserts, dairy products (e.g. yoghurt), confectionery, beverages, dressings and meat/fish products for its emulsifying, stabilizing, foaming and gelling properties. WPI, which contains a substantial amount of β-lactoglobulin, can sometimes be used as a less expensive alternative.

Properties

The most distinct property of β-lactoglobulin is its gelling behaviour. This property depends on protein concentration, ionic strength, pH and heating procedure. At a pH close to the isoelectric point and at a high ionic strength, heat-induced, coarse gels are obtained. Fine-stranded, heat-induced gels are formed at pH values far from the isoelectric point and at a low ionic strength. Prolonged heating at pH 2 results in a fibrillar structure. In cold-set gelation the proteins are first converted into aggregates during heating at a protein concentration below the critical gelation concentration. In the final application, gelation can be induced by, for instance, adding calcium ions at room temperature.

Literature

Bolder SG (2007) Fibril assembly in whey protein mixtures. PhD thesis, Wageningen University.

Etzel MR (1999) Isolating β-lactoglobulin and α-lactalbumin by eluting from a cation exchanger without sodium chloride. US 5,986,063 [patent].

Strohmaier W (2004) Chromatographic fractionation of whey proteins. In: *Advances in Fractionation and Separation: Process for Novel Dairy Applications*. Bulletin of the International Dairy Federation no. 389, pp. 29–35.

Glycomacropeptide

Description

This glycomacropeptide (GMP), or caseinomacropeptide (CMP), is formed by enzymatic cleavage of κ-casein and is released in the cheese whey. GMP can be produced by membrane processes or by ion-exchange chromatography. The latter is given here as the example. GMP can be manufactured simultaneously with WPI. GMP is isolated by anion exchange. Use is made of the low isoelectric point of GMP (pH 2.8). At pH 5 GMP is negatively charged and the protein is adsorbed. After loading, the charge is reversed and GMP is desorbed. In addition to elution, concentration and spray drying are carried out to obtain the final product.

Product analysis[a]

Chemical

Protein (N × 6.47; % DM)	≥ 82.5%
GMP (N × 7.07; % DM)	≥90.0%
Fat	≤1%
Lactose	≤1%
Ash	≤7%
Moisture	≤7%
pH	≤7.0

Nutritional (g per 100 g of protein)

Alanine	5.5
Arginine	0.5
Aspartic acid	8.6
Cysteine	0.1
Glutamic acid	20.5
Glycine	1.1
Histidine*	0.3
Isoleucine*°	10.1
Leucine*°	2.6
Lysine*	5.9
Methionine*	1.8
Phenylalanine*	0.5
Proline	12.5
Serine	6.1
Threonine*	15.8
Tryptophan*	0.0
Tyrosine	0.1
Valine*°	8.0

*Essential amino acids.
°Branched-chain amino acids.

Microbiological

Standard plate count	≤10,000/g
Coliform	≤10/g
E. coli	neg./g
Salmonella	neg./375g
Staphylococcus CP	<10/g
Listeria	neg./25 g
Yeast and moulds	≤10/g

[a] Davisco Foods International.

Applications

The unique amino acid composition of GMP makes it useful for the treatment of individuals with phenylketonuria (PKU). This is a rare hereditary disorder in which the amino acid phenylalanine cannot be metabolized by the human body. GMP also has potential as a multifunctional compound with promising biological applications, for instance, in oral hygiene and clinical nutrition.

Properties

The three aromatic amino acids (tyrosine, phenylalanine and tryptophan) are hardly present in pure glycomacropeptide products. This is useful for some nutritional applications. As far as health-promoting properties are concerned, GMP interacts with toxins, viruses and bacteria. This activity is strongly mediated by the 'gluco' fraction. This fraction contains sialic acids, which are also found in the human brain. The acidic peptide is highly soluble and heat stable.

Literature

Doultani S, Turhan KN & Etzel MR (2003) Whey protein isolate and glycomacropeptide recovery from whey using ion exchange chromatography. *Journal of Food Science* **68**: 1389–95.

Holst HH & Chatterton EW (2002) Process for preparing a kappa-caseino glycomacropeptide or a derivative thereof. US 6,462,181 [patent].

Thomä-Worringer C, Sørensen J & López-Fandiño R (2006) Health effects and technological features of caseinomacro peptide. *International Dairy Journal* 16: 1324–33.

Lactoferrin

Description

Lactoferrin (LF) is a minor protein in bovine milk. This glycoprotein can be captured from skim milk, but predominantly whey is used as a starting material. The main operation for its isolation is cation-exchange chromatography. Use is made of the net positive charge of LF (isoelectric point 7.8–8.0) at the pH of sweet whey. At this pH the major whey proteins have a negative charge and are rejected, while lactoferrin is adsorbed. Subsequent desorption is carried out, for instance, with a salt solution at a low concentration. In combination with microfiltration and ultrafiltration, LF is obtained with a low bacterial count and a high protein content. After spray drying, a powder with a salmon pink colour is collected.

Product analysis (typical)[a]

Chemical

Protein (N × 6.38)	96%
Lactoferrin (% of protein)	98%
Ash	0.5%
Iron saturation	120 ppm
Moisture	3.5%
pH (2% solution, 20°C)	6.0

Microbiological

Standard plate count	<100/g
Enterobacteriaceae	<1/g
Salmonella	neg./10 g
Staphylococcus aureus	neg./g
Yeast and moulds	<10/g

[a] FrieslandCampina.

Applications

Lactoferrin belongs to the category of the highest priced well-known minor milk ingredients. It is applied in a variety of products, for instance in infant formulas in Asian countries, nutritional supplements, and oral and skin care products. The ingredient has GRAS status (Generally Recognized As Safe) for food products (adults) in the USA.

Properties

Lactoferrin is a protein with several physiological functions. Protection against a variety of pathogens is the best known. Its antimicrobial activity is related to its ability to bind Fe^{3+} ions. Moreover, LF induces immunity and is reported to have anticarcinogenic activity. A critical aspect in the production of LF is its low heat stability.

Literature

Burling H (1994) Isolation of bioactive components from cheese whey. *Scandinavian Dairy Information* **8**: 54–6.

Gonzalez-Chavez SA, Arevalo-Gallegos S & Rascon-Cruz Q (2009) Review: Lactoferrin structure, function and applications. *International Journal of Antimicrobial Agents* **33**: 301.e1–301.e8.

Kussendrager KD, Kivits MGC & Verver AB (1997) Process for isolating lactoferrin and lactoperoxidase from milk and milk products, and products obtained by such process. US 5,596,082 [patent].

Steijns JM & van Hooijdonk ACM (2000) Occurrence, structure, biochemical properties and technological characteristics of lactoferrin. *British Journal of Nutrition* **84**(Suppl. 1): S11–17.

Wakabayashi H, Yamauchi K & Takase M (2006) Review: Lactoferrin research, technology and applications. *International Dairy Journal* **16**: 1241–51.

Lactoperoxidase

Description

Lactoperoxidase (LPO) is another minor milk protein. The enzyme can be obtained from skim milk, but cheese whey is mostly used. It is isolated in the same process scheme as lactoferrin. The main unit operation is cation-exchange chromatography. LPO has a very high positive charge with an isoelectric point above pH 9. At the pH of sweet whey the major whey proteins are negatively charged and are rejected, while LPO is adsorbed by the cation-exchange column. Subsequently, desorption of LPO is carried out. During elution a higher salt concentration is applied compared to lactoferrin. A combination of microfiltration, ion-exchange chromatography and ultrafiltration yields a pure ingredient with a low bacterial count.

Product analysis (typical)[a]

Chemical

Protein (N × 6.38)	95.5%
Ash	0.5%
Moisture	4.5%
pH (2% solution 20°C	6.0
Activity:	1050 U/mg
ABTS method, pH 5.0, 37°C	

Microbiological

Standard plate count	≤100/g
Enterobacteriaceae	<5/g
Salmonella	neg./50 g
Staphylococcus (CP)	neg./g
Yeast and moulds	<10/g

[a] FrieslandCampina.

Applications

Lactoperoxidase can be used as a natural preservative in a variety of applications. With the current pressure on the use of antibiotics, isolated bovine LPO appears to be an attractive antibacterial compound for supplementation in calf starters and fish feeding. Toothpaste, mouthwashes and cosmetics are other applications. Lactoperoxidase can even be used in pesticides in the control of fungi and bacteria.

Properties

The enzyme LPO is another example of the enormous diversity of milk components. Next to xanthine oxidase it is the most abundant enzyme in milk. Its concentration is around 30 mg/L, constituting about 0.5% of the whey proteins. LPO has an antimicrobial effect, which is based on the concomitant existence of three components, namely the enzyme, thiocyanate and hydrogen peroxide. This system forms the antimicrobial hypothianate.

Literature

van Hooijdonk ACM, Kussendrager KD & Steijns JM (2000) *In vivo* antimicrobial and antiviral activity of components in bovine milk and colostrum involved in non-specific defence. *British Journal of Nutrition* **84**(Suppl. 1): S127–S134.

Kussendrager KD & Hooijdonk ACM (2000) Lactoperoxidase: physico-chemical properties, occurrence, mechanism of action and applications. *British Journal of Nutrition* **84**(Suppl. 1): S19–S25.

Kussendrager KD, Kivits MG & Verver AB (1997) Process for isolating lactoferrin and lactoperoxidase from milk and milk products, and products obtained by such process. US 5,596,082 [patent].

Ravensberg WJ, Van der Pas RK, Kussendrager KD & Maas JAM (2002) Pesticide against plant-pathogenic microorganisms. US 6,447,811 [patent].

Osteopontin

Description

Osteopontin (OPN) is one of the numerous minor constituents in bovine milk. It is preferably produced from clarified acid whey. The advantage of acid whey is that it does not contain glycomacropeptide (GMP). Isolation by anion-exchange chromatography is used here as an example. In this procedure a product with a high purity is obtained. OPN is a highly acid glycoprotein with an isoelectric point of about pH 3.5. At the pH of acid whey (4.5), OPN is negatively charged. The protein is adsorbed to the strong anion-exchange column. Desorption takes place with a specific salt solution. The protein solution is purified by ultrafiltration and spray dried.

Product analysis[a]

Chemical

Protein (N × 7.17)	≥88%
OPN of protein (N × 7.17)	≥89%
Fat	≤1%
Lactose	≤1%
Ash	≤9%
Moisture	≤5.5%

Nutritional (g per 100 g of protein)

Alanine	3.3
Arginine	2.2
Aspartic acid	23.4
Cysteine	0.1
Glutamic acid	15.8
Glycine	1.5
Histidine*	3.4
Isoleucine*°	2.6
Leucine*°	5.7
Lysine*	8.2
Methionine*	0.2
Phenylalanine*	4.1
Proline	7.4
Serine	10.4
Threonine*	7.2
Tryptophan*	0.9
Tyrosine	1.8
Valine*°	4.3

*Essential amino acids.
°Branched chain amino acids.
Total amino acids >100 due to water-containing molecules.

Microbiological

Standard plate count	≤5000/g
Enterobacteriaceae	≤10/g
Salmonella	neg./125 g
Staphylococcus aureus (CP)	neg./g
Yeast and moulds	≤100/g
Bacillus cereus	≤30/g

[a]Arla Foods Ingredients.

Applications

Osteopontin is a multifunctional bioactive protein that is used as a dietary supplement, for instance in clinical nutrition. An important application is 'humanized' infant food. Addition of purified bovine OPN to infant formulas increases the concentration to that of human milk (average 2.1% of total protein). Its aim is stimulation of the immune system.

Properties

Osteopontin is a sialic acid-rich, highly phosphorylated glycoprotein. In bovine milk the concentration is on average 0.05% of total protein. OPN is involved in diverse biological processes. It is believed to be involved in, for instance, wound repair, cancer biology, inflammation and renal stone disease. Its strong calcium-binding properties play an interesting role in bone growth. Calcium-binding is related to the high contents of aspartic acid, serine and threonine.

Literature

Azuma N, Maeta A, Fukuchi K & Kanno C (2006) A rapid method for purifying osteopontin from bovine milk and interaction between osteopontin and other milk proteins. *International Dairy Journal* 16: 370–8.

Mazzali M, Kipari T, Ophascharoensuk V, Wesson JA, Johnson R & Hughus J (2002) Review: Osteopontin – a molecule for all seasons. *Quarterly Journal of Medicine* **95**: 3–13.

Schack L, Lange A, Kelsen J *et al.* (2009) Considerable variation in the concentration of osteopontin in human milk, bovine milk and infant formula. *Journal of Dairy Science* **92**: 5378–85.

Lactose monohydrate edible

Description

Edible, or food-grade, lactose is produced from whey or permeate. The liquid is concentrated and during the cooling process the oversaturated lactose solution crystallizes. The crystals are separated, washed and again separated. The refined edible lactose has a slightly higher purity than the regular type. Subsequently, the lactose is dried, milled and sieved to various particle sizes. An example of a medium-ground type is given here.

Product analysis

Chemical

Protein (N × 6.38)	<0.5%
Ash (sulphated)	<0.3%
Lactose monohydrate	≥99.0%
Total moisture	4.5–5.5%
Loss on drying	<0.5%

Physical

Appearance:	
Powder – slight yellow to white	
Solution – opaque, slightly yellow to white	
Particle size distribution[a]:	
% <75 μm (200 mesh)	38
% <150 μm (100 mesh)	57
% <250 μm (60 mesh)	79

[a]Mesh indicates the number of apertures per linear inch (25.4 mm) in a sieve.

Microbiological[b]

Aerobic plate count	≤10,000/g
Coliform	≤10/g
E. coli	neg./g
Salmonella	neg./1500 g
Staphylococcus (CP)	<10/g
Yeast and moulds	≤10/g

[b]Davisco Foods International.

Applications

Edible lactose can be used in a range of products, for instance, in infant foods, confectionery, baked goods, processed meats, soups, sauces, flavourings and as a protein standardizing agent in milk powders. For infant foods the most stringent standards are set. The coarse grades are applied where a good product flow is required. The finer grades are usually selected when other fine ingredients are used. In these cases increased wettability and minimal stratification are also desired.

Properties

Lactose is a carbohydrate found almost exclusively in milk from mammals and is therefore of importance in the nutrition of newly born mammals, including humans. Edible lactose is the milk sugar in its most common form, the so-called alpha-lactose monohydrate. It is a reducing sugar and is known for its influence in developing characteristic golden-brown colours, which may be caused by the Maillard reaction, or caramelization. Lactose differs from other carbohydrates due to its low sweetness and solubility. Furthermore, it is not fermented by baker's yeast. Edible lactose is available in unground, medium-ground or fine-ground varieties. The mesh number is generally an indication of the particle size distribution.

Literature

Association of Lactose Manufacturers (1988) *The ALM Guide to Lactose Properties and Uses.*

Reimerdes EH (ed.) (1990) *Lactose as a Food Ingredient.* Association of Lactose Manufacturers/Expoconsult Publishers.

Schaafsma G (2008) Review: Lactose and lactose derivatives as bioactive ingredients in human nutrition. *International Dairy Journal* **18**: 458–65.

Lactose monohydrate pharma

Description

Pharma lactose monohydrate is isolated from whey or whey permeate. The evaporative concentration is followed by a pre-set crystallization procedure. To produce pharmaceutical grade lactose, a second crystallization step, including a treatment to remove impurities, is needed. The resulting highly pure white crystalline lactose is dried on a fluid bed and is subsequently milled/sifted. A 200-mesh lactose is given as an illustration.

Product analysis

Chemical

Proteins and light-absorbing impurities:	
Absorbance (210–220 nm)	≤0.25
Absorbance (270–300 nm)	≤0.07
Ash (sulphated)	≤0.1%
Heavy metals (mg/kg)	≤5
Acidity (mL 0.1 N NaOH/6 g)	≤0.4
Specific optical rotation	+54.4° to +55.9°
Total moisture (KF method)	4.5–5.5%
Loss on drying (80°C/2 h)	≤0.5%
Initial solubility (20°C)	7.4 g/100 g water

Physical

Appearance:	
Powder	White
Solution (10%)	Clear and colourless
Absorbance (400 nm)	≤0.04
Particle size distribution (air jet sieve)[a]:	
% <45 μm (325 mesh)	50–65
% <100 μm (150 mesh)	min. 90
% <150 μm (100 mesh)	min. 96
% <250 μm (60 mesh)	99–100

[a]Mesh indicates the number of apertures per linear inch (25.4 mm) in a sieve.

Microbiological[b]

Aerobic plate count	≤100/g
E. coli	neg./10 g
Salmonella	neg./100 g
Yeast and moulds	≤10/g

[b]DMV-Fonterra Excipients.

Applications

The use of lactose in the pharmaceutical industry is a highly sophisticated outlet for whey products. Lactose monohydrate is the most common of the various lactose ingredients. It is used as a diluent, filler or bulking agent. The finer grades are mostly used in the traditional wet granulation process for the production of tablets. The coarser grades are used as a flow improver, for instance, in capsules and sachets and as a filler in direct compression tablet formulations. Besides the pharmaceutical industry pharma lactose is also used in the veterinary industry and as a raw material for the production of chemical lactose derivatives.

Properties

For use as a pharmaceutical ingredient (excipient), lactose monohydrate has to fulfil the strict requirements of the various pharmacopoeias – Eur-Ph (European Pharmacopoeia), USP-NF (United States Pharmacopeia and The National Formulary), JP (Japanese Pharmacopoeia). A range of milled and sieved varieties are available, each with a specific particle size distribution profile. The finer grades can be found among the milled forms. The advantages of lactose are its bland taste, good compatibility with other ingredients and low hygroscopic nature.

Literature

European Directorate for the Quality of Medicines and Healthcare (2002) Lactose monohydrate monograph. In: *European Pharmacopoeia* 4th edn. EDQM, pp. 1444–5.

Lactose anhydrous pharma

Description

Pharma anhydrous lactose is obtained from whey or whey permeate. The concentration process is followed by a double crystallization procedure, which is required for lactose of pharmaceutical quality. After the second crystallization step the lactose monohydrate crystals are dissolved. They are purified again and the temperature is kept above the crystallization temperature of anhydrous beta-lactose (93.5°C). This high temperature is maintained during the drying step and a lactose with no crystal water is obtained. Roller driers are commonly applied as a final step in the process.

Product analysis

Chemical

Proteins and light-absorbing impurities:	
Absorbance (210–220 nm)	≤0.25
Absorbance (270– 300 nm)	≤0.07
Ash (sulphated)	≤0.1%
Heavy metals (mg/kg)	≤5
Acidity (mL 0.1 N NaOH/6 g)	≤0.4
Specific optical rotation	+54.4° to +55.9°
Total moisture (KF method)	≤1.0%
Loss on drying (80°C/2 h)	≤0.5%
Initial solubility (20°C)	48 g/100 g water

Physical

Appearance:	
Powder	White
Solution (10%)	Clear and colourless
Absorbance (400 nm)	≤0.04
Particle size distribution (air jet sieve)[a]:	
% <45 μm (325 mesh)	≤20
% <150 μm (100 mesh)	40–65
% <250 μm (60 mesh)	≥80

[a]Mesh indicates the number of apertures per linear inch (25.4 mm) in a sieve.

Microbiology[b]

Aerobic plate count	≤100/g
E. coli	neg./10 g
Salmonella	neg./100 g
Yeast and moulds	≤10/g

[b]DMV-Fonterra Excipients.

Applications

Anhydrous lactose is another value-added ingredient obtained through the dairy industry. In the pharmaceutical industry it is a useful ingredient in the efficient production of tablets by the direct compression method. It acts as an excipient, applied as a non-drug filler-binder.

Properties

The relatively coarse roller-dried powder is composed mostly of agglomerated small crystals of anhydrous beta-lactose. Due to this production method, about one-fifth is present as anhydrous alpha-lactose. The anhydrous roller-dried product can be easily compacted into tablets without the use of a solvent. The product is sweeter than the lactose monohydrate form and is also more soluble in water. In solution at room temperature, an equilibrium is formed in a ratio of 38% alpha- and 62% beta-lactose.

Literature

European Directorate for the Quality of Medicines and Healthcare (2002) Lactose monohydrate monograph. In: *European Pharmacopoeia*, 4th edn. EDQM, pp. 1443–4.

Fonterra (2005) *Introduction to lactose New Zealand – Pharmaceutical lactose*.

Vromans H (1987) Studies on consolidation and compaction properties of lactose. PhD thesis, Groningen University.

Lactose spray-dried pharma

Description

Lactose spray-dried pharma is produced from whey or whey permeate. Concentration of this process stream is followed by double crystallization, including a treatment to remove impurities. This is the common procedure for the production of pharmaceutical lactose. In addition, the milled crystalline alpha-lactose monohydrate is suspended in a saturated lactose solution at an elevated temperature. The suspension is cooled at a preset temperature, after which seed material can be added and the product is spray dried.

Product analysis

Chemical

Proteins and light-absorbing impurities:	
Absorbance (210–220 nm)	≤0.25
Absorbance (270–300 nm)	≤0.07
Ash (sulphated)	≤0.1%
Heavy metals (mg/kg)	≤5
Acidity (mL 0.1 N NaOH/6 g)	≤0.4
Specific optical rotation	+54.4° to +55.9°
Total moisture (KF method)	4.5-5.5%
Loss on drying (80°C/2 h)	≤1.0%

Physical

Appearance:	
Powder	White
Solution (10%)	Clear and colourless
Absorbance (400 nm)	≤0.04
Particle size distribution (air jet sieve)[a]:	
% <45 μm (325 mesh)	≤15
% <100 μm (150 mesh)	30–60
% <250 μm (60 mesh)	≥98

[a]Mesh indicates the number of apertures per linear inch (25.4 mm) in a sieve.

Microbiological[b]

Aerobic plate count	≤100/g
E. coli	neg./10 g
Salmonella	neg./100 g
Yeast and moulds	≤10/g

[b]DMV-Fonterra Excipients.

Applications

This pharma lactose spray-dried variant is used as a filler-binder for tablets produced by the direct compression method. It can also be used in sachet and capsule formulations due to its good flow properties.

Properties

The spray-dried particles consist of spherical agglomerates of small monohydrate crystals bound together by glassy amorphous lactose. The amount of amorphous lactose is about 10%. The amorphous part improves its compactibility, which is an essential functional property. The spherical form gives good flow properties. However, due to the hygroscopic nature of amorphous lactose it needs to be stored in sealed containers in a cool environment.

Literature

European Directorate for the Quality of Medicines and Healthcare (2002) Lactose monohydrate monograph. In: *European Pharmacopoeia*, 4th edn. EDQM, pp. 1444–5.

Kussendrager KD, Van den Biggelaar HAM & Vromans H (1989) Spray dried lactose and process for preparing the same. US 4,802,926 [patent].

Lactose inhalation pharma

Description

This alpha-monohydrate is obtained from whey or whey permeate. Concentration of these process streams is followed by two crystallization steps, which includes a refining treatment. Subsequently, extensive drying takes place. The highly purified crystals are gently milled or sieved with special techniques in a carefully standardized environment. An example of a sieved variant is given here.

Product analysis

Chemical

Protein (Bradford)	≤0.02%
Proteins and light-absorbing impurities:	
Absorbance (210–220 nm)	≤0.25
Absorbance (270–300 nm)	≤0.07
Ash (sulphated)	≤0.1%
Heavy metals (mg/kg)	≤5
Acidity (mL 0.1 N NaOH/6 g)	≤0.4
Specific optical rotation	+54.4 to +55.9
Total moisture (KF method)	4.5–5.5%
Loss on drying (80°C/2 h)	≤0.5%

Physical

Appearance:	
Powder	White
Solution (10%)	Clear and colourless
Absorbance (400 nm)	≤0.04
Particle size distribution (air jet sieve)[a]:	
% <32 μm (425 mesh)	5–10
% <63 μm (230 mesh)	≥70
% <100 μm (150 mesh)	100

[a]Mesh indicates the number of apertures per linear inch (25.4 mm) in a sieve.

Microbiological[b]

Aerobic plate count	≤10/g
E. coli	neg./10 g
Salmonella	neg./50 g
Staphylococcus aureus	neg./10 g
Yeast and moulds	≤10/g

[b]DMV-Fonterra Excipients.

Applications

Pharma inhalation lactose is a common ingredient for use in therapeutic aerosol powders. Dry powder inhalers (DPI) include a portable device and its content. The powder – drug and its carrier lactose – is administered via the lungs. The drug is micronized and attached to the larger lactose particles. The latter serve to provide the drug in a manageable quantity and facilitate the detachment of the drug in the respiratory system.

Properties

Inhalation lactose is available in various particle size distributions in milled or sieved forms. Mostly it is a customized ingredient. The choice is influenced by particle size distribution, shape and surface roughness. Optimal delivery in the lungs is in general achieved by a mixture of coarse and very finely ground lactose. A regular shape and smooth surface result, for instance, in good flow properties. Also important are a very low percentage of amorphous lactose as well as a low surface moisture content.

Literature

European Directorate for the Quality of Medicines and Healthcare (2002) Lactose monohydrate monograph. In: *European Pharmacopoeia*, 4th edn. EDQM, pp. 1444–5.

Hickey AJ, Mansour HM, Telko MJ *et al.* (2007) Physical characterization of component particles included in dry powder inhalers I. Strategy review and static characteristics. *Journal of Pharmaceutical Sciences* 96: 1282–01.

Te Wierink H, Diepenmaat P & Damhuis R (2002) Formulation of lactose for inhaled delivery systems. *Pharmaceutical Technology Europe* **11**: 1–4.

Galacto-oligosaccharides syrup

Description

Galacto-oligosaccharides (GOS) are produced from high quality lactose. The polymerization reaction is carried out in a concentrated solution by means of the enzyme β-galactosidase. In addition, the solution is decolorized/purified, and a heat treatment takes place in which the enzyme is inactivated. The final product – a syrup – is obtained through an additional concentration step.

Product analysis[a]

Chemical

Moisture	25%
Nitrogen	0.01% on TS
Ash (sulphated)	0.01% on TS
Nitrite	0.03 ppm on TS
Carbohydrates:	
Galacto-oligosaccharides	59% on TS
Lactose anhydrous	21% on TS
Glucose	19% on TS
Galactose	1% on TS
pH	3.2

Physical/sensory

Appearance	Clear to slightly yellow
Flavour	Slightly sweet

Microbiological

Total plate count	<10/g
Enterobacteriaceae	Absent
E. coli	Absent
Salmonella	Absent
Staphylococci CP	Absent
Yeast and moulds	<1/g

[a] FrieslandCampina Domo (GOS-Vivinal: VVGS-EU.03).

Applications

The main application of GOS is the fortification of infant and follow-on formulas. Partly due to its heat and acid stability, it is also used in many other products such as acid beverages, dairy products, chocolate bars, bakery products and geriatric nutrition.

Properties

GOS belongs to the group of non-digestible carbohydrates with a dairy origin. It is a complex mixture of species having polymerization degrees between 2 and 8. The syrup consists partly of non-reacted lactose and monomer sugars. The longer chains comprise several galactose units and one terminal glucose unit. They are generally considered as dietary fibre components, or so-called prebiotics. GOS can replicate some of the functional attributes of the third largest component of mother's milk, namely the heterogeneous oligosaccharide fraction. These oligosaccharides have a positive effect on the growth of the favourable bifidobacteria and lactobacilli in the intestine. Additional effects include enhanced mineral absorption, stool improvement and weight management. Being partially indigestible, the caloric value of GOS is estimated to be 1.7 kcal/g. The product has GRAS status in the USA and FOSHU (Food for Specified Health Uses) status in Japan.

Literature

Environ (2007) GRAS notification for galacto-oligosaccharides.

Lamsal BP (2012) Production, health aspects and potential food uses of dairy prebiotic galactooligosaccharides. *Journal of the Science of Food and Agriculture* **92**: 2020–8.

Macfarlane GT, Steed H & Macfarlane S (2008) Bacterial metabolism and health-related effects of galacto-oligosaccharides and other prebiotics. *Journal of Applied Microbiology* **104**: 305–44.

Torres DPM, Gonçalves M, do Pilar F, Teixeira JA & Rodrigues LR (2010) Galacto-oligosaccharides: Production, properties, applications, and significance as prebiotics. *Comprehensive Reviews in Food Science and Food Safety* **9**: 438–54.

Milk mineral concentrate

Description

Milk mineral concentrate is usually produced from pre-concentrated whey permeate by applying a specific heat treatment at a slightly elevated pH. Calcium salts precipitate during this treatment. Moreover, they are separated, refined, dried and ground into a calcium supplement.

Product analysis[a]

Chemical

Protein (N × 6.38)	≤3%
Fat	≤1%
Lactose	±8%
Minerals	73%
Calcium	≥24%
Phosphorus	±12.5%
Chloride	±2%
Sodium	±1%
Potassium	±0.6%
Magnesium	±0.6%
Zinc	±0.06%
Citrate	±4%
Moisture	≤6%
pH (10% solution)	6.5–7.3
Calcium/phosphorus ratio	±2

Physical/sensory

Appearance	White to cream
Flavour and odour	Clean/light milky
Particle size:	
% <5 μm	50
% <15 μm	99

Microbiological

Standard plate count	≤10,000/g
Enterobacteriaceae	≤10/g
Salmonella	neg./125g
Staphylococcus CP	neg./g
Yeast and moulds	≤100/g

[a] Arla Foods Ingredients.

Applications

Milk mineral concentrate can be used particularly for the calcium enrichment of a wide range of food products. Examples are fruit juices, drinks, nutritional bars, breakfast cereals, milk products, bakery products and supplements.

Properties

Milk mineral concentrate contains approximately 25% calcium, mainly as calcium phosphate and calcium citrate complexes, and a wide range of natural trace elements. It distinguishes itself from other calcium sources by its dairy origin, good bioavailability and neutral flavour/odour. Due to its favourable particle size distribution, it is suitable for applications in which suspension stability and mouthfeel are essential factors.

Literature

Hunt CD & Nielsen FH (2009) Nutritional aspects of minerals in bovine and human milk. In: Fox PF & McSweeney P (eds), *Advanced Dairy Chemistry, vol. 3: Lactose, water and mineral constituents*. Springer, pp. 391–456.

Whey permeate powder

Description

Permeate powder is produced from clarified sweet cheese whey from which most of the proteins have been separated by ultrafiltration. The clear liquid is concentrated by an evaporator and the lactose is partially crystallized. Subsequently, the resultant concentrate is spray dried.

Product analysis

Chemical

Protein (N × 6.38)	2–6%
Fat	<0.5%
Lactose	75–85%
Ash	<10%
Calcium	0.6%
Magnesium	0.1%
Potassium	2.5%
Sodium	0.7%
Phosphorus	0.7%
Citric acid	2.6%
Moisture	<5%
pH	6.0
Protein/ash ratio	0.4

Microbiological[a]

Standard plate count	<1000/g
Coliform	<10/g
Salmonella	neg. in 375 g
Yeast and moulds	<30/g

[a] Agri-Mark.

Applications

Whey permeate powder is a replacer for whey powder. It is a less expensive ingredient and is particularly a source of lactose. It can be used in a variety of processed and bakery products. In the latter it contributes to an attractive golden-brown colour. The milk-related composition of whey permeate powder is also of value in animal feed applications such as for calves and piglets.

A permeate powder can be produced directly from milk. It is called milk permeate, and is utilized, for instance, in standardization of the protein level in milk powders.

Properties

As opposed to whey powder the whey permeate powder only contains the non-protein-nitrogen fraction. For this reason, this ingredient is more difficult to dry than whey powder. Partial crystallization of lactose reduces the hygroscopic nature of the powder. The lower protein content and its low protein/ash ratio limit its use in food products. The lactose in whey permeate powder has a low sweetness level compared to other carbohydrates, which is in some cases advantageous. For animal feeds it is – like most whey products – highly palatable and digestible.

Literature

Durham RJ (2009) Modern approaches to lactose production. In: Corredig M (ed.), *Dairy-Derived Ingredients – Food and Nutraceutical Use*. Woodhead, pp. 103–44.

European Whey Products Association (2007) Whey in Animal Nutrition – a Valuable Ingredient. EWPA, pp. 1–19.

O'Doherty JV, Nolan CS & McCarthy PC (2005) Interaction between lactose levels and antimicrobial growth promoters on growth performance of weanling pigs. *Journal of the Science of Food and Agriculture* **85**: 371–80.

Delactosed whey permeate liquid

Description

Delactosed permeate is the residue left after the processing of cheese whey. Most of the proteins are separated by ultrafiltration and a major fraction of the lactose is harvested by means of crystallization. The remaining part is mostly available as a liquid, which may show some variation in composition. The example given here illustrates its composition after applying a high-yield lactose process.

Product analysis[a]

Protein (NPN)	2.1% on TS
Carbohydrates	55.2% on TS
Lactose	55.2%
Galactose	<0.4%
Glucose	<0.4%
Acids	9.2% on TS
Citric acid	3.5%
Lactic acid	5.7%
Minerals/ash	17.0% on TS
Calcium	0.7%
Magnesium	0.2%
Potassium	6.9%
Sodium	2.3%
Phosphorus	1.6%
Chloride	5.3%
Moisture	74.1%
pH	5.2
Protein/ash ratio	0.1

[a] Liang *et al.* (2009).

Applications

Due to the high moisture content the delactosed permeate is difficult to incorporate into food products. Its use as a natural preservative in fresh-cut tomatoes is reported as well as its salt-enhancing properties. Delactosed permeate is commonly utilized in animal feed.

Properties

Delactosed permeate is a side-stream of a series of processes such as separation of cream, cheese manufacture, removal of whey proteins and lactose crystallization. The processing including the time needed for storage can last more than a week before the delactosed whey is ready for delivery. The liquid still contains various interesting substances which might be currently underestimated. Due to its stickiness and hygroscopicity caused by high mineral content and enhanced lactic acid levels, delactosed permeate manifests drying problems and post-drying instability. Careful processing, use of special drying equipment or application of some additional substances makes conversion into powder possible.

Literature

Ahmed L, Rico D, Martin-Diana AB & Barry-Ryan C (2011) Optimization of application of delactosed whey permeate treatment to extend the shelf life of fresh-cut tomato using response surface methodology. *Journal of Agricultural and Food Chemistry* **59**: 2377–85.

Bund RK & Hartel RW (2010) Effect of protein on sorption characteristics of delactosed permeate. *Journal of Food Engineering* **96**: 388–93.

Burrington KJ (2011) USDEC-Application Monograph: Permeate for Sodium Reduction. Available at: www.innovatewithdairy.com

Liang B, Bund RK & Hartel RW (2009) Effect of composition on moisture sorption of delactosed permeate. *International Dairy Journal* **19**: 630–6.

Index

From Milk By-Products to Milk Ingredients: Upgrading the Cycle, First Edition. Ruud de Boer.
© 2014 John Wiley & Sons, Ltd. Published 2014 by John Wiley & Sons, Ltd.